FUNDAMENTALS
OF
PLANT SCIENCE

A home made beautiful by the plants which surround it.

FUNDAMENTALS
OF
PLANT SCIENCE

By
John Gaylord Coulter

DISCOVERY PUBLISHING HOUSE
NEW DELHI-110002

Reprinted - 2021

ISBN: 978-81-7141-213-6

Fundamentals of Plant Science

Published by:
DISCOVERY PUBLISHING HOUSE PVT. LTD.
4383/4B, Ansari Road, Darya Ganj
New Delhi-110 002 (India)
Phone: +91-11-23279245, 23253475; 43596065
E-mail: discoverybooksindia@gmail.com
discoverypublishinghouse@gmail.com
web: www.discoverypublishinggroup.com

Printed at:
Infinity Imaging Systems
Delhi

PREFACE

This book is for boys and girls who study about plants. It is a book about the fundamentals of plant life, and about the relations between plants and man more than it is a "textbook of botany." Yet it presents, as fully as the author believes to be desirable in required courses, those large facts about plants which form the present basis of the science of botany. These facts also form, it is believed, a minimum of knowledge about plants to which every high school student is entitled. To present this minimum adequately, rather than lengthily to cover a maximum has been the aim.

Appreciation. — The book seeks to give its reader a certain appreciation of plants and of the relationship of plant life to his own life. The study of "botany" may or may not yield such appreciation. Boys and girls by mere accumulation of "organized knowledge about plants" may never come to that appreciation of plants as a part of life which is believed to be very desirable, and one of the proper ends of the study of plants in high schools.

Delimitations. — The book presents the large essentials of plant life. It emphasizes their significance to man. But it does not pile up specific illustrations and applications which may not illustrate or apply in what is common to the lives of high school students in general. The effort is to include what has proper place in the education of *all* young people, and to exclude special information which properly has required place only in the education of *some* young people. Thus, as to agriculture, it is believed that such general study of plants as is presented herewith should precede the special study of that subject, but it is doubted whether such study of plants as is required of all students should be a study primarily of agriculture. In so far as is possible, plant life is presented in this book in terms of its largest relations to human life, but the treatment has in view preparation for life in general, and not preparation for any particular kind of calling.

Forestry, Plant Breeding, Weeds, Plant Enemies and Diseases, Plant Culture, Decorative Plants, and Economic Bacteria are topics which are discussed where such discussion seems pertinent to the general theme, but special chapters are not devoted to these topics. Certainly such topics should form a part of the course, especially if it is a year course, but it is questionable whether special chapters on such topics should form a part of the basic text. Such topics, treated in separate chapters, seriously impair that unity of organization which should characterize a foundational text. Also they form that part of the course which the teacher needs to organize largely in terms of the locality. A treatment of the fundamentals of plant life may fit all cases, but any treatment of the topics given above will not fit all cases, and the misfits may do the cause of practical plant study more harm than good. Thus town children and parents have rebelled against plant studies which are purely agricultural. Finally, government and state bulletins of local application and works of reference have been found to be of much better service in the study of such topics than any brief miscellany of facts put together in "special chapters."

Coherence. — The pupil's facts should hang together for him. In this sort of course he needs to see relationships. This outcome may hardly be expected when there is an overload of facts included for their own sake and at the expense of the course as a whole. Like plants themselves, courses dealing with plants are often in need of pruning in the interest of the fruit desired.

It is a pruned-back course which is presented herewith, even though the informal style of presentation may lead the reader sometimes to think it otherwise. Omissions have been weighed equally with inclusions. In the interest of a certain end in view, whatever seemed dispensable has been dispensed with. That certain end in view is a broad and a true conception of plants and of their principal relations to mankind. If our subject is to be a science, and not a mere mass of information, then such a conception is the first claim upon plant knowledge which we are called upon to satisfy, as well as the necessary foundation for additions of special knowledge. All that is herein included is meant to be contributory to that conception; bricks, as it were, in that foundation.

Style. — The manner of the book has been determined by a desire to make what is important seem interesting to young readers. If the author has been successful in what he has attempted, the intellectual effort per page needed to comprehend the text will be considerably less than if the book had been written with the idea of a formal textbook in mind. The teacher, in making assignments, should take this matter into consideration.

Introduction. — The introduction aims at appreciation. It rests on the principle of Herbart that "the pupil must know from the beginning what is aimed at if he is to employ his whole energy in the effort of learning." It is a frank attempt at motivation. It may be included or it may be omitted in the assignments. It falls into two parts. The first is devoted to appreciation of the subject matter to be studied; the second, to appreciation of the method of study to be used. The idea of this second part is to get the student interested at the very outset in scientific method, and to lead him to see that this method may be of great value to him in life as well as in the laboratory. Trial of this introduction has indicated that more of interest and effort is aroused when such material is included than when it is omitted. Many teachers have been interviewed as to their opinion of such an introduction. It is evident that the idea meets with approval, whatever opinion may prove to be as to the manner in which it is carried out in this book. If the introduction is used as lessons, it should be used before laboratory work is begun.

CONTENTS

CONTENTS

CONTENTS

CHAPTER FIVE: STEMS

CONTENTS

CONTENTS

CONTENTS

INTRODUCTION

Part I — Plants and Ourselves

1. **Our Dependence upon Plants.** — Plants are a part of life. Without them all life, our own included, would soon come to an end. Our lives depend upon their lives.

You are very much interested in life, but the importance of plants as a part of it may not have occurred to you. You know that food is one of the necessities of life, but you may not know that all food, directly or indirectly, comes from plants. You know that the country in summer is covered with plants. You know that their leaves hide the soil and make the hills and valleys green. But you may not know that green leaves are the food factories of the world.

When you eat an apple you are eating food that comes from a plant directly. When you eat an egg you are eating food that comes from plants indirectly. The hen which laid the egg lived on plants and on small animals which got *their* food from plants. All food comes thus from plants. *Other living things can change foods from one kind to another, but only plants can change substances which are not food into substances which are food.*

It is not for food alone that our lives depend upon the lives of plants. There are many other things which make them important to us. In this study you are to learn of the important part which plants have in your own life, as well as of the nature of plant life itself.

Think of the many uses which we make of plants besides their use as food. From plants we get the principal material out of which homes are built, and it is plants that we use to beautify the grounds around our homes. When a man is able to make the sort of home he wants, he is sure to surround it with grass and trees, with flowers and shrubs. They seem a necessary part of the sort of home we all want. Think, too, of the use of plants for clothing. Cotton and linen are both made from materials taken from plants. Wool and silk come from animals which live on plants.

Food is the first need of man, and this need could not be supplied at all except for plants. Shelter and clothing come next in order of our needs, and for these too we depend very largely upon plants. Houses are for shelter, but they would be poor shelter in winter if they were not heated. Coal, wood, and gas are the substances principally used in producing heat, and all of these are derived from plants. Coal is composed of the transformed bodies of plants which lived many thousands of years ago, and gas is derived from coal. Even when we use electricity we are usually depending indirectly upon plants, for the power-houses which generate the electricity use steam engines, and coal is used to generate the steam. The steam engines, in turn, run the machines which generate the electricity. Think of all the factories, and locomotive engines, and steamships which also use coal. Evidently one of the principal uses of plants to man is that they provide him, or have provided him, with fuel. Gasoline has come to be one of the important kinds of fuel, especially in connection with automobiles. It is derived from petroleum, and petroleum, like coal, was formed chiefly from the dead bodies of plants which were buried deep under the earth thousands of years ago. Lubricating oils are also derived from petroleum, while some oils, like cottonseed and linseed and castor oil, are derived from plants directly.

Wood is principally used in the building of houses and furniture, but there are other important uses of it. Think, for example, of the great quantities needed for telegraph and telephone poles, for railroad ties, and for fence posts.

Sometimes medicine is needed, and this need also is principally supplied by plants. Most of the drugs of which medicines are made are derived from plants. Rubber and gutta-percha, paper and perfumes, and many other much used substances are also derived from plants.

Each day you discover things which are new to you; things which make life seem to you more interesting and important. Just now you are to discover that knowledge of plant life will make your own life more interesting and more important. You are anxious to find out what you must know and what you must do in order to make your own life a success. The more you understand about life in general, the more likely you are to judge wisely for your own life in particular. To understand plant life and its relationships to human life will help you to understand your own life and its relations to the great world of which it is a part. You will learn that plants, besides furnishing food which makes life possible, may also furnish thoughts and feelings which make life pleasant; besides furnishing what is necessary to the development of our bodies, they may also furnish something needed for the development of our minds.

Plants are partners in life with us. They lived in the world before men lived in it, and they made the world habitable for men. The laws of life which control them are the laws of life which control us. To understand our own lives we must understand their lives. To understand their lives is to improve our own.

2. **Our Enjoyment of Plants.** — We enjoy the country chiefly because of the plants which cover it. We enjoy spring chiefly because then the grass turns green, flowers begin to bloom, leaves come out on the trees, and young plants grow up in the gardens. We may find it hard to explain this pleasure that we get from plants, but we feel it none the less. Something in us seems to respond to them. In fine weather we are glad to get out of the city, to go for a holiday into the country, to lie on the grass in the shade of trees, to look at what is green and growing.

The surface of the land is hidden by plants. They form the natural covering of the soil. Suppose you were up in an airship on a summer day. Green would be the color of the earth beneath you, the green of plants. The roads would make a few white lines across the green, and here and there you would see a house or catch a glimpse of a town. But nearly everywhere the earth's surface would be green, — green with the millions of plants which grow upon it. Of animal life hardly any evidence at all would be visible.

Suppose it was a July day and you were looking down upon farm lands in the Mississippi Valley. Corn fields would be below you, miles on miles of them, deep green and growing, shining in the sun. Here and there you would see fields of wheat and oats, pale green or turning yellow, through with their growth and ready for the harvest. In some fields you

View across a fertile valley in New York State.

would see men at work, the grain already piled in shocks. On hills and near streams you would see cattle grazing in the pastures, gathering their food from plants.

That great view of fertile country would delight you. Perhaps it would set you thinking. You might think of the miracle of nature by which this huge harvest has come from tiny seeds. You might think of the thousands of acres needed to yield food for one small city. You might think how small man is in the midst of the millions of plants which sustain him. You might think of the soil, that wonderful and complex layer which covers the rocks and holds, firmly and nourishingly within itself, the roots of plants; it is of the soil we think when we say "Mother Earth," for it is the soil which permits Earth to be "mother" of us all, plants and animals alike. As you looked down upon the great green carpet of plants below, you might remember that your own life depends upon this green life of plants, and perhaps you would feel grateful. At least you would enjoy its beauty, and might wish to know more about it.

Or suppose you were on a ship coming near to the steep shore of some tropical island. There you would see no sign of men, but high above you would stretch hill and mountain sides, covered with the deep green of tropical forests, only the bare white trunk of a tree showing here and there. All the land would be hidden, the year round it is hidden, under this heavy mantle of plant growth. Trees you would see everywhere, trees even growing out to sea, as is a habit of some trees in the tropics. Coming nearer, you might see little houses along the shore, but all the work of men would be a trifle compared with the mass of vegetation surrounding it. And so wherever you go in

Tropical vegetation. The U. S. Naval Station in Samoa. The harbor is an extinct crater.

the world, even to the very edge of the arctic circles, you will find plants in the growing season covering the earth with green, making it beautiful and fruitful, making it habitable for man.

Evidently plant life is one of the big things in the world in which you live. It is so big a thing and so important that you need to know about it. You need to know about it whatever your business is going to be. Even though you are not going to be a farmer, or a florist, or a forester, or in any other way make your living directly out of knowledge of plants, none the less it will pay you to have such knowledge. It will pay you to be able thereby better to appreciate the work of the farmer, who works for us all. Though knowledge of plants may not pay you in money, it will pay you in something else which is just as important as money. Money helps us enjoy life. Knowledge of plants may also help us enjoy life, and even more certainly than money does.

Every boy who likes to fish or swim or tramp through the woods is a lover of nature. Every girl who likes to gather flowers or watch the sunset is a lover of nature. All of us, whether we call it that or not, have some love of nature. It seems to be born in us. The pleasure that we take in plants is a part of it, a very considerable part of it. Nothing will more certainly add to our enjoyment of life than cultivation of this love of nature, and one way to cultivate it is to learn *about* nature. The more you know about nature (of which plants form such an important part) the more you care for the country. Knowledge of plants and animals makes it a different place for you from what it was before. With your eyes opened by knowledge, you can enjoy the country more than most people enjoy the city. To those who are interested in nature and understand her ways, life in the country is the most agreeable of all kinds of life. The farm boy who learns the ways of plants and animals finds more to interest him on the farm than he does anywhere else.

So, through knowledge of plants, something may be gained which makes the world, wherever we go, a far more interesting place than it was before. We gain something that gives every acre of ground and every plant growing on it a new meaning for us. We gain something that makes us enjoy finding out the names and habits of all the plants we meet; we wish to become personally acquainted with them and to watch their changes through the seasons. We gain something that makes us see in a grove of trees or in a field of corn things that we did not see before — things that change the trees and the corn to us; we see them with new eyes; their growth and their behavior become matters of the greatest interest; we know that they have a history just as we have a history and that as we keep changing so they too keep changing. This *something* that we gain may be called *appreciation of nature*, and those who possess it are very fortunate, for there is nothing in the world that is more certain to make their lives enjoyable. They can find pleasure in watching even a dusty patch of weeds, while woods and fields have for them a meaning like fine music.

You yourself in your study of nature in school can begin to get this thing which we have called the appreciation of nature, and the sooner you get it the more it will be worth to you. It is one of the finest things in life, and the study of plants, if you are truly interested in them, can lead you straight to it.

3. Plants in Spring. — It is in spring that we wonder most about plants. Whether you go into fields and woods or stay in town, it is the same. You see things growing. You watch the trees and shrubs turn green. You see green things sprouting from the cold and seemingly lifeless ground. Sometimes you seem almost to feel things growing.

Who can see, day by day, this fresh color coming back into the trees and over the ground, and not find pleasure in it? Who can find the first violets, or see the first green coming among willows along brooksides, and feel no thrill?

Have you ever wondered how this plant awakening comes to pass? Have you ever felt that you would like to know the plants, somewhat as you know your friends, — know their names, know how they live, and know how man, whose life depends upon them, can, through knowledge, improve their beauty and their usefulness? Have you ever thought that when you come to travel or have days free for roaming in the fields and woods it would add much to your pleasure to know the trees and flowers you meet? Have you ever wished to be able to judge why plants grow in the manner and in the places that we find them? Have you ever thought you would be glad to know how best to beautify with grass and shrubs and trees your home, or the home that some day may be yours? Have you ever wished to have a garden or a farm of your own?

Have you ever suspected that the lives of plants are like our lives, at least in that they, too, must have food and air and water? Has it ever occurred to you that there is a kinship between all living things, both plants and animals, and that you, the most intelligent of all living things, should understand all you can about this kinship?

Have you ever felt, while things are growing about you, something in yourself that commands *you* to grow; something that makes you know that to grow you must have knowledge, just as plants to grow must have air and water; something that makes you hungry for learning just as you might be hungry for food?

It is quite possible that you do not recall having had such a feeling, and, even if you do, you may not see that it and the study of plants are connected. Indeed this very study of plants may seem to you now one of the dull things from which you would like to escape. But you may not be yet a fair judge of such things. Whether you see it or not, there is in the study of plants a chance to get the very thing you want most, yet hardly know you want. That thing is a large true view of life.

Two things are sure. One is that if you have not yet felt a "thirst for knowledge," then, sooner or later, more or less, you are going to feel it, for it comes to us all as surely as green sprouts come from good seeds. And the other is that the study of plants is necessary if we are to know about life. To learn of plant life is one way to gain an understanding of all life, and nothing is more important to us than that. For this, perhaps, you must take our word. We cannot prove it to you now. We can only promise it as one of the rewards of interest and of attention.

4. **Plants and History.** — We think of history as the record of mankind, a record of how cities and nations came to be what they are to-day. We do not find much in history about plants, yet plants have had a great deal to do with making cities and nations what they are. They

have had a great deal to do with making your own life what it is.

Plants make little stir in the world. They are silent. Where they grow, there they remain. Yet they have had even more to do with the making of history than kings and armies have had with all their wars and commotion. This is not just because men have depended upon plants for food and shelter and clothing; it is not just because the abundance of plants has made plenty, and the lack of plants has made famine. Even more than all that, it is because plants have changed. As men have changed, so plants too have changed. There is a history of plant life as well as a history of human life. These two histories have gone along together, and each has had a great effect upon the other.

The changes in plant life have been of two great kinds. First, there are those which have been caused by nature alone, and the history of these changes is a very ancient history. It is far older than the history of man; it runs down to to-day, and it is still going on. Men change from generation to generation, and so do plants.

The second great kind of changes in plants is found in those which have been wrought by man working with nature. The history of these changes is that of the cultivation of plants, and it is this kind of change which has had most to do with affecting human affairs. By means of cultivation the plants which now you see in the fields have been derived from ancestors which, in ancient days, grew wild upon the plains and in the forests. And in those ancient days our own ancestors roamed over the plains and through the forests. Civilization has made us very different from our savage forefathers. In like manner cultivation has made the plants of the fields and of the orchards very different from those wild plants which once they resembled. It is this cultivation of plants which is so important a part of human history. This, more than anything else, has brought into the world such peace and prosperity as we have. This, more than anything else, has made man's life more than a mere struggle for existence. If the whole history of the cultivation of plants could be written, it would be a sort of history of man himself.

Some thousands of years ago all men lived a good deal as the Indians lived when Columbus discovered America. They were hunters and fishers. They roamed over the land. Their food came from the animals which they killed and from the wild fruit and roots which they gathered as they went about. The only advantage which they had over the wild beasts was that they had better brains. They learned to remember and they learned to reason. They learned which plants were good for food and which were not. They remembered when and where

A native home in the tropics. The buildings are composed entirely of materials obtained directly from plants which grow near by. In the foreground is a patch of *taro*, a plant whose underground parts are much used for food.

were to be found the fruits and roots which were good to eat. They learned that the plants which produced food would grow better if other plants around them were cut away, and probably this was their first step toward plant cultivation. They observed how plants reproduce themselves, and they began to clear the ground of useless plants and to save those which produced food. Thus the first little farms appeared.

The coconut palm grows in almost all parts of the tropics. It is one of the most valuable plants in the world. No plant exceeds it in variety of uses. Food, oil, soap, rope, and building material are some of the things derived from it. It is also one of the most beautiful of plants.

Think how farming must have changed the life of our ancestors. It was foolish to plant crops and then go away and leave them. Those who planted began to wait for the harvest. Gradually they began to give up free roaming of the forests, and began to live near their little fields. They tended and guarded for the harvest which was to come. They began to make homes. They began to settle on the lands which were most fertile. They began to band together with their neighbors for mutual protection, and to help each other with the crops. Men living together form what we call society. Thus you can see the truth of the statement that "society grew out of waiting for the crops." It was plant life which made men cease their roaming and settle down. It was farming that created society, and it is farming that sustains it to-day. It was from farming principally that cities grew and nations arose.

Thus it is that the cultivation of plants has made our own lives what they are. It is the farmer, the man who works with plants, whose work is the most necessary of all. It is upon what he produces and does not use himself that all the rest of us live. It is the land and the plants which they bear which have made our history what it is, and they are to-day far more necessary to our existence than are cities and all the rest of civilization.

5. **The Improvement of Crops.** — After farming began, men learned how to improve their crops. They learned to do more than simply clear the land and plant. They learned to plow. They learned that the stirring of the soil improves the growth of plants upon it. They learned to save for seed the better ears of grain, and to transplant cuttings from the better vines and fruit trees. They learned to make the soil more fertile by putting certain substances upon it. They learned that it is best not to grow the same crop upon the same field season after season; that it is better to change in some seasons to other crops. They learned to cross one plant with another, and thus they produced new kinds which were of value. In these and in many other ways they improved their crops, and, as farming went on through some thousands of years, the

cultivated plants became more and more different from the wild plants which were their ancestors until now the fine fruits and grains are very different from those dwarfish forms from which they came. Yet even to-day the wild plant-ancestors survive. Even to-day a wild and hardy plant from which wheat was derived grows among the sterile hills of Palestine, while in tropical America thrives the ancient parent of the corn.

A corn field in the Mississippi Valley.

These are wonderful results which in the past have been attained in agriculture, yet even more wonderful results are being attained to-day. Agriculture has become a science. In the past men found that certain things they did to plants changed them; some of these changes were improvements, and so the crops were improved. Nowadays men are finding out more than that. They are finding out *why* plants change. They are learning just what to do in order to get the results they want. They are learning how to breed plants just as they have bred horses and cattle. They are finding just what to do in order to make the fields yield more than they did before, and they are finding how to make farm plants grow where they never grew before.

It is the knowing *why* as well as *how* that makes agriculture a science. Success in farming depends very much on knowing why plants behave as they do; why the doing of certain things produces good results, and why the doing of other things produces poor results. Farms cannot be run best by rules alone. Each field is a problem in itself, and the farmer needs to know how to solve his problems for himself. To do this he must understand the principles of plant life. He must understand the conditions which are most favorable to plant growth, and learn to recognize

Corn standing in shocks.

what conditions are unfavorable to it. He must understand why it is that crop plants gradually poison the soil for themselves, and why it is an advantage to change the crops. He must understand why plants of the clover family increase the fertility of the soil, and why it is that deep plowing and frequent crumbling of the surface also increase it. He must understand the relation of water to plant life, and why drainage increases fertility. He must understand the principles which should guide him in the choice of the seed which he plants. He must understand how to encourage that invisible plant life in the soil which helps his crops, and how to combat that parasitic plant life above the soil which injures his crops by causing crop diseases.

Wheat in Montana ready for threshing.

Such an understanding of plants is one of the necessary first steps in scientific agriculture, and it is also one of the principal things which this book seeks to give you.

PART II — THE STUDY OF PLANTS

6. **Reasons for It.** — You are entitled to know why you are to study this subject, just as you are entitled to know why you study any subject. Much of your time in the next few months is to be devoted to it. You have the right to know and should know why the study of plants is considered a good investment of this time. We are

usually careful to see that we get our money's worth, but it is even more important to make sure that we get our time's worth.

Your teacher is as much interested as you are in your understanding the reasons for studying plants. If you understand the aims of this course and keep them clearly in mind, you are certain to be interested in it. If you are interested, you are pretty sure to get your time's worth. If not, you are pretty sure to fail to get it. It is a simple matter. You must become interested if only to keep from wasting time, which is about the most valuable thing you possess.

So here are set down some of the reasons for studying about plants. Consider them carefully. If they seem to you good, the hours you spend should yield much profit. If they seem to you insufficient, there may be others which will appeal to you more. But by all means or by any means *get interested if you possibly can*, if only for the sake of getting a proper return for your precious time.

Reasons for studying plants: —

a. Because without plants we should all die. We depend upon them for food, for clothing, and for shelter. These are the three fundamental needs of life. Whoever is interested in life needs knowledge of the means which support life.

b. Because knowledge of plant life may help us keep well and prosperous; it may keep us from being sick or hungry. Even such a brief course as this will give you a chance to understand the principles of plant cultivation. This is the art upon which our food supply chiefly depends. Knowledge of its principles is necessary for good gardening or good farming. Certain very simple plants called ***bacteria*** affect our lives very much, both helpfully and harmfully. Some knowledge of them is highly important to enable us to avoid disease.

c. Because knowledge of plants may increase our enjoyment of beauty. This applies especially to the decoration of our homes. Knowledge about lawns and shade trees, shrubs and garden flowers, is almost sure to be of service to any one sooner or later.

d. Because in the study of plants we come to a better understanding of life as a whole. We find that the life of plants is in many respects like our own. We observe the working of laws which control our lives as well as theirs. We learn of facts which indicate that all life is constantly changing and being modified by new conditions. Plant behavior throws direct light upon what our own behavior should be.

e. Because (and this is very important) in the study of plants you may study the subject itself, and not merely what some one else has written about the subject in a book. This means that you must observe closely and do some reasoning for yourself. You cannot get your lessons about plants properly by memory alone, or by using some one else's formula in order to solve the problems. This method of study is not very important if you use it only to study plants, but it is very important and very valuable if you use it in solving many of the serious questions of life. It is called the *scientific* method.

These five reasons are by no means all the good reasons for studying plants. Very likely you can think of others for yourself; perhaps others may interest you more than these. If so, they are more important to you at present than the five which have been given. The important thing just now is not *what* reason seems to you important, but that *some* reason seems to you important. Try to find some reason or reasons why the study of plants is important *to you personally*. Try to find a reason that makes you willing to do hard work in order to get from this study that which is important to you. Certainly there are such reasons. It is for you to find them out and apply them to yourself. If you can do this, the work is sure to be interesting and profitable. If you cannot do it, the work will probably seem difficult and tiresome. It is possible for each new exercise to be to you like a new scene in a play which pleases you. It is also possible for the whole course to be disagreeable. Which it is to be for you depends principally upon yourself.

7. **Ways of doing It.** — You will study about plants in two quite distinct ways. It is important to distinguish between them and to realize the value of each.

With one of these ways you are already familiar. It is the way in which you have already been doing most of your studying. It is the way in which you will learn most of the facts which you are expected to learn. When studying in this way about plants you do not learn from the plant itself, but you learn, from a book or from your teacher, what some one else has learned from the plant itself. This method is important because it is a great time saver, but to use this method alone would lead you to depend too much upon your books and teacher and too little on yourself. You must respect and use the knowledge of others, but you must also respect and use your own powers to gain knowledge for yourself; otherwise you lose them. This indirect method is called the *didactic* method.

(*Didactic* means that something is *told*, rather than learned by first-hand observation.)

In the other way you will learn but few of the facts you are expected to learn, but you will learn them from the plant itself, or from experiments which concern plants. This method is not important for the number of facts it teaches, but it is of great importance for the way in which it teaches them. It requires you to rely on yourself, to make your own observations, raise your own questions, and draw your own conclusions. This way of study makes you think for yourself. It keeps you from being a mere follower of some one else's thought. It requires you to use other powers than the power of memory alone. It requires you to get your information directly from the thing itself. This direct method of learning is what we have already called the scientific method.

In the preceding section you read something about this scientific method, but not enough. It is too important a matter to be dismissed with a few words, especially since it is a method which you may not have used in school before. If you do not realize at the outset its great importance to you, you may fail to put your best efforts into acquiring it and so miss one of the most valuable things which school can give you. Like all other really valuable things, this method of study and thought is not to be gained easily. It calls for the very best you can give of attention and of effort. You may have formed the habit of using little besides your memory in learning your lessons. It may seem strange and difficult at first to use this other method. But you can learn to use it. You are able to observe and to decide for yourself if you are willing to make the effort, and only by means of such efforts are strong minds developed.

Most of the facts about plants which you learn, unless you continue the study, you will presently forget. But if you learn this method once, and appreciate its importance, and practice it, you never will forget it. You will use it all your life. It will serve you best when you come to settle the most important questions of your life, it will lead you to the right answers in the great questions of behavior, and you will come to esteem it among your most precious possessions.

Knowledge changes as new facts are found, but this method does not change. It is one way to knowledge. Learning is knowledge of facts, but unless we know how to use the facts all our learning is of no more use than a dusty encyclopedia which is never taken off the shelf. Wisdom tells us how to recognize facts, how to understand them, and how to act in accord with them. Indeed wisdom is just this scientific method by another name. Though named with the name of science, this method applies to far more than just the things you think about when you say science. It applies to all the affairs of men. It is not a thing which concerns just astronomers and botanists and the like. It is a thing which concerns every man, and which is of as much importance to every man as it is to the scientist. It is called the scientific method only because it is through the study of science that man has developed it. It was by the study of stars and of plants, by working out the laws of life and of light, by seeking new knowledge in every field of nature — by such means the scientific method came to be what it is. It is more valuable to man than all his discoveries and achievements, for it is the method by which all the best of these discoveries and achievements were accomplished.

By means of it, in the yesterday of history, a few men discovered much to make us wonder. By means of it, in the to-day of history, many men are discovering much which makes us think and act. And by means of it, in the to-morrow of history, all men will learn to work together to make life best worth living.

8. Your First Plant Study. — Already you have had a good deal to do with plants. They have furnished you shelter and food and clothing. You have seen them alive and at work and have taken pleasure in their beauty. Consciously or unconsciously you have already acquired some miscellaneous knowledge about them. Presently you can begin to put that knowledge in order. Knowledge of plants put in order is the thing called *botany*. It is chiefly in books. Plants, however, live regardless of books and often break rules laid down for them.

Your first plant lesson is not concerned with botany or with books. It concerns only two living things, you and a plant. It is a little matter between yourselves. Let botany and books be for the time forgotten, but let your own good powers for getting information for yourself be well remembered. Perhaps those powers have been idle for a good while — idle in school, at least, where one can get along pretty well without using them. Now it is time to arouse them; to use them for all they are worth. Once aroused they will be worth far more to you than the power to learn lessons out of books. This is to be more than your first lesson with a plant. Possibly it is your first school lesson in getting knowledge for yourself — your first lesson in the scientific method.

In this lesson your teacher cannot help you much, must

not help you much. The important part of the lesson is what you find out for yourself. It may not be easy for you to find out things for yourself. You have been in the habit of finding things out in books. Now you must also learn how to find things out without books. You must use your own good powers of observation and of reasoning. Your brain may have formed the lazy habit of letting other brains think for it. It is time to make it begin to think for itself. Perhaps the best thing about the study of plants is that it does make you think for yourself, at least, it ought to, and this first lesson with a plant may be the most important lesson in your life. It may be your first lesson in using your brain *for all it is worth*, and not for only a part of what it is worth. For you it is a serious matter and calls for the best effort you know how to make.

Any plant will do for this first study. A plant, a piece of paper, and yourself awake, and we have materials for the first lesson which cannot be excelled. It is easy to supply the plant and paper. Any teacher can supply these, but no teacher can supply *yourself awake*. That rests with you. By *yourself awake* is meant awake in all your senses, attentive to the matter in hand, ready to appreciate its importance to you and to accept its invitation to self-improvement. You are to be an investigator, and the little plant before you is to be the subject of your investigation. The first step is to observe it carefully. Nothing makes us more sure to observe carefully than to draw the thing we are observing. So you are to make a careful outline drawing of this plant. Remember all the time that it is a living thing just as much as you are, and that each tiny part or position of a part may be important in this life you are investigating. Make a clear outline drawing, showing each part in its natural position. While you are drawing, think of the life and growth of this plant; think of such questions about its life as you would ask it if it could answer. Your drawing is not to be shaded or indistinct, such as it might be in the art department. Each line is to stand for something you see, and everything you see is to have a line to stand for it — so far as that is possible. Draw as well as you can, but remember this is not an exercise in drawing. It is an exercise in observation and careful record.

When the outline is completed, you will record with words such of your observations as you could not record in the drawing. You have been observing carefully. Sight and feeling have both been working on this investigation. Few features of the structure or color or texture of that plant should have escaped your keen senses. You have noted also its general relations to its surroundings. You are now to make a list of these observations which will go along with the drawing to complete the record. The writing will help explain the drawing, and the drawing will help explain the writing. You made them for just the same purpose. They are to record your observations. Observations are worth little unless they are recorded.

With that finished, the questioning begins. You have made yourself think about that plant while you observed it closely. Your thoughts raise questions. You will not be satisfied now until some of your questions are answered, and it is just here that one of the most excellent characteristics of plant study appears. That excellent and to you most valuable characteristic is that for some of the questions you have raised *you can find the answers for your-*

self, not in a book, nor by any figuring process which some one else devised, but right here in nature, exactly as the pioneers of science discovered them. You are to play at being a pioneer of science yourself, and, if you play the game well, you can fairly feel the powers of your mind waking up and stirring within you. If you play the game well, you will let no one supply you with answers or explanations which you can find for yourself. That "finding for yourself" is the whole object of the game. To let some one else do this for you is as foolish as to steal a game from yourself. The object of any game is to try fairly to win. You win in this game, as in the game of life, by doing your own thinking. You lose in this game, as in the game of life, when you let other people do your thinking for you.

It is by such work in school as this work with plants that your powers to think are to be aroused. If you solve thoughtfully for yourself some little questions about plants, you may by the same method solve wisely for yourself some of the big questions of life. In this lesson it is not important that your solutions be right, but it is immensely important that they be your own and as right as you can make them. Observations and questions too must be your own; they are important even if you cannot find the answers.

So if you give your best efforts to this task and others like it, the powers of your mind will begin to work for you as perhaps you never knew they could work. "After all," you will begin to think, "this thing of solving real problems, especially the problem of my own living, is not going to be so very difficult, if I only have the sense to apply myself diligently to it. There's nothing to be afraid of except myself. I have the power to work independently and to get results. The great thing is to keep that power at work. It is not at work whenever I let others do thinking for me that I should do for myself." You will go out from your study of plants, if you study them as we would have you, feeling new power growing within you. You have your own clear brain, and you are learning how to use it. It is the most valuable thing in the world when rightly used, and nobody has two.

But to go back to your questions. Dreams of power must never make us miss the next careful step. If we do miss it, we never shall reach our goal. Perhaps the little plant before you is a geranium in a pot. You will look it over. You will remember what you already know about it. You know that the parts you see are leaves and stems and that in the soil there are roots. What are they for? You know that plants grow, and hence you conclude that they are alive. What are the surroundings in which they will grow and what are those in which they will not grow? You know that living things require food. Where does the plant get its food? How does it grow? You know that this plant may presently produce a cluster of flowers. What are they for? You know that it reproduces itself. Though it dies, its kind does not die. Thus you already know that plants, like all other living things, have two great kinds of work to do. One is to keep themselves alive, the other is to keep their kind alive. One is *nutrition*, the other *reproduction*.

It seems fair and helpful to give you this much of a start. But now the suggestions stop, and you must go on alone. You will observe and draw the different parts of the plant. You will note other facts and other questions will occur to you. Your own common sense will help you tell what questions are sensible and what are not. You will make a list of what seem to you sensible ones. Your teacher will discuss these lists, and select for your further study one or two questions whose answers it is quite possible for you yourself to find.

The first lesson now is over. The drawing and the writing are the only visible evidences of what you have done. But the most important part of the lesson is the invisible part, the part that goes on in your brain. However excellent the drawing and the writing may appear, the lesson is a failure unless this drawing and this writing have been produced by your brain alone. Have you borrowed ideas from any one else? Then, for your own sake, do the lesson over with some other plant.

It may be that your first plant study will not be the study of a whole plant. It may be the study of a part of a plant, such as a flower or a seed. But whatever the material used in this first study, whether it be the whole plant or only one part of it, the method of study you are to use is the same. It is the method which has just been described.

9. **Later Studies.** — Your first study will result in a collection of facts and of questions. Your later studies will result in explanations as well as in facts and questions.

As a part of your first study you are to prepare a list of questions. Those chosen from your list for further study may be like these: What is the work of roots? of stems? of leaves? A complete answer to any one of these questions has not been worked out by botanists themselves, yet the principal kinds of work done by roots and stems and leaves are not difficult for you to discover for yourself.

In seeking the answer to any one of these questions you are sure to meet facts which help supply the answers to the others. So it is well to have a number of questions in mind even when you are giving particular attention to one. You will not solve your problems one at a time. You will solve them together. That is because everything in life depends so much on everything else. All the parts of living things work together. They do not work independently. They help each other. Thus all plant work, as you have seen, may be divided into nutrition and reproduction. Roots, stems, and leaves are sure to be working together to accomplish these great common tasks. So in finding out what share a particular part has in the whole work, you are sure to have light thrown on what the other parts are doing. Similarly, for example, you may have studied an automobile engine. In studying one part, you learned something also about the other parts.

In trying to explain facts you have observed about plants, you will need advice on several points, and one principal point is this — avoid haste. Hasty thought may be worse than no thought at all, but clear thought is the secret of wisdom. You cannot think clearly of these matters and think in a hurry. Don't let the idea bother you that some one else may find an explanation before you do. What you want to do is to forget everything but the problem. Forget even yourself. You are a true scientist only when you put yourself and your feelings entirely out of mind and consider the question only in the light of the facts.

Here is the plant body which grows. How do the materials which are added to it get into it? Certainly not through a mouth. as in animals. How then? Your own clear thought working on facts you can observe will furnish you an explanation for this, but hasty thinking is almost sure to lead you into error.

Thinking attentively, you are almost sure to find some explanation of what you have observed. The next step is to test the explanation. See whether any other will do as well. For example, you might explain the refreshed look of plants after a rain by saying that the leaves absorb water as it falls on them. But another explanation of this fact will do as well, and other facts, discovered later, will show which is more nearly true.

New facts will constantly become your property as you proceed, and help you form explanations of such matters. Your later studies will thus be lessons in one of the great uses of facts — the use of them in forming explanations.

10. **The Notebook.** — The notebook is usually the most unpopular feature of high school botany; this is natural, since boys and girls of your age usually find writing and drawing tasks disagreeable.

Though the task may seem small and insignificant, to do your best is not small or insignificant, and to do less is to lose much, for it is to lose respect for the sincerity of your own effort. A sentiment like that might be useful on every page of a science notebook. Nothing in the whole course is better training for you than the keeping of the right sort of notebook, keeping it carefully, and keeping it up to date. It deserves your best efforts, and will reward them. It is the best device yet found for insuring careful observation and careful record, two things indispensable to science.

As to the drawing, against which so many rebel, you

will be interested in what was said by Huxley, one of the greatest of science teachers: —

> In addition [to reading and writing] I should make it absolutely necessary for everybody, for a longer or shorter period, to learn to draw. Now, you may say, there are some people who cannot draw, however much they may be taught. I deny that *in toto*, because I never yet met with anybody who could not learn to write. Writing is a form of drawing; therefore if you give the same attention and trouble to drawing as you do to writing, depend upon it, there is nobody who cannot be made to draw, more or less well. Do not misapprehend me. I do not say for one moment you would make an artistic draughtsman. Artists are not made; they grow. You may improve the natural faculty in that direction, but you cannot make it; but you can teach simple drawing, and you will find it an implement of learning of extreme value. I do not think its value can be exaggerated, because it gives you the means of training the young in attention and accuracy, which are the two things in which all mankind are more deficient than in any other mental quality whatever. The whole of my life has been spent in trying to give my proper attention to things and to be accurate and I have not succeeded as well as I could wish; and other people, I am afraid, are not much more fortunate. You cannot begin this habit too early, and I consider there is nothing of so great a value as the habit of drawing to secure those two desirable ends.

Notes, to be of much value to you, must be original with you. They must be the record of your own thoughts and observation, not a record of the thoughts and observations of others. If the notebook is not largely original with you, the time spent upon it is largely wasted. Notes which are not original may be valuable memoranda, and good practice in handwriting, but they are not the kind that rouses the power of your brain.

The study of plants is in your course and the notebook is a part of that study largely in order to induce you to do your own thinking. The structure and the behavior of plants you can observe for yourself. You can think about what you see, and you can write down your thoughts and observations in the notebook. Most of the important things, even in science, you will learn from books or from other people. You do not have time to find out very much for yourself. You must take "short-cuts" to knowledge. Yet it is of extreme importance that, with respect to a few things at least, you do *find out for yourself.* Your power to find out for yourself and to judge for yourself must be developed, for it is a power of which you will have great need. So, in the study of plants, you are to find some things out for yourself. You are to have an excellent opportunity to observe, to reason, and to come to conclusions. But all your observing and your thinking will be of much less value if you do not make careful record of it in a notebook. Thoughts come to us as we write them down, and they are of value only as we express them. Thus the notebook is an opportunity you must not neglect. If you force yourself to observe, and to think, and to record your observations and your thoughts, that work will put power into your brain as starch puts stiffening into a limp rag. If you neglect such an opportunity, you are in a fair way to have a limp-rag sort of brain all your days.

It is when you get to notebook work that you begin to get the most important results from your observations and your thought. Pen and paper are mechanical aids without which good thinking loses much of its value. It vanishes like a fog and is lost unless you make a record of it. Thinking is only a misty sort of thing at best until you set it down in black and white. Once you learn to do that, you have begun to realize on the product of your mental machine instead of letting it run to waste. All the good thinking that has ever been done would have amounted to little except that a few good thinkers took notes.

So the science notebook is really one of the valuable opportunities of high school life, and one of the least appreciated. The value of your notes does not depend upon your reading them again. That may have little to do with it. The value lies in the process of making them, and making them clear-cut and honest expressions of your own observations and conclusions; it lies in fixing clearly in mind ideas which otherwise would be lost; it lies in training you in careful expression. The drawings also serve these purposes. Drawing enforces accuracy of observation; it will lead you to notice many things which otherwise you would overlook.

11. **The Textbook.** — After all that has been said about making observations, and about thinking, and about taking notes and making drawings, it may seem to you that the study of this book is not going to be a very important part of the course. That is about right. The book is perhaps the least important part of the course, and yet it is a necessary part. It is necessary because your time is so limited. The great facts about plants are to be learned, and you can learn but few of them by your laboratory work alone. So you will turn to the book. The book is for information. Your laboratory work is not so much to give you information as it is to show you how to get information, how to get it without books and without being told. The laboratory is principally for learning the method of science, while books principally give us the facts of science.

In the laboratory and in the field many questions about plants may occur to you. The answers to many such questions are in this book, but the answers to some of them may not be in this book or in any other book. These unanswered questions may be of more value to you than the answered ones. They may keep you thinking, while an answer might stop your thinking. Some time you may find the answer for yourself. The most fascinating thing about botany to the botanist is that there *is* such a great deal yet to be found out. What you learn for yourself from the plants rather than what you learn from the book will be the pleasantest part of your little journey into the plant world. Any journey may be dull with every mile of it planned in advance, and all adventures eliminated. That is what your plant journey would be if it were a journey into books alone. But instead of that it is to be a sort of voyage of discovery.

To learn about plants by print alone is like reading of travels in a rocking chair when you might be roaming the world yourself. So this book is a sort of necessary evil. Its principal excuse for existence is that in the few months that you have for your botanical journey you can discover only a few things for yourself. The making of those discoveries will be the best part of your work, but you are also entitled to a general view of the subject. Hence this book. It should not come between you and the plant. Its place is after the plant, to supplement.

QUESTIONS AND SUGGESTIONS

SECTION 1. 1. Explain the statement that all the food of animals comes from plants. 2. What are the principal plants used by men for food? 3. What are the principal plants used by animals for food?

4. Explain the use of plants in clothing. 5. Explain the use of plants as fuel. 6. Explain the use of plants in connection with the production of electricity. 7. Explain other uses of plants to man in addition to the uses for food, clothing, and fuel. 8. Mention the substances derived from plants which you yourself have used to-day. 9. Mention the kinds of business in your community which depend directly upon plants and the products of plants. 10. What foods sold by the grocer come from our own country and what from foreign countries? 11. Mention the plants principally used for timber. 12. What are some of the ways in which plant life and animal life are alike?

SECTION 2. 1. What kinds of plants principally form the natural covering of the soil? 2. What are the most abundant plants in the country round where you live? 3. How do plants get started on soil which is bare? 4. What is the principal cause which may prevent soil from having its natural covering? 5. Describe the appearance in summer of the farm land round where you live as you might see it from a balloon. 6. What are the principal crops in the county in which you live? 7. Explain why knowledge of plant life may be of value to you apart from its value in making a living. 8. Seeds are the principal means by which plants are started. Do you know any other ways in which they are started? If so, describe them.

SECTION 3. 1. Describe some of the things you have noticed about plants in spring. 2. About what time do the leaves begin to appear, and what are the plants whose leaves appear first? 3. Do you know any plants whose flowers appear before the leaves, and can you think of any advantage in this habit? 4. What are the first wild flowers you have noticed? 5. What are the first cultivated flowers to appear? 6. What are three of the things necessary to plant life as well as to your own life?

SECTION 4. 1. What are the two great kinds of changes which have occurred in plants? 2. Describe the relation of primitive men to plants. 3. Describe the beginnings of the cultivation of plants. 4. Explain the statement that "society grew out of waiting for the crops."

SECTION 5. 1. What are some of the ways in which farmers learned to improve their crops? 2. Where are plants like the ancestors of wheat and corn found? 3. What is it that makes agriculture, or any other subject, a science? 4. What are some of the things which a farmer must understand in order to be scientific?

SECTION 6. 1. What is meant by "getting your time's worth"? Does this seem to you more important than getting your money's worth? Why? 2. What are some of the reasons for studying plants? 3. Do you feel a personal interest in plant life? Why? (Note: One excellent reason for your being interested in anything is simply *because* you are interested in it. It is not necessary to understand an interest in order to possess it. But our interests are of more value to us when we understand them than when we do not.) 4. What do some of the facts about plant life indicate with reference to all life? 5. State the advantages which the study of plants has over any study which requires memory alone.

SECTION 7. 1. What are the two ways in which you are to study about plants? Describe them. 2. Which of these methods is more useful in the acquirement of knowledge? Why? 3. Which of these methods is more useful in acquiring the habit of thinking for yourself? Why? 4. Give an example of learning by the scientific method in your life outside of school.

SECTION 8. 1. Write a brief statement of your present ideas of how plants live. 2. What is botany? 3. Describe the method of plant study suggested in this section. 4. Have you used this method in other kinds of school work? Have you used it outside of school? Give examples. 5. What materials are necessary for the first lesson? 6. What is meant by *yourself awake?* 7. What are the things you are to do in this lesson? 8. What is the purpose of the drawing and writing? 9. What are the two great kinds of work which all living things have to do?

SECTION 9. 1. In what ways will your later studies differ from your first studies? 2. Why is it that in studying one part of a plant you learn about other parts?

SECTION 10. 1. What two things, indispensable to science, do we aim to secure by the keeping of a notebook? 2. State the point of the quotation from Huxley. 3. What are the principal qualities which notes must have in order to be of value to the maker? 4. Have you ever found that writing helps you in thinking? Give an example.

SECTION 11. 1. Why is a textbook necessary in this course? 2. Compare purpose of study of book with purpose of study in laboratory.

CHAPTER I

THE PLANT: A GENERAL EXTERNAL VIEW

12. **Seed Plants and Other Plants.** — Of the many kinds of plants we shall consider first those most familiar to you. One simple way in which to divide all plants into two groups is to divide them into those which produce seeds and those which do not. Ferns and mosses, seaweed and pond scum, toadstools and mushrooms, are examples of plants which do not produce seeds. But all the familiar trees and shrubs and herbs, and the plants which the farmer grows in his fields, are seed plants, and we shall consider seed plants first. *Whenever the word plant is used in this part of the book the thing referred to is a green plant which produces seeds.* We shall consider how seed plants are constructed, how they live, and how they reproduce their kind.

13. **Plant Life and Ours.** — You know that a plant is a living thing. You may have wondered how its life compares with your own. You may have wondered how it is constructed and what is going on inside of it.

You and a plant are, of course, very different in appearance, but as to what must be done in order to keep alive, you and a plant are a good deal alike. To understand the ways in which you and a plant are alike is even more important than to understand the ways in which you are different. It is the likeness between all living things which gives a real basis for that feeling of kinship to nature to which we have referred.

Our life depends absolutely upon certain things. Chief among these things are food and air and water. Without them we die. Plant life also depends upon food and air and water. It has the same needs for them that we have. Our whole body is mainly an equipment for securing these things and using them. Our bodies are chiefly composed of bone and muscle, of blood and the digestive organs, and the principal work of all these parts is to secure food and use it. Similarly a plant is composed principally of roots, stems, and leaves, and the principal work of these parts is to secure food and use it.

Plants are much like ourselves, then, in their relations to food and air and water; here are three life relationships necessary to living things whether they be plants or animals. But there is another life relationship of plants in which they are different from animals. That is the relationship to light. Animals can live in the dark, but green plants cannot. They must have light, and it is their need for light which explains the form of their stems and leaves more than anything else explains it. Look at *Figures 10* (page 55), *73* (page 210), *74*, *76*, and *78*. Also consider the forms of green plants wherever you see them. Do they not all appear to be seeking the light?

To get light, to get food, to get air, to get water, and to reproduce, appear to be the chief ends of plant life; herein we have the principal secret of plant structures; herein we have a sort of formula by which we may explain the parts of plants which we study. These are the great purposes, and, in some way or other, each part serves these purposes. You will see many kinds of plant bodies, just as you see many kinds of houses in which men live. In appearance they are very different, but in general purpose they are all alike.

Nutrition and Reproduction. — Every living thing has two great tasks to perform. If any kind of living thing fails to perform these two great tasks, and perform them well, that kind soon disappears from the face of the earth. One of these tasks is to maintain the life of the individual, and, to perform this task, food and air and water are among the things necessary to all living creatures. This task is what we call *nutrition.* The other great task is to maintain the life of the race, and it we call *reproduction.* These, then, are the two great and fundamental similarities of all living things — they must keep themselves alive and, by means of reproduction, they must keep their kind alive. The many kinds of living things all differ more or less from each other in the ways in which they solve these two great life problems, but the problems themselves are the same for all, and in the ways in which they are solved you will find that there is more of similarity than there is of difference. From the most powerful man to the lowest plant, the life of all of us is determined by the fulfillment of the laws of nutrition and of reproduction more than by anything else. These are the fundamentals of life. They cannot be omitted. All other things are built upon them. They are the foundation.

14. **Nutrition.** — By nutrition we mean *all that has to do with the maintenance of the life of the individual.* It is something that you already know a good deal about. At least, you have had a good deal of experience with it.

You have been practicing it every instant of your life. You know that in order to keep alive you must take into your body certain substances from the world about you. You know that you must get rid of certain wastes. You know that health depends upon this income and this outgo, and that loss of health results when these processes are not properly performed. You know that too much heat is bad for you, and likewise too little heat. All these are the outward evidences of nutrition. As they are true for you, they are true also for plants.

You know that you live at the bottom of a great ocean of gas we call the air, and on the surface of a pleasant solid we call the earth, and that both in the air and in the earth and on it we find an agreeable liquid called water. You know that from the sun there come to us forms of energy called light and heat, which are of much comfort to our lives. You know that if air, earth, water, light, and heat, or any one of these should be lacking, our lives would soon come to an end. All this is likewise true of plants. Their life, like ours, depends upon the presence of these things, and the great task of their lives is to secure such portions of these things as are essential to their well-being. Your task, just now, is to find how they do this.

A. Food. — You know that, to live, you must have what we call food. Plants, too, to live, must have what we call food. You know that you get your food from the world outside of you. Plants, too, get food, *but they do not get it from the world outside of them.* From the world outside they get materials from which food is made, but food itself they make within their own bodies.

Evidently, to understand what has just been said, you

need to understand the difference between food and food materials. Food is a general term. You may think you know what you mean by it, but you would find it hard to define. Just now we need a definition. We may define as food *all substances which contribute directly to the energy, growth, or repair of living bodies.* You may have been thinking that plants get their food from the soil and from the air. It is true that they do take in substances from the soil and from the air, but it is also true that these substances are not food. They are substances out of which food is made, but they are no more food itself, strictly speaking, than bits of gold and iron ore are a beautiful and complicated watch. Gold and iron ore may furnish the materials from which the watch is made, but surely they are not the watch itself. Here, then, at last, we have found a fundamental difference between the nutrition of plants and the nutrition of animals. Plants manufacture food, animals do not. Plants get from outside themselves the crude materials from which food is made, but these materials are not the food itself. The food itself is made inside the body of the plant. All this is true only of green plants. Some plants have no green parts. A toadstool is an example. Such plants get their food from outside their own bodies, just as you and I do.

B. Digestion and Assimilation. — You have probably learned already that the food which we eat, in order to become of use to us, must be digested and assimilated. *Digestion is the transformation of food into such liquid form as may go to all parts of the body. Assimilation is the transformation of food into living parts of the body.* Plants, like ourselves, must digest and must assimilate their food. These are parts of the great work of nutrition, and they are common to all living things. They are among what are known as the " general " life processes.

C. Photosynthesis. — The process by which plants manufacture food out of substances which are not food occurs only in the presence of light. (You see now why green plants cannot live in the dark, and why leaves turn to the light.) This process is called *photosynthesis.* Except for its length, photosynthesis is a very good word. It is a word which explains its own meaning. The *photo* in it means light. *Syn* means together. *Thesis* means putting. So the whole word may be translated *putting together in the light*, and it refers to the power of the plant *to put together in the light certain substances in such a way that they form food.* Photosynthesis is not all that there is to food making. The food which is made by photosynthesis may be afterwards transformed into other kinds of food. *But photosynthesis is the only process by which food is made out of materials which are not themselves food.* We ourselves have the power to transform foods from one kind into another, but we do not have the power to make food out of materials which are not food. This is done only by green plants. In the foods which they manufacture they store up energy which is derived from sunlight.

Thus we see that, so far as the general features of nutrition are concerned, there is but one in which plants are dissimilar from ourselves. They are dissimilar from us in that they possess one more nutritive power than we possess, and that power is extremely important. It is a power which makes us absolutely dependent upon them. It is a power which green plants alone of all living things possess.

FIG. 1. — The young plant-body of a common weed called lamb's quarters or pigweed. It is composed of roots, stems, leaves, and the buds of flowers. The scientific name of this plant is *Chenepodium album*.*

It is the power of transforming certain crude and lifeless substances of the soil and air into those complex and life-giving substances on which all life depends. It is the power of using the energy of the sun and thereby sustaining the life of the world.

D. Organs and Functions. — The main divisions of the plant body are root, stem, and leaf. These are the *organs* of nutrition. The work of any living body is divided among its parts, and these parts are called organs. The word organ implies that work is done; thus when we call roots, stems, and leaves the organs of nutrition, we mean that the work of nutrition is subdivided among them. The work which an organ does is called its *function*. We

* The scientific names of all plants and animals have two parts. The first part names the *genus* to which the form belongs, while the second part names the particular kind or *species*. Thus *Chenepodium* is the *generic* name of this plant, and *album* is its *specific* name. The plural of genus is *genera*. Species has the same form in both singular and plural. Genera are subdivided into species, while groups of genera form what are called the *families* of plants. When only one name is given for a plant, it is the name of the genus.

are now to consider the structure and principal functions of those organs called roots, stems, and leaves. (See *Figure 1*.)

E. Protection. — The great functions of living things are sometimes spoken of as nutrition, reproduction, and *protection.* Thus, for example, your skin, and what may be called the skin of plants, are organs whose function is largely to protect the more delicate organs which lie beneath. The teeth and claws of animals are organs which they often use to protect themselves and their young as well as to procure food. However, since nutrition and reproduction are properly used to refer to the whole maintenance of life, as well as to mere nourishment and the production of offspring, these words include protection. Protection is thus a subdivision of both nutrition and reproduction.

15. **Roots.** — By roots we generally mean that part of the plant which is related to the soil. Roots branch freely and penetrate the soil in all directions.

Roots of the same kind of plant are much alike if they are growing in soil of the same nature, but if they are growing in soils of different nature, they may be quite different. Thus they tend to extend farther in poor soil than they do in rich soil; they grow deeper in dry soil than they do in moist soil; they seem to be seeking for what they need. They appear to be modified to suit their surroundings. Such modifications are called *responses.* The modified organ appears to have *responded* to the conditions which surround it.

The roots nearest the stem may be large and strong, but the last branches are always soft and slender. These soft and slender branches are usually very numerous. The total length of the root is often much greater than the total length of the stem and all its branches. The total length of all the roots of an ordinary corn plant is several hundred feet.

Think of the delicate root of a young plant as it first begins to burrow into the soil. Pull it up gently, brush off the soil grains carefully, and you find a tip as soft and tender as a baby's finger. Little here that suggests great power to dig! The smooth surface is broken only by a delicate fuzz of fine hairs. They are just behind the tip. Unless you have been very careful, these *root-hairs* have been broken off. (See *Figure 2*.) This delicate root tip has a remarkable power. It has the power to penetrate stiff soil, to burrow deep below the surface. Sometimes even hard rocks are found broken by the gradual work of roots.

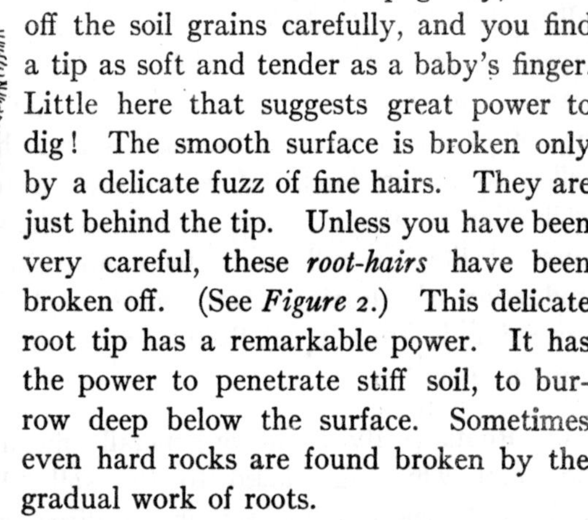

FIG. 2. — A young root tip of corn.

The structure and the position of roots suggest to us the things which they do. These principal functions of roots are *to get from the soil those things necessary to the life of the plant which are found in the soil, and to hold the plant firm.* The entire root system holds the plant firm, but only the young and tender parts do the work of absorption. By means of all its roots the plant is anchored so that its above-ground parts can grow up into the sunlight, and not be upset by the first breeze that comes along. By means of the tender tips of the roots and the hairs which they bear (see *Figure 3*) the plant absorbs materials which are to be used up there in the sunlight in the making of food. The older and tougher parts of the roots cannot do this work for the reason that, as they grew older, their surface became changed in such a way that it no longer admits water.

The amount of water which enters the roots of a plant is surprising. On a warm dry day more than a quart enters the roots of a sunflower of medium size. Think, then, of the thousands of gallons which enter the roots of a forest. All this water moves through the roots to the upper parts of the plant; the roots are the paths to the stem. So, *to anchor the plant, to take in water*, and *to furnish the path to the stem* are three functions of roots.

FIG. 3. — A single root-hair of wheat, showing its close contact with grains of soil.

Another comes to mind when we think of all the edible roots which find their way to our table. Of what advantage is a fleshy root to a radish? The advantage to us is plain enough, but what is the advantage to the radish? If left to itself the radish will use the food stored in its fleshy root to erect a stalk which will bear flowers and fruit. Similarly many other kinds of roots are used for the storage of food for later use. Thus a fourth function of roots is *storage*.

The roots are of use to the plant in still other ways. Like some other parts of the plant, they seem ready as need arises to do other than their regular work. Roots sometimes do work usually done by stems, and stems sometimes do work usually done by roots. In machinery each part has a certain function and can perform no other. But in this matter plants are not like machines at all.

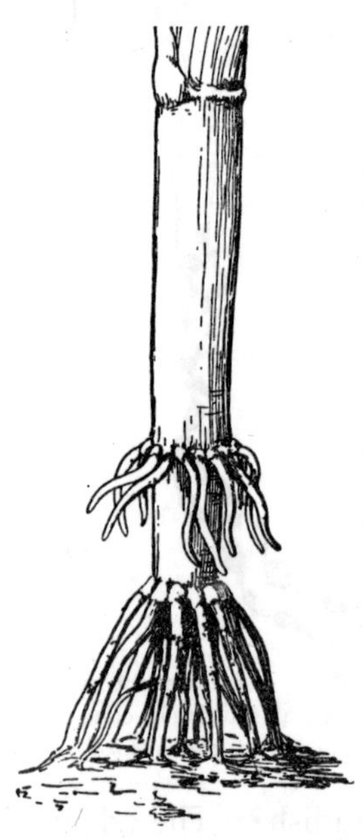

FIG. 4. — The prop roots of corn.

Often we find organs doing work which may not be their usual function. Thus in corn you may have noticed roots which act as little props at the bottom of the stem. (See *Figure 4.*) They are more in the air than they are in the ground. In greenhouses you may have seen those strange plants, the orchids, many of which live without touching the ground. Their roots dangle in the moist air of their tropical homes and absorb moisture directly from it.

The use of underground parts of plants for reproduction may be already familiar to you. The production of plants from *bulbs* is an important part of gardening, onions being an example. *Tubers* may be cut up and the pieces planted to get a new crop, potatoes being an example. But neither onion bulbs nor potato tubers are roots, as you shall see. A tuber is a swollen part of an underground stem. A bulb is an underground stem and leaves, most of it being leaves. They are mentioned here to exclude them rather than to include them, to avoid confusion rather than to hold to the subject very strictly.

A sweet potato, however, is truly a root, and it does give rise to new plants. The dahlia is another example of a plant whose true roots are used for reproduction. The silver poplar, the osage orange, and some kinds of willows give off new plants which rise as branches from their roots. In general, however, true roots are not much used for direct reproduction. Indirectly, of course, they have an important part to play, for without the work of roots no part of the plant could reproduce.

16. **Stems**. — Stems are the great helping organs of the plant. They appear to be more concerned in helping other organs to work than they are in doing a special work of their own. They appear to be chiefly concerned with getting other organs into positions in which these other organs can work to best advantage. Stems determine the positions of the leaves and the flowers. Flowers to accomplish their work successfully require positions of a certain sort just as leaves do. Fruits, as you know, follow the flowers, and for them too, certain positions are much more advantageous than others. The stems serve the fruit in this matter just as they serve the flowers which precede them.

The stems are the great intermediate organs of the plant. Roots must be in the soil. Leaves must be in the light. Both must be separated and yet connected. It is the stem which separates them and yet connects them. It furnishes channels through which move the foods and the materials from which foods are made. Often food is stored up in stems to be used later, perhaps next season, before the leaves have begun their work. Underground stems especially are used by the plant in this way.

Stems tend to grow up as roots tend to grow down. Roots place the root-hairs in positions favorable to the absorption of water. Stems place the leaves in positions favorable to the absorption of light.

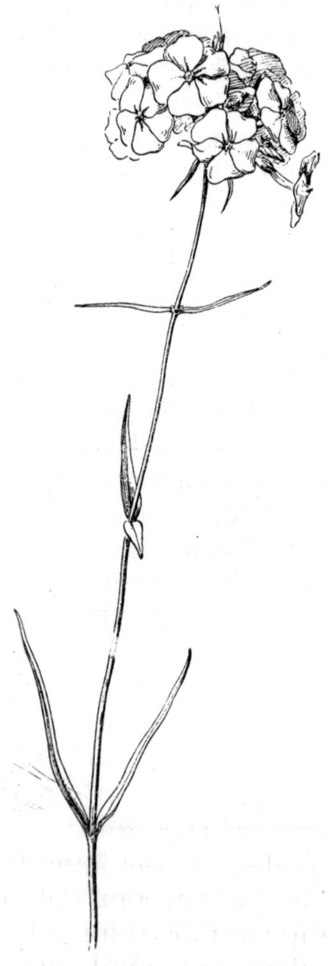

FIG. 5. — Phlox, showing nodes and internodes and a flower cluster.

Stems are jointed. Roots are not. In old stems the joints are often hard to see. They often disappear as the stem enlarges. In young stems, however, they are very plain. These joints are called *nodes*. The lengths of stem between the nodes are called *internodes*. Corn stems show nodes and internodes very plainly. It is at the nodes that leaves and branches regularly arise from stems. (See *Figure 5*.) It is because they possess nodes and internodes that some underground structures, like the potato, are known to be stems and not roots.

In the section on nutrition you learned that green plants make their own food. By green plants we mean those plants which have some green parts. Leaves, of course, are usually the principal green parts of plants. In them food is made. Some stems, however, are also green, and in them too food is made. Perhaps you have noticed how very green some young stems are. Under the thin bark there is sometimes a layer which is quite as green as the leaves themselves. In this green layer of stems photosynthesis occurs just as it does in green leaves. Thus we find that stems sometimes do the ordinary work of leaves. You recall that we found they also sometimes do the ordinary work of roots.

FIG. 6. — The giant cactus. These leafless stems are green and contain a great deal of water.

There are some green plants which have no food-making leaves at all, and such plants perform all of their photosynthesis in the stems. Cactuses are the best known plants of this kind. Perhaps you have seen them under cultivation. They are commonly found in conservatories where they are cultivated on account of their strange forms and on account of the beautiful flowers which they produce at long intervals. It is in deserts that cactuses are found growing wild in greatest abundance. Their fleshy, water-containing bodies permit them to live where other plants would perish. (See *Figure 6*.)

Have you ever noticed an asparagus bed in the summer? The old asparagus plants have numerous slender branches which are green and do the work usually done by leaves. The true leaves of asparagus are simply little brown scales.

Uses of Stems to Man. — Apart from their value as being necessary to the plant as a whole, the woody stems of plants, especially the trunks of trees, are of great value to man. Probably their first value to him was in affording refuge. Fortunately for our remote ancestors, they were better climbers than many of their enemies.

The first rude huts were probably built of plant stems and, even in this age of steel and concrete, the wood of plant stems remains the most important material in the building of houses. The furnishings of the house, as well as most of the houses themselves, are made of it. You are to learn of the method of nature in producing wood and the principles which underlie proper care of the wood supply. You have probably heard of *forestry*. Forestry is the art of proper care and cultivation of the forests; it is important because the stems of trees are important. The aim of forestry is to secure as much timber as possible without injuring future timber production. Forestry is based upon the science of plants. The government employs many foresters, all of whom must have a thorough knowledge of the principles of plant life.

Stems contribute to man's food and clothing as well as to his shelter. Cane sugar comes from the stems of sugar cane; maple sugar from the stems of maple trees. The stems of asparagus, cauliflower, and cabbage are eaten; of celery we eat the stem-like part of the leaf. The potato, as you know, is part of an underground stem. Corn and grass stems are fed to stock. As to the contribution of stems to clothing, linen is the most important item. Linen is principally composed of fibers derived from the stems of a plant called flax.

Rubber is another important product of stems. It is derived from the milky juice of certain tropical trees. This juice in appearance is like the juice of the common milkweed.

17. Leaves. — Often you have had a leaf in your hand. You have noticed the stem-like part. That is the *petiole.* You have noticed the broad green part. That is the *blade.* You have noticed that the blade is supported by a sort of fine framework. That framework is composed of what are called the *veins.* The largest vein, if there is a largest, is the *midrib.* (See *Figure 7.*) Extensions of the veins run from the leaf down through the stem of the plant clear to the ends of the roots. Inside the stem and the root these extensions are known by other names than veins, but they have the same structure. The veins and their extensions form the paths along which movement inside the plant principally occurs. On the

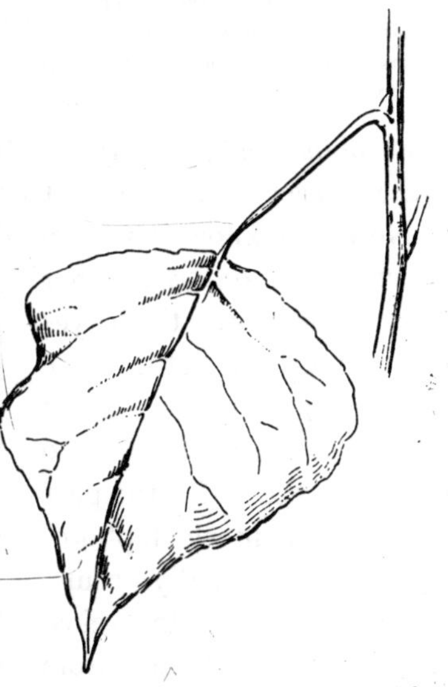

FIG. 7. — Leaf of poplar showing petiole, blade, veins, and midrib. Also a bud in the axil.

under surface of the leaf are many *stomates.* (The Greek word *stoma* means *mouth.*) A stomate is an opening in the skin of a leaf. It is so small that it requires a microscope to see it; there may be thousands of stomates on a single leaf. (See *Figure 8.*) Gases pass in and out through the stomates. The upper angle between the base of the petiole and the stem is the *axil.* If there is no petiole, the upper angle between the base of the blade and the stem is the axil. It is in the axils of leaves that buds regularly arise. Buds develop into branches, either flower bearing or leaf bearing or both.

FIG. 8. — A little of the under surface of the leaf of *Coleus,* very highly magnified. Two stomates appear.

Now you know very well that more light falls on the leaves than on any other part of the plant; the leaves seem to seek the light. (See *Figures 9* and *10.*) So it does not surprise you to find that the leaf is the part of the plant in which photosynthesis principally occurs. Up to it, through the veins, come those materials from the soil which it uses in this work. Into it, through the stomates, come those substances of the air which are necessary. Down to it,

FIG. 9. — A nasturtium with light coming to it from above.

from the sun, comes streaming the light which gives it power to do this work. Out from it, down the veins and into the stem, passes the stream of manufactured food, descending and moving slowly to other parts of the plant.

Fig. 10. — The same plant as in *Figure 9*. For six hours the light has come to it from one side.

A green leaf is a perfect symbol of the service of nature to man; a symbol of the last great change in the world before he appeared on it; a symbol of the final adjustment of nature's forces which made his existence possible. It is more than a symbol. It is the thing itself.

The substances which the leaf uses in photosynthesis are inorganic substances. All substances are either *organic* or *inorganic*. *Organic substances are substances which are of the bodies of living things or have been derived from the bodies of living things. All other substances are inorganic.* A shell is organic. Though never itself alive, it was once a part of a living thing. Pure clay and pure sand are inorganic, but dark soil owes its darkness to the organic substances which have resulted from the decay of plants and which are mingled with sand or clay.

The green leaf is nature's one great link between the inorganic and the organic. Out of inorganic things, parts of the world since it began and parts of other worlds than ours, green parts of plants, and they alone, can build up those organic things which make life possible and presently become the substance of life itself. Incombustible substances from which no energy can be extracted are transformed by leaves into combustible substances, stored richly with that energy which the leaf has caught from the light. All food, as we have defined it, is organic, and photosynthesis is the one great process of nature by which organic substances are made out of inorganic ones.

18. **Reproduction.** — Under reproduction we may include all those processes which have to do with the maintenance of the life of the race. This would include the care of the young. In man it might be taken to include the education of the young, and all those other things which are designed to make the next succeeding generation better than the present one.

No individual, plant or animal, can avoid responsibility for reproduction. Nature makes reproduction the highest duty of the individual; in fact she seems to design the individual life as a means to that end even more than as an end in itself. When we think of the life of the race, individuals seem almost like insignificant beads upon a precious golden chain. The golden chain is the unbroken thread of life which passes through the individuals and on to other generations. For a time this precious thread is in the keeping of the individual; his highest duty to nature is to pass it on untarnished.

As we grow older, we realize more and more the small importance of our own life as compared with the life of the race. We find that even to ourselves we are not so important as the good we may be able to do is important. We find that even to ourselves we seem valuable only as we contribute in some way to the value of others. We find that within ourselves there is something which forever keeps us from being happy unless we are conscious of sincere effort to help others. We find that the happiest lives are those which are most devoted to the good of other lives. All this seems to be connected in some way with the great law of nature that the individual shall be devoted to the good of the race; it seems a part of nature's provision for reproduction. This truth that the highest fulfillment of the individual is in what it does for the race is evident in plant life as well as in human life.

Plants reproduce *asexually* and *sexually*. Asexual reproduction is accomplished when a portion of the body of one individual produces another individual. Sexual reproduction is accomplished when, portions from the bodies of two individuals having united, a new individual is produced as a result of this union.

Examples of asexual reproduction are familiar. Every one who has helped make a garden is familiar with onion sets and bulbs, with "seed" potatoes which are not seed, but are the potatoes themselves, and with the cuttings whereby grapevines and rose and berry bushes are propagated. Even to those of you who live in large cities the bulbs from which hyacinths and Chinese lilies grow are quite familiar. New geranium plants are made by planting slips, which are simply short pieces of stem bearing a bud or two. The bright tulip beds of the parks are made by planting bulbs late the fall before. All these are asexual methods of plant reproduction.

As to sex reproduction, the results of it are familiar in seed plants though the process itself may not be. The seeds themselves are results of a sex process. They contain a structure which develops into the new plant. That structure is the *embryo*. It is produced as a result of the union of portions from the bodies of separate parents. The explanation of this process, and of the structures concerned with it, will have to wait until after you have studied the flower and its parts. It will then be much easier to understand. It is in the center of the flower that this process occurs, and the portions which unite are much too small to be seen with the naked eye.

Flower, fruit, and seed all have to do with sex reproduction. Though no one of them is itself a sex organ, they are all devoted to helping in one way or another in the fulfillment of reproduction by this method.

19. Flowers. — Stems and roots and leaves we take for granted. There is not much about them to make us exclaim; not much that we see in them surprises us. But it is not so with flowers. We exclaim at their beauty. Their form and fragrance and color constantly surprise and delight us. Even when we were little children, flowers delighted us. As we grow older we seem to enjoy them more and more.

When a child, your first consciousness of plants was a consciousness of the bright colors of flowers, was it not? You were conscious of flowers first, and of the plants which bear them afterwards, were you not? The other parts of plants gratify primarily our physical needs, but flowers gratify a love of beauty which savages and children have as well as educated people. We are impelled to learn more of them by our enjoyment of their beauty, as well as by our desire to know more about an organ upon which the production of new plants depends.

You may have heard it said that people who pick flowers apart to study them do not enjoy them so much as people who do not. You may have heard that to know the names of the parts of flowers and what they do will not add to the pleasure you get from them. Do you believe it? Here is a structure which arouses our admiration and interest as no other part of the plant does. When we journey about we shall find flowers, new and strange to us, which will excite new admiration and interest. Now shall we say: "Here is a wonderful and beautiful thing which interests me very much. I know that it has something important to do with the reproduction of plants, and I know that my life depends on the reproduction of plants. But I do not care to know anything more about it. I can enjoy it more if I do not understand it." The truth is, however, that your enjoyment of flowers will be very greatly increased by learning as much as you can about them. Their beauty, which pleases you now, will please you far more when you understand their usefulness, and the various forms of their structures.

The Structure of Flowers. — The first step toward understanding flowers is to understand what are the parts of which they are usually composed. Look at *Figure 11.* It shows three flowers of syringa, a shrub whose white, sweet-scented blossoms are seen in June on many lawns and in the parks. The white, leaf-like parts (marked by *p* in the picture) are *petals;* taken together they form what is called the *corolla.* The uppermost flower in the picture is the oldest; its petals have dropped off. Under the petals are five green parts (*s*). These are *sepals;* taken together they form the *calyx.* It is the calyx which is the covering of the buds. The flowers of syringa show plainly the inner parts. Of these there are two kinds. The numerous slender structures with knobs at the tops (*st*) are *stamens*. The knobs themselves are *anthers*, and they produce a yellow powder called *pollen*.

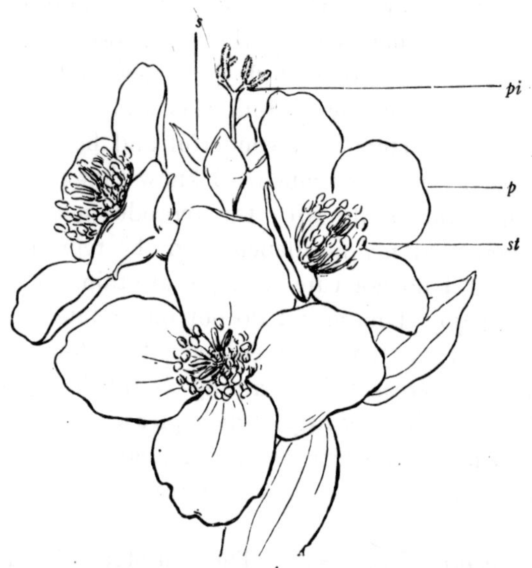

FIG. 11.—Flower bearing twig of syringa or mock-orange (*Philadelphus coronarius*). *p*, petals. *st*, stamens. *s*, sepal. *pi*, pistil of old flower whose petals and stamens have fallen off; note the four branches of the stigma.

The central structure (*pi*) with four branches is the *pistil*. The enlarged base of the pistil is, in syringa, covered by the bottom part of the calyx. This hidden base of the pistil is the *ovary*. It contains the structures which later develop into seeds. The production of seeds is a complex process. One of the things necessary in this process is that pollen should reach the top of the pistil. We find that the topmost part of pistils is a part well suited to catch and hold pollen. This part is called the *stigma.* In syringa, as you can see, the stigma consists of four branches. That part of the pistil which connects the stigma with the ovary is the *style.*

Many flowers do not have all the parts which have just been described, but in most of the flowers with which you are familiar you will find them. In order to understand descriptions of flowers and their functions it is quite necessary for you to understand the meaning of the terms to which you have just been introduced.

There is a great variety of flower forms, an even greater variety than there is in the forms of the plants which bear them. Notice *Figures 12* and *13* and think of the differences in the forms of flowers with which you are familiar. Yet, just as it is with plants as a whole, so with flowers — they all have the same task to perform. The devices are extremely various, but the purpose to be accomplished is the same. Roots, stems, and leaves, as you have seen, secure certain supplies and use them, and so maintain the life of the individual. The flower produces seed and so

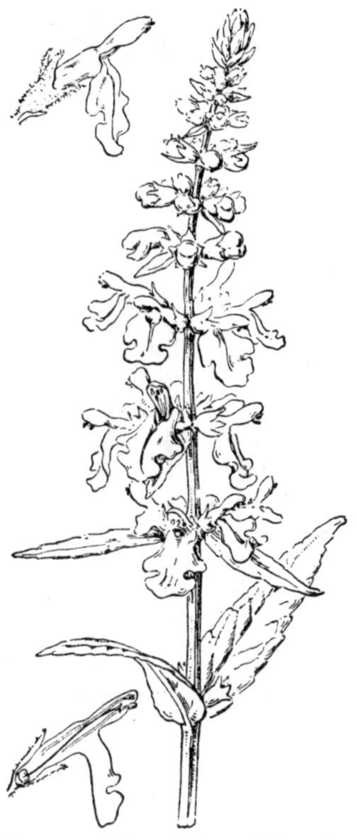

FIG. 12. — Flowers of one of the mint family (*Stachys*).

helps maintain the life of the race. Roots, stems, and leaves are organs of nutrition; the flower is an organ of reproduction. Leaves are principally for photosynthesis. Flowers are exclusively for seed production. The way in which they produce seed is described in a later chapter.

FIG. 13.—"Brown-eyed Susan" (*Rudbeckia*). The part commonly called the "flower" of this plant is composed of many small flowers such as are shown, in three stages, at the left.

20. Fruits.—At the grocer's we buy fruits and vegetables. The grocer has no difficulty in telling a fruit from a vegetable, but he would have great difficulty in telling what fruits and vegetables are. Try it yourself to see. These two terms are convenient, but not exact. An exact term stands for a definite idea, and only exact terms can be accurately defined. You can make lists of fruits and vegetables, but a list is not a definition.

Exact terms may not be necessary in the grocery business, but they are very necessary in science. So in science the word fruit is used to denote *that structure, whatever its appearance, which follows the flower and contains the seeds.* Thus a tomato, though a vegetable to a grocer, is a fruit to a botanist.

Fruits, beside being seed containers, are often devices for securing the distribution of seeds, a process which is called *dissemination* or *seed dispersal.*

Two questions naturally arise: (1) What are the advantages in having the seeds widely scattered? (2) What are the ways in which fruits aid in securing this scattering? The answer to the first your own thought will furnish. In its effort to maintain the race it is not enough that each plant produce one other. No kind of plant or animal would survive long at that rate. Life is too uncertain. For one seed that produces a new plant, hundreds perish. To make sure of a new generation as numerous as its predecessor, each plant must produce many seeds. So each plant appears to be seeking, not only to perpetuate itself, but to populate the world. Its seeds, traveling far from the parent, may find lodgment where none of its own kind compete with it, and, under these easier conditions, it may establish a vigorous new colony.

FIG. 14. — Winged fruit of the maple.

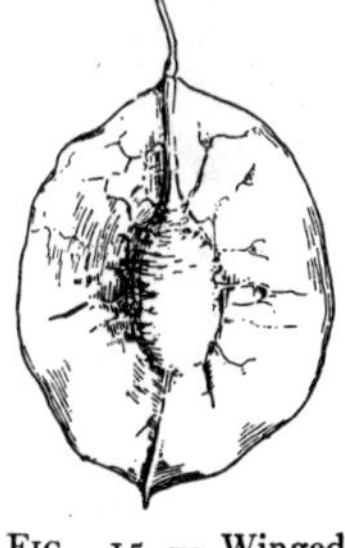

FIG. 15. — Winged fruit of the hop tree.

In answer to the second question, you may remember devices you have seen which secure the scattering of seeds. Perhaps you have noticed the winged fruits of the maple or of the elm. (See *Figures 14* and *15.*) Think of the wind-blown fruit of the

FIG. 16. — Fruit of the cocklebur.

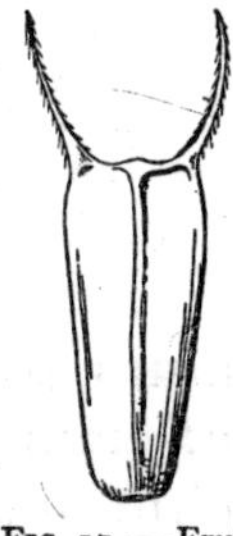

FIG. 17. — Fruit of the sticktight.

dandelion (see *Figure 19*), or of the sticktights you bring home on your clothes from an autumn ramble. (See *Figures 16* and *17*.) Think of the burs in a cow's switching tail, or of the seeds in a bird's crop which pass uninjured from its body. (See *Figure 18*.) Think of buoyant fruits that float over leagues of ocean uninjured, or of the mud carried on the feet of wading birds; mud of the kind from which Darwin got five hundred and thirty-seven seedlings out of three tablespoonfuls. Thus we see some reasons why buoyancy, edibility, and the clinging power of fruits are of advantage to the kinds of plants which produce them.

FIG. 18.—Fruit of the mulberry. The seeds of such fruits pass uninjured through the alimentary tract of birds.

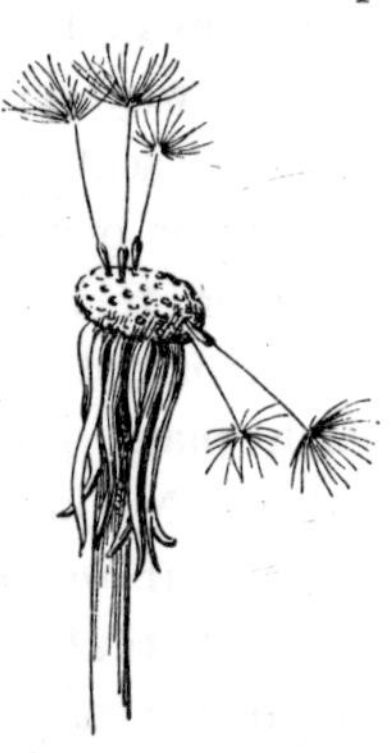

FIG. 19. — The plumed fruits of dandelion.

21. Seeds. — Sometimes the seed itself bears the device which secures a journey before germination. You may recall the beautiful plumes of the milkweed seed by which it floats gracefully away when the pod bursts in autumn, or in early spring. (See *Figure 20*.) It is from the soft and abundant fibers that grow from the seeds of cotton that the cotton of commerce is spun.

But the main business of the seed is not dissemination. That is only an important aid. The main business of the seed is to protect and nourish that structure within it which develops into the new plant. That structure, as you have learned, is called the embryo. The embryo begins with that joining together of portions of its parents which occurs in the very heart of the flower, hidden from the naked eye. After that joining together, the seed begins to develop around the embryo, and the fruit begins to develop around the seeds. A supply of food is deposited around the embryo ready to be used up by it as it grows. Even while it remains in the seed, and even while the seeds remain in the fruit, and even while the fruit remains in the flower, the embryo grows. After it has reached a certain size, it may stop growing for a long time. You know that seeds may be kept a long time and still be good for sprouting. But when the seed does begin to sprout, there is the little plant within it, quite large enough to be seen with the naked eye, and ready to develop rapidly the parts which will enable it to make its own food and begin an independent existence. (See *Figure 21.*)

FIG. 20. — A pod of milkweed with the winged seeds escaping. — *After Hunter.*

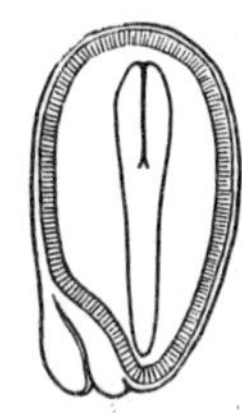

FIG. 21.—Diagram of a seed of violet. The embryo lies in the center of its food supply.

Open a seed to see. A bean or a peanut kernel will do. Within the protecting coats you find an oval body which separates easily lengthwise into halves. These halves are parts of the embryo itself. In this kind of seed the embryo has grown to fill the whole of the inside. But these fat halves, filled with food, are not that part of the embryo which develops into the new plant. These fat halves are *cotyledons*, and at the end of one of them, on its inner face, you will find a tiny structure, slightly bent. This tiny structure is the part of the embryo which will develop into the new plant. As it grows, it will draw into itself the food which is in the cotyledons, and they will gradually wither and die. (See *Figure 22.*)

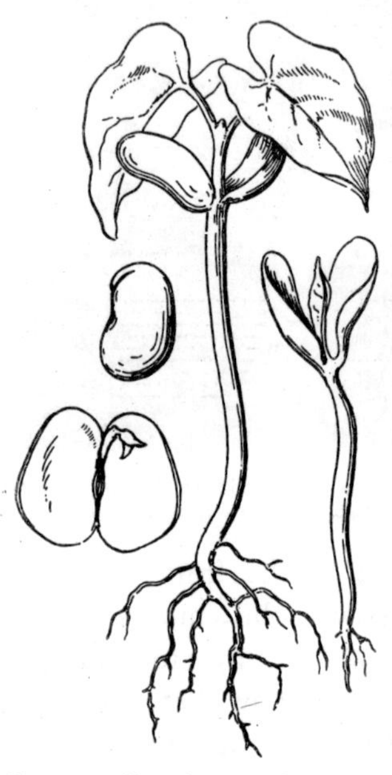

FIG. 22.—Two views of a bean, and two young stages of the plant which grows from it.

But not all seeds are formed on the same plan as the bean or the peanut. As you find variety in flower forms, so you find variety in seed forms. Thus in corn you find that the embryo does not fill the whole inside of the seed. It occupies only a small part at one end. The rest is filled with food, but the food part of this kind of seed is called *endosperm*, not cotyledons. The embryo in seeds like corn has but one cotyledon instead of two, and it is but a small structure, not enlarged and swollen with the food supply. (See *Figure 23.*)

But, whatever the structure, the end to be secured by the seed is always the same. Seeds always contain a supply of food stored for the use of the little plant when it begins to grow again. And this food is also the principal source of food for mankind. It is for the food in their seeds that men cultivate rice and wheat, corn and oats, barley and rye. So in seeds we have not only the means to create crops of value; we have also the most valuable crop itself.

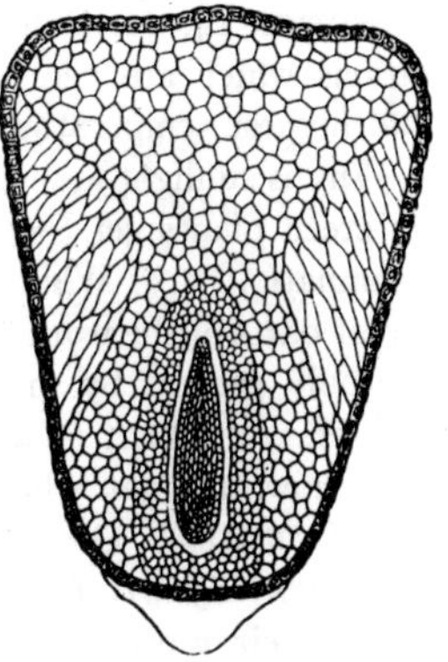

FIG. 23. — Diagram of a section taken through a grain of corn near one of its broad surfaces. The oval structure in the lower part is a slice through the embryo.

QUESTIONS AND SUGGESTIONS

SECTION 12. 1. State your idea of what a seed is. 2. Of what uses are seeds to plants? 3. Name the kinds of seeds you can identify when you see them. 4. Name seeds which are used for food. 5. Do you know any uses of seeds to man except for food and planting? If so, describe them. 6. What plants do you know, other than those named in the text, which do not produce seed? 7. In what month do most seeds ripen? 8. Where you live when does corn ripen? When does wheat ripen? 9. What are some of the crops whose seeds are not allowed to ripen? 10. Does the size of a seed seem to bear any relation to the size of the plant which produces it? 11. Find out what you can about how seeds are stored in order to keep them in good condition. 12. How is seed corn stored? How is it tested before planting? 13. Will young seeds grow as well as seeds which have been kept over winter?

SECTION 13. 1. What are the things which you and a plant must both do in order to keep alive? 2. Describe a life relationship of plants which is not found among animals. 3. Does grass grow where the shade is dense? Describe the appearance of grass after it has been covered for a few days. 4. Describe the shape of a plant which is kept indoors with the same side constantly toward the window. 5. What are the principal things which plants seem to try to do?

6. What are the two great tasks of all living things? Describe them. 7. Do you know any structures, other than seeds, by means of which plants are reproduced? If so, describe them.

SECTION 14. 1. Define nutrition. 2. What are the external conditions which are necessary to your life? 3. What are the lowest and highest temperatures we can endure without great discomfort? 4. Why is sunlight necessary to your life? 5. What is food? 6. Where do green plants get their food? 7. What is digestion? 8. What is assimilation? 9. What is photosynthesis? 10. What do the different parts of the word photosynthesis mean? 11. Is photosynthesis the only process by means of which food is made? Explain your answer. 12. What is an organ? 13. What is a function? 14. Explain how an organ which protects may be properly considered an organ of nutrition. Give an example. 15. Define species, genus, family.

SECTION 15. 1. Describe the way roots grow. 2. What things affect their growth? 3. Compare them with stems. 4. What is meant by response? 5. Compare as to length the above-ground and underground parts of a plant. 6. Describe the appearance of a young root. 7. What are the principal functions of roots? 8. What parts of roots do the work of absorption? 9. Why cannot all parts of the roots absorb? 10. Give some idea of the amount of water absorbed by the roots. 11. Of what advantage is it to the plant to use roots for storage? 12. What advantages have roots over stems as places of storage? 13. Name some plants other than radish which have fleshy roots. 14. Are the organs of living things limited to certain functions? Illustrate your answer with examples of roots. 15. Give and describe examples of the use of underground parts of plants in reproduction. 16. Define tuber and give an example. 17. Define bulb and give an example. 18. Are there any roots which give rise directly to new plants while continuing their ordinary functions? Give examples.

SECTION 16. 1. Compare stems with roots as to the way in which they grow, and as to the work which they do. 2. What are nodes? 3. What are internodes? 4. What stems do you know, besides that of corn, which show nodes and internodes plainly? 5. Describe leaf work done by stems. Give examples. 6. What sort of positions do you consider of advantage to fruit? Why? 7. Describe uses of stems to man. 8. What is forestry?

SECTION 17. 1. Describe the structure of an ordinary leaf. 2. What are the largest leaves you have ever noticed? The smallest? 3. Do leaves appear before the flowers in spring or the flowers before the leaves? 4. Do leaves fall off before the fruit ripens or afterwards? 5. What plants have you noticed that keep their leaves on all winter? 6. Have you noticed any plants whose leaves turn brown in winter but do not fall off? 7. Define petiole, blade, vein, midrib, axil, stomate. 8. Give an example of a kind of leaf which has no petiole? 9. Have you noticed any striking difference in leaves as to the way in which their veins are arranged? Describe it. 10. Observe the veins in a blade of grass by holding it between you and the light. How are they arranged? How are the veins of a maple leaf arranged? 11. What are some of the plants with which you are familiar whose leaves are used by man for food? For other purposes? 12. What plants and what parts of them are principally used as food for cattle and horses? 13. What tropical plants do you know whose leaves are used by us for food or for other purposes? 14. Find from an encyclopedia or otherwise from what parts of plants the following plant products are derived: sugar, coffee, quinine, opium, chocolate, manila hemp, olive oil, licorice, peppermint, vanilla, pepper, cinnamon, cloves, cotton, linen, rubber. 15. What things go into living leaves and what things come out of them? 16. Explain the difference between organic and inorganic substances. Give examples. 17. What is meant by the statement that the green parts of plants form "the link between the inorganic and the organic"?

SECTION 18. 1. Is the highest "fulfillment" of the individual in what it does for its own benefit or in what it does for the benefit of the race to which it belongs? Explain. 2. A sense of physical comfort results from satisfying hunger. Does any similar sense of comfort result from aiding others than ourselves? Have you ever experienced it? Have you ever felt a sort of "hunger" to aid others? 3. What is meant by asexual reproduction? Give examples of it other than those given in the book. 4. By what process is the embryo of seeds formed? 5. What are some of the organs of plants that have to do with sex reproduction?

SECTION 19. 1. Write a list of the cultivated plants which you know by their flowers. 2. Write a list of the wild plants which you know by their flowers. 3. Define petal, corolla, sepal, calyx, stamen, anther, pollen, pistil, ovary, stigma, style. 4. Do all flowers have the same function? If so, what is it? 5. Do you know any kind of plant which produces more than one kind of flower? If so, describe it. 6. Have you ever noticed flowers on trees? If so, describe some kind you have noticed. 7. Does corn have flowers? Wheat? Grass? 8. Why do insects visit flowers? 9. Are the visits of insects to flowers of any advantage to the plants? 10. Are the bright colors of flowers of use to the plant? Is the fragrance of the flowers of use to the plant? 11. Do the commonest plants have bright and fragrant flowers? 12. At what season are wild flowers most abundant? 13. Are wild flowers in the fall found most abundantly in the places in which they are found most abundantly in the spring, or in different places?

SECTION 20. 1. What is the botanical definition of the word fruit? 2. Is it of advantage to a plant to produce a fruit which is eaten by animals? 3. Write a list of edible fruits produced in temperate regions. 4. Write a list of edible fruits produced in the tropics. 5. What is meant by dissemination? Of what advantage is it to a plant? 6. Give examples of fruits which are scattered before they open and set free the seeds. 7. Give examples of fruits which open before the seeds are scattered. 8. What are the first wild fruits which you have noticed in spring? 9. What plants have you noticed whose fruit hangs on through the winter? 10. Describe various ways in which animals aid in scattering fruits.

SECTION 21. 1. What is the chief function of seeds? 2. What are the three principal parts of a seed? 3. Define cotyledons and endosperm. 4. What plants have you noticed which produce seeds in great numbers? 5. Do you think it better for a plant to produce a smaller number of large seeds or a larger number of small seeds? 6. When we speak of the "most successful" plants we mean those plants which, without the aid of man, are the most common. That is, it is by abundance that we judge a plant's "success" in a general way. Do the most successful plants that you have noticed produce great numbers of small seeds or small numbers of comparatively large seeds? What other properties besides size have you noticed to be characteristic of the seeds of the most successful plants? 7. Make a list of seeds which are used by man for food and for other purposes. 8. What conditions are necessary to cause a seed to sprout?

CHAPTER II

THE PLANT: A GENERAL INTERNAL VIEW

22. **The Inside of Plants.** — A boy in the country broke the stem of a milkweed. Following a country boy's tradition, he rubbed a wart with the thick, sticky juice. He looked at that slowly oozing, milk-white fluid and wondered if it was the blood of the plant. He wondered what was going on in that broken stem. He wondered if it hurt a stem to break it. He wondered what was going on inside the trunks and roots of the trees at the roadside, or in the leaves rustling over his head. Though no one had told him, he knew that these things were alive, like himself or the turtle in his pocket. Their growth proved that. He knew that parts of plants are good to eat. He had proved that by eating them. He knew that many plants possess powers of healing. Was not the milkweed to cure his wart? So, just now, it occurred to him that very interesting things were probably proceeding inside of plants, and that important things were to be discovered if one only had eyes like microscopes. He wished that he had them.

Just now such eyes would be useful to you too, for the inside of plants is your next lesson. The boy was right about important secrets to be discovered there. But because we can discover them only through a microscope, they remain secrets. Their importance is not generally appreciated.

23. Cells and Protoplasm. — Suppose we were to take the stem of some common weed, cut a very thin cross slice of it with a razor, and then examine it under a microscope. It would look a good deal like *Figure 24.* We would see a great number of very small compartments of different shapes and sizes. These little compartments and what they contain are called *cells.*

But this discovery that plants are composed of cells is only the first step toward learning about the life which goes on inside of these cells. What you see in *Figure 24* might be called the framework or skeleton of plant life. This framework is not itself alive. That which is alive is a colorless, transparent substance inside the cells. This living substance is called *protoplasm.* You will find protoplasm to be an exceedingly important word in this or in any other study of living things.

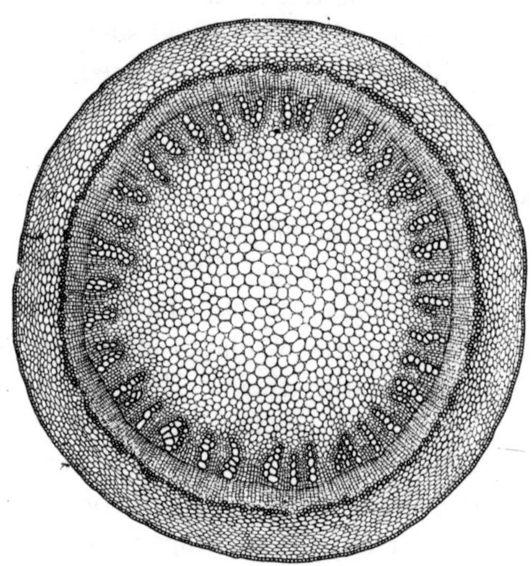

FIG. 24. — Cross section of the stem of the castor-oil plant. This is not a woody plant. It dies in winter.

Huxley called protoplasm *the physical basis of life.* No better definition has been given to it. Just what it is no one knows. The most striking thing about it appears to be that it is constantly changing. But, as long as it is alive, whatever its nature may be, it is called protoplasm. Whether in plants or in animals, whether active, as in a growing leaf, or dormant, as in the embryo of a seed, whether in a loose mass on decaying wood or at work in the brain of the wisest man, this life stuff is called protoplasm. If we knew its secrets, we should know the secrets of physical life. All other parts of living bodies serve it; it manufactures them and uses them for its various purposes. The cell walls which you see in *Figure 24* were all manufactured by protoplasm within them.

We can understand cells and protoplasm best by considering a single cell and its contents. Look at *Figure 25*. It shows the principal parts of a living cell. That portion of protoplasm which is contained in a single cell is called a *protoplast*. You may think of protoplasts as the *units of function* of plants. It is by their action together that everything which a plant does is accomplished. The protoplast is the essential part of a living cell, but it is not easily seen. The other parts, which it manufactures, are more easily seen. They are usually opaque and have definite form, while the protoplast is transparent and does not have definite form.

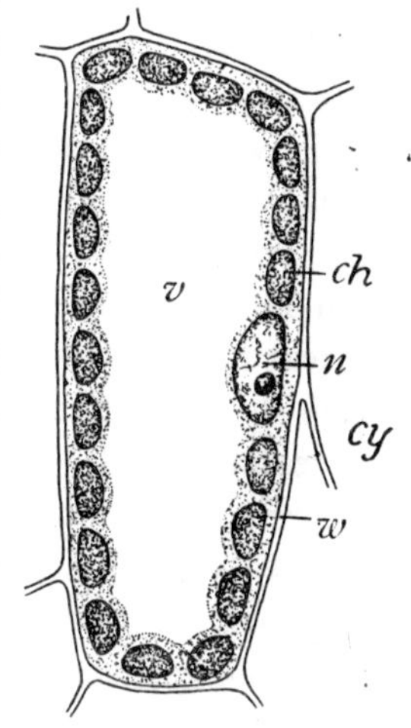

FIG. 25. — Diagram of a cell from the inside of a leaf; *cy*, cytoplasm; *n*, nucleus; *w*, wall; *ch*, chloroplast; *v*, vacuole.

Protoplasm has the consistency of very thin jelly. In studying its structure stains are used. The dots which you see inside the wall in *Figure 25* crudely represent the appearance of protoplasm and its contents after they have been stained. Protoplasm has more fluid and less fluid parts. The more fluid part is called *cytoplasm.* The cytoplasm streams slowly about in active cells. Through exceedingly small spaces in the wall it connects one protoplast with its neighbors. A denser part of the protoplasm which has rather definite shape is called the *nucleus.* The nucleus is indicated by *n* in *Figure 25.* So far as we know, every living cell has a nucleus. The cytoplasm appears to be concerned with the nutritive work of the cell, and the nucleus with the reproductive work of the cell. Cells reproduce by division, but the power to divide is not possessed by all cells.

Nucleus and cytoplasm are the only things which occur in all living cells, but there are other things which occur in many living cells. Some of these other things are of great importance to plant life. *Chloroplasts,* for example, occur in the green cells of leaves and stems, and they are of great importance to plant life. They are the structures which do the actual work of transforming inorganic substances into food. That is, they are the organs of photosynthesis. The numerous chloroplasts in *Figure 25* are indicated as *ch.* (See also *Figure 26,* in which the chloroplasts are scattered through the cell.) The chloroplasts are principally composed of protoplasm. In density they are similar to the nucleus. They contain *chlorophyll,* a substance manufactured by the protoplasm. (*Chloro* means green, *phyll* means leaf.) It is the chlorophyll which gives the green color to the chloroplast; without chlorophyll photosynthesis does not occur.

Another important thing usually present in cells is the *vacuole.* (Indicated by *v* in *Figure 25.*) There may be one vacuole or many. (See *Figure 26.*) Vacuoles are spaces filled with *sap.* In young cells the protoplast fills the interior, but as the cells enlarge much of the interior comes to be filled with sap. Sap is the most fluid part of plants; it is simply water with various substances dissolved in it. You may be familiar with the sap of maple trees whose sweetness is due to the sugar dissolved in it.

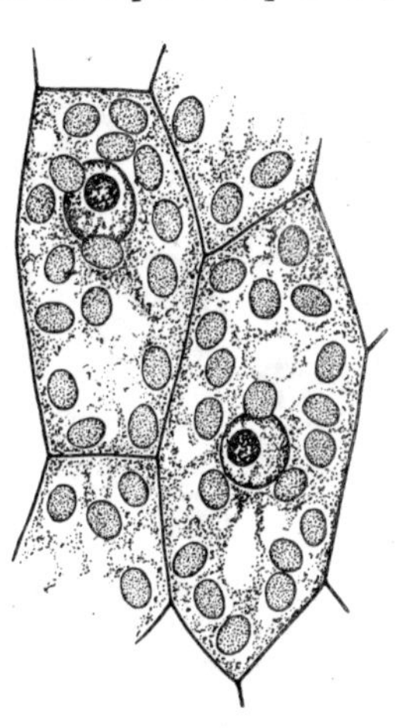

FIG. 26. — A few cells from a leaf of moss, showing nuclei, chloroplasts, cytoplasm, and scattered vacuoles.

Sap enters cells more freely than it passes out of them. The protoplasm appears to encourage its entrance and to discourage its exit. It enters living cells until there comes to be a pressure outwards, like the pressure of gas against the inner walls of a balloon. The walls of active cells are elastic and they tend to curve out as the sap presses against them. The cells press upon each other, thereby helping make the whole plant body rigid. This swollen state of the cells is called *turgidity*. (See *Figure 27.*) The cells of a fresh leaf are turgid. The cells of a wilted leaf have lost their turgidity. Water has evaporated from them more rapidly than it was replaced.

Plant cells when mature have many different forms and functions, but when young they are all much alike. They become different as they grow. This process of becoming different is called *differentiation.* Mature cells may be very dissimilar in appearance, as the pictures show. Yet there are certain life processes which occur in all living plant cells. When you understand what is known of these life processes in one plant cell, you understand what is known of them in all plant cells. Thus the living cells are the plant's stomach, its lungs, and its heart, in so far as a plant has stomach, and lungs, and heart. In truth, of course, the plant has no such organs, yet what those organs do for you must also be done in some manner by a plant. Though the plant does not have the same organs that animals have, it does have the same general functions to accomplish. Digestion and respiration are examples of these functions. You have learned of them in your study of physiology. In plants these functions are accomplished by each cell for itself.

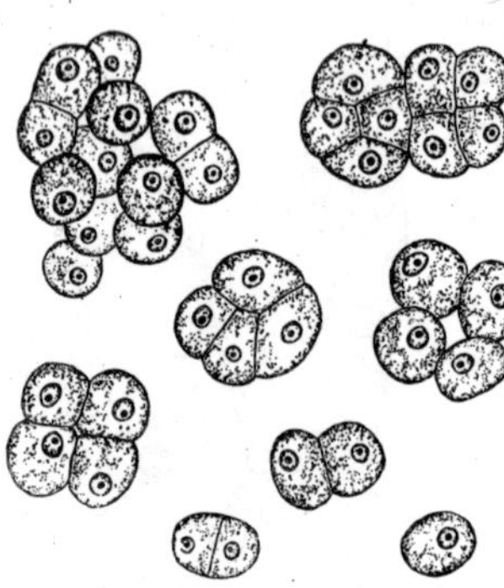

FIG. 27. — Turgid cells of a very simple, one-celled plant as seen under a microscope.

All this about the internal structure of plants, about cells and protoplasm and the work they do, may not interest you, for it is about things with which you are not familiar and cannot see with your naked eyes. In order to make it interesting, you need to use your imagination. There is need for imagination in connection with the study of science as well as in the study of anything else. For the sake of this lesson it is perhaps a pity that a plant cell is not as big as a tree and as common as a tree in your experience. In importance it is certainly as big as a tree, and you cannot begin to understand a tree until you begin to understand the cells which compose it. Your mind can readily picture a tree. Similarly, let it picture a plant cell, not as it appears in a picture, but as it exists in a plant, one of millions working together and completing the life of the plant. You will remember that this life is of great importance to you. Just as your own life goes on unceasingly in your body, you will think of the life of a plant going on unceasingly in those millions of busy cells which are hidden within it. It is a life which words and pictures and microscopes cannot portray. They can only reveal certain dead evidences of it. The thing itself you must imagine. The cells are the seat of real life like your own. You must examine them dead, but you must think of them alive.

24. **Tissues.** — A tissue is a group of similar cells. There are scores of different kinds. *Epidermis* is the name of one. It is the tissue which is on the outside. *Wood* and *pith* are other kinds. The milk of the milkweed stem is contained in the cells of a certain tissue which manufactures it. When this tissue is cut, the milk flows as blood flows from a cut blood vessel. There is one tissue through which the water absorbed by the roots ascends to the leaves, another by which the food manufactured in the leaves goes to other parts of the plant. There are tissues which strengthen, like the tissue which strengthens the wheat stalk. There are tissues which protect, like the outer tissues of buds. There are tissues in which food is stored, as in the potato. There are tissues in which food is made, and these tissues are green, as in leaves. Thus the plant body has its work subdivided, but all the tissues coöperate, each making some contribution to the two great ends of nutrition and reproduction.

The pictures (see *Figures 24, 29, 30*) show the appearance

of thin slices of tissues, stained, and magnified by the microscope. But they can give you little idea of the life that goes on in these tissues. Nor can words give you this idea very well. Life cannot be explained in any such limited medium as language. Your own protoplasm has powers which we cannot explain. So has a plant's.

25. **From Root to Leaf.** — A plant stands at work. Its roots are in the rich moist earth. Its leaves are bathed by air and sunlight. Its cells are filled with the complex, ceaseless workings of protoplasm. All its energies are engaged in the great tasks of life. Always it strives to sustain itself and to multiply its kind.

The tender root-hairs cling closely to the soil grains. Around each grain there is a film of water. (See *Figure 28.*)

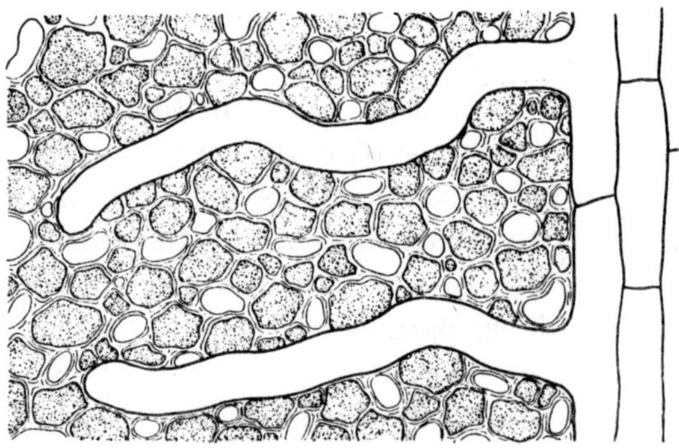

FIG. 28. — Diagram of root-hairs and soil. The clear spaces among the soil grains indicate air. The lines of shading around the grains indicate films of water. Note how closely the root-hairs are pressed to some of the soil grains.

The smallest possible particles of some substances of the soil are dissolved in the water. The root-hair draws water away from the grains and into itself. The grains renew the water film. Physical laws operate to keep moisture spread evenly through the soil. As the moisture spreads it carries with it particles which the plant uses in making food. Thus, in part, the fertility of the soil is renewed.

The movements through the wall of the root-hair are not all inward. Substances pass out of it into the soil as well as into it out of the soil. Substances which pass out of living bodies are called *excretions*. It is known that these plant excretions affect the fertility of the soil. This is a matter of much importance in agriculture. It has been found that plant excretions may have injurious effects upon other plants of the same kind, and this may be one of the important causes of decrease in the fertility of soil.

From the root-hair the water and the soil substances dissolved in it move into the young root. Only the young roots do this work of absorption. Upon leaving the root-hair, the water enters first a tissue of thin-walled cells called the *cortex*. (Study *Figure 29*.) The cortex is much softer than the central part of the root which it surrounds. This central part is called the *stele* (*stē'lē*). In many young roots you can easily separate the cortex from the stele with your fingers. The stele is composed of several different kinds of tissue. The movement of substances in the plant is principally through the cells of the stele. The water which has come from the soil moves through the cortex and enters that part of the stele called the *xylem*. (See *Figures 29* and *30*.) The cells of the xylem are much longer than they are broad. Through them the water travels toward the leaves.

These xylem cells have thick walls. They give rigidity to the plant. The trunks of trees are principally composed of xylem. As you probably have already guessed, the common name for xylem is wood.

Another tissue of the stele is the *phloëm*. (See *Figures 29* and *30*.) Through the phloëm the manufactured food principally moves, having descended from the leaves.

The advantage to roots in having the harder tissues in the center rather than on the outside is evident. Think of the way they must bend and burrow in the soil. But

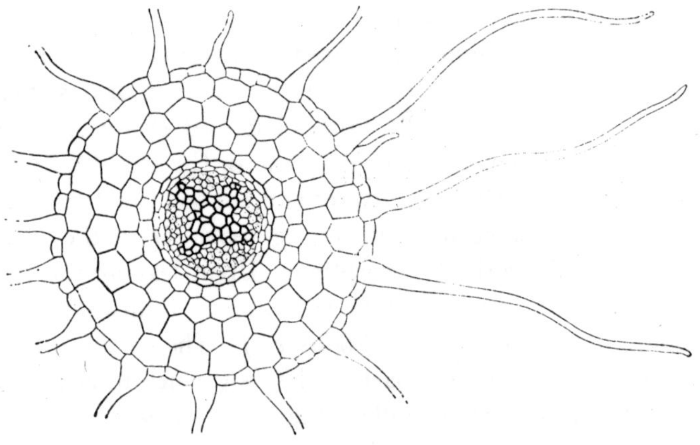

Fig. 29. — Cross section of a young root showing root-hairs. The outermost layer of cells is the *epidermis*. Note that the root-hairs are really prolongations outwards of certain cells of the epidermis. The thin-walled tissue of larger cells is the *cortex*. The central region of smaller cells is the *stele*. The thick-walled cells are *xylem*. The cells with shaded contents are *phloëm*.

with the stem it is different. In it the tissues of the stele are differently arranged. Here rigidity is an advantage. The leaves must be held firmly up in the light. The stem must have strength to withstand winds. The pliability of the root would be a disadvantage to the stem, unless it is the stem of a creeping or climbing plant. Evidently it is an advantage to most stems to have the tissue-arrangement of the root reversed; it is better to have the stronger tissue outside and the weaker tissue within. If you examine young stems, you will find that this arrangement is quite common. Evidently in such plants there is a change in the arrangement of the tissues of the stele as they pass from the root to the stem. Where the root joins the stem, the stele spreads out. In stems it is much larger in pro-

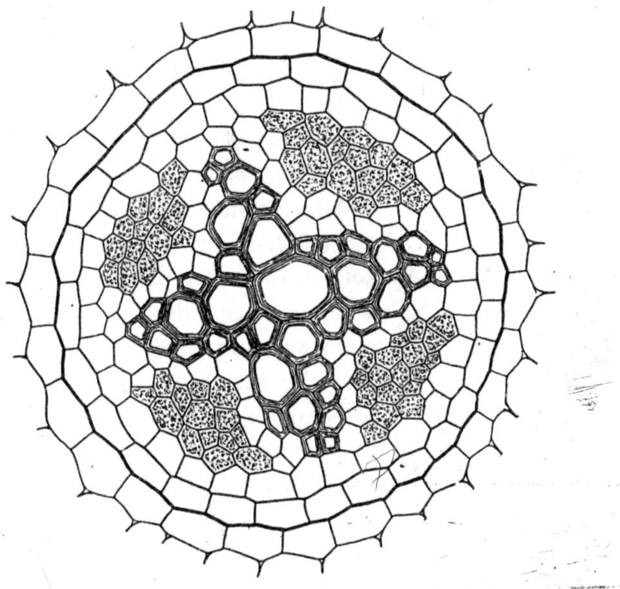

FIG. 30. — The stele of a root in detail. (Compare with *Figure 29.*)

portion to the whole diameter than it is in roots. The center of stems, at least of young stems, is occupied by soft-walled cells which form the tissue called pith. The pith is a part of the stele. The xylem and the phloëm form a cylinder around the pith. This cylinder is called the *vascular cylinder*. In cross section this cylinder forms a ring of groups of xylem and phloëm. (See *Figures 31* and *24.*)

These groups or bundles of xylem and phloëm are called the *vascular bundles*. You have frequently noticed vascular bundles even though you did not recognize them by that name. When you eat celery which is a little too old to be tender, it is the stringy vascular bundles of it which get in your teeth. When you peel a banana you are sure to see vascular bundles. They either come off with the skin or stick to the pulp. If they stick to the pulp you strip them off, for they are too tough to eat.

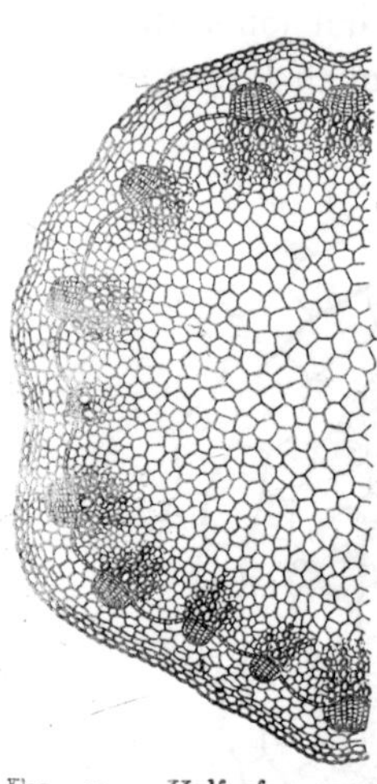

Fig. 31. — Half of a cross section of the stem of a common plant. The vascular bundles are quite distinct in this stem, ten of them being shown. Compare with *Figure 24* in which the bundles lie closely side by side. The tissue lying within the ring of the bundles is pith. (So also in *Figure 24*.)

Many stems, however, are not distinctly hard in the outer part and soft in the inner part like the stems we have been describing. In some stems you find no pith. Woody stems, like the stems of trees and shrubs, are hard all through. Such stems last over winter and form new layers of wood year by year. They do have vascular bundles which are arranged in a cylinder, but the arrangement is made complex by the addition of new layers year by year. The arrangement of woody stems is discussed in the chapter devoted to stems.

The stem of corn has still another arrangement. In it the hard parts, the vascular bundles, are scattered through the body of the stem. (See *Figure 32*.) They look in the picture like islands lying in the sea of softer tissue. Stems having this arrangement are common, but they never become so large and strong as the stems of trees which have their wood in layers. It is not an arrangement which is so well suited to great size and strength.

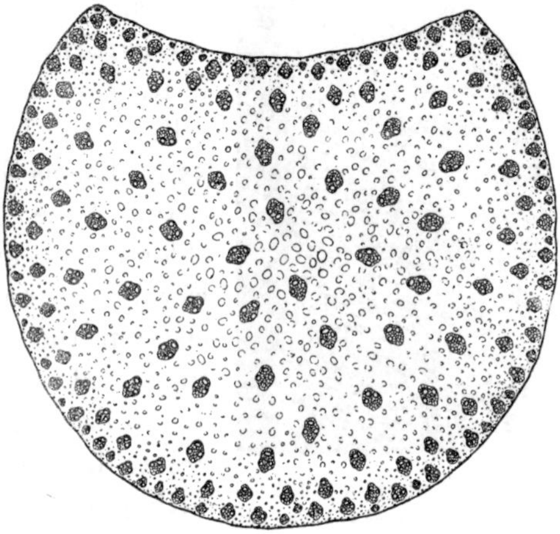

The water ascends through the xylem of the stems as readily as through the xylem of the roots. It becomes indistinguishable from the sap of the plant. It is the sap. It continues to move up.

Fig. 32. — Cross section of corn stem showing the scattered arrangement of bundles.

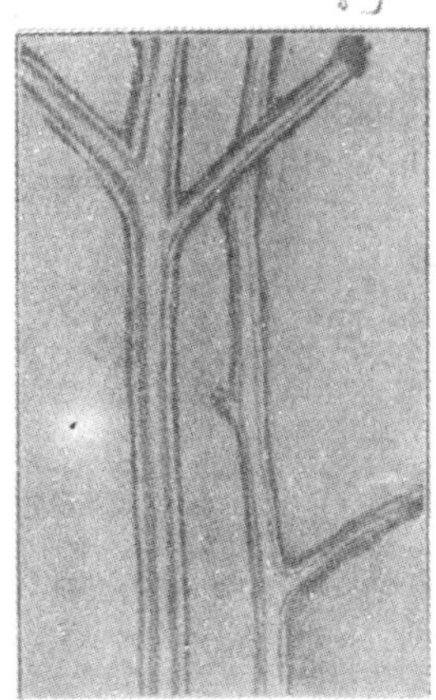

Fig. 33. — Apple twigs split to show the course of colored water up the stem. — *After Hunter.*

If the cut ends of vigorous young branches are immersed in red ink and left in the sunlight for a few hours, the ink will enter the stems. Its color reveals the path which the ascending sap follows. (See *Figure 33.*)

The vascular bundles branch out from the stems and enter the leaves. Here they are called veins. Hold a thin leaf to the light and see the delicate tracery of the veins. (See *Figure 34.*) Into every part of the blade they penetrate.

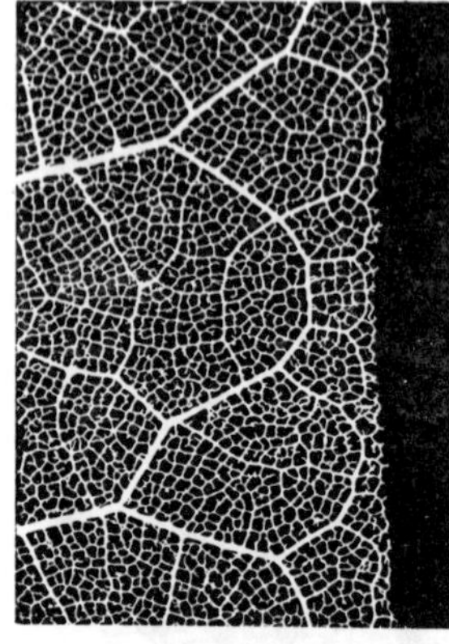

FIG. 34. — Part of the leaf skeleton of a rubber plant. The soft parts of the leaf have been removed. The smallest veins of all do not show in the picture. — *Photo. by Land.*

Through them the water, and the particles of other substances in it, reach at last the soft green cells of the leaves. (See *Figure 35.*) This is the end of the journey we have followed. Here the chloroplasts, food makers for the world, are busy while daylight lasts in the work of photosynthesis. Here inorganic substances are to become organic. Particles are to be rearranged and joined with other particles. Here, in the new food particles, is to be stored the force which sustains life, the energy on which all living substance feeds and from which it is renewed.

26. Inside the Leaf. — Here are dim, green spaces between loose, thin-walled cells. The light filters through. The tissue under the epidermis is called the *mesophyll.* (*Meso* means in the midst of; *phyll* means leaf.) Under the upper epidermis a layer of the mesophyll forms what is called the *palisade* tissue. (See *Figure 36.*) The cells of the palisade stand closely side by side, like logs

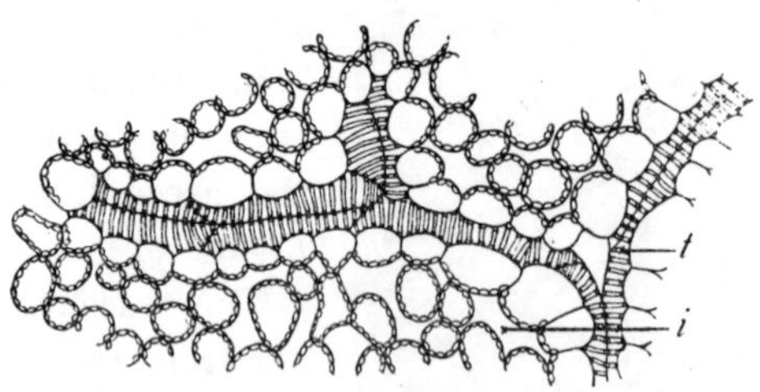

FIG. 35. — The ending of one of the smallest veins in the mesophyll; *t*, a water-conducting vessel; *i*, an intercellular space.

in the palisades which surrounded the blockhouses of the early settlers. This palisade tissue is firm, but the mesophyll beneath it is soft and spongy. In the spongy mesophyll the cells are loosely arranged; there are open spaces between them. The stomates of the under epidermis open directly into large intercellular chambers. (See page 54.)

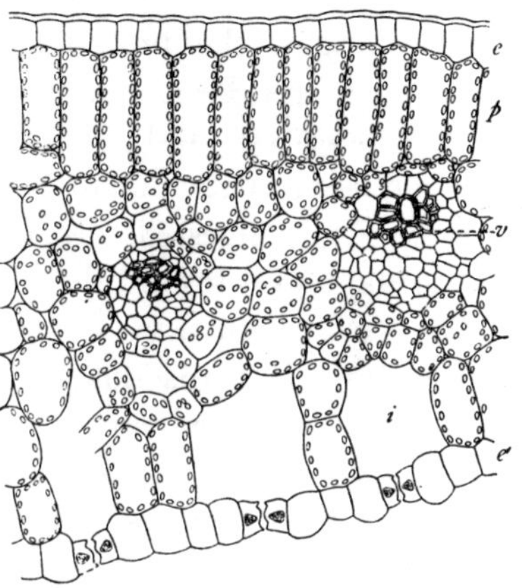

FIG. 36. — Cross section of a leaf of a lily; *e*, upper epidermis; *p*, palisade tissue; *v*, vascular bundle or vein, showing xylem and phloëm; *i*, an intercellular space with loose mesophyll above it; *e'*, lower epidermis showing three stomates opening into intercellular spaces.

Air passes through the many stomates of the under epidermis. It diffuses freely among the intercellular spaces. From it the cells of the mesophyll absorb the gases that they need: oxygen for respiration and carbon dioxide for photosynthesis. To it they yield other gases, chiefly water vapor, for the water brought from the soil evaporates. The substances which came with it do not evaporate. They remain and are used by the leaf in its work.

Thus together in the leaf we find these three things necessary to its work — the soil materials, the air materials, and the chloroplasts. And through the leaf the sunlight strikes, yielding to it that energy whereby its work is done, that force which is itself transformed and stored in the finished product. For from sunlight green plants derive energy which is stored in the food they make.

27. The Working Together of the Whole Machine. — Now perhaps you are ready to picture to yourself the working together of the whole machine. Think of a seed, and observe one as you think. With moisture and temperature right, the protoplasm in the embryo becomes active. It has been dormant long enough. The seed absorbs moisture and swells. Its cells become turgid. The old seed coats burst. The stored food changes its nature. It is digested. It is dissolved by the entering water, and moves toward the embryo, which begins to grow.

The seed sprouts. One part of the sprout seeks soil and water. It turns down. It is the root. The other part seeks light and air. It turns up. It is the *shoot*. By the term shoot we refer to all of the plant except the root.

As the young leaves unfold and chlorophyll appears, the tender root-hairs sprout, and begin to clutch the soil grains. The great effort of the plant appears to be to place its root-hairs favorably in relation to the soil and its chloroplasts favorably in relation to the light. And now the cells, whose growth unerringly placed root tip where it belongs and green leaf where it belongs, stand ready to be the paths of movement. The root-hairs absorb water. The fresh, new chloroplasts absorb light and air. The currents of delivery and exchange move through the little plant. Its life as an independent organism has begun.

It continues to grow. All day long it makes food. All night long it grows, more rapidly than in the daylight, unless it is too cold. Its internodes elongate. At the nodes new leaves appear. At their axils new buds appear. The buds expand into little branches. The branches expand and bear more leaves. Presently a little branch assumes a form different from its predecessors. Its unfolding leaf-like parts do not all turn green. Some of them show white, or delicate shades of blue or pink or yellow. This new kind of little branch does not continue to elongate. In the center of its tip appear organs which do not look like leaves at all, though they follow the rule of leaves in the manner of their birth. From the swollen tips of some of these a soft powder escapes. It is the pollen. The flower has appeared, and the plant has entered upon the second stage of its existence. It is ready for the task of reproduction.

QUESTIONS AND SUGGESTIONS

SECTION 22. 1. Make a list of the wild plants whose fruits, roots, or other parts you have eaten. 2. What wild plants do you know which are used for medicine? What parts are used? 3. At what time of year are maple trees tapped? How is this done? 4. What plants do you know which have hollow stems? 5. What plants do you know which have pith in the stems? 6. What plants have stems which are good to eat? 7. Of the trees that you know, which kinds have the largest stems? Which have hard wood? Which have soft wood? Which are most valuable for lumber? 8. What kind of wood is easiest to whittle? Why? What kinds are best for floors? Why?

SECTION 23. 1. What are cells? To examine them, how would you proceed? 2. Describe protoplasm. 3. What is a protoplast? 4. Define cytoplasm and nucleus. What appears to be the division of work between them? 5. Define chloroplast and chlorophyll. 6. Define vacuole and sap. 7. What is turgidity and how is it produced? 8. What is meant by differentiation? How is it illustrated in plants? 9. What are the principal processes performed by stomach, lungs, and heart? Are these processes also performed in the body of a plant? If so, by what parts?

SECTION 24. 1. What is a tissue? Give examples. 2. How many different kinds of tissue can you observe in a banana? In an orange?

SECTION 25. 1. Describe some of the relations between root-hairs and soil. 2. Why is it easier for roots and root-hairs to grow and work in soil which has been spaded or plowed? 3. Have plant excretions anything to do with fertility? Explain your answer. 4. Describe the inner structure of a root, defining cortex and stele. 5. Describe xylem and phloëm. 6. Can you, with a knife, separate the cortex of a young root from the stele? 7. Why is rigidity an advantage to stems and not to roots? Why is too much rigidity a disadvantage to stems? 8. What plants have stems which stand erect, and yet bend easily without breaking? 9. What is the vascular cylinder and where is it found? 10. What are vascular bundles? Give examples of vascular bundles you have seen. 11. How do trees increase in diameter? 12. Describe the way in which the vascular bundles of a corn stem are arranged. 13. What difference is there between water and sap? 14. Examine a leaf and describe the way in which its veins are arranged. 15. Describe the movement of sap from roots to leaf.

SECTION 26. 1. Describe the structure of the inside of a leaf as shown by *Figure 36*. 2. What do the cells of the inside of the leaf take from the air, and what do they give up to it? 3. What is an advantage to the plant in having its stomates on the under rather than on the upper surface of a leaf? 4. Where would you expect the leaf of a water lily to have its stomates? Explain your answer. 5. Is there more chlorophyll on the upper or under side of a leaf? What advantage in this? 6. What four things are mentioned as necessary to the work of the leaf? 7. Does it appear from their structure that leaves absorb water which falls on them or shed it? 8. Is it better to sprinkle the leaves of plants or the soil over the roots? Explain your answer.

SECTION 27. 1. What conditions are necessary to the sprouting of a seed? How would you arrange experiments to prove just what conditions are necessary? Are light and soil necessary? 2. What happens when a seed sprouts? 3. Define the term shoot. Can you think of any exceptions to the rule that shoots grow up? Could you place a sprouting seed under such conditions that its shoot would grow down and its roots grow up? 4. What is pollen? Name some plant whose pollen you have observed. What color is it? 5. Describe "the working together of the whole machine."

CHAPTER III

NOT ALTOGETHER ABOUT PLANTS

28. A Break in the Story. — We cannot very well go straight ahead with the story of plant life. Like other stories, it must be interrupted for explanations. Certain matters, though not exactly a part of the story itself, must be explained so that the story itself may be understood.

Already some things have been mentioned which you may not understand. Evaporation and solution have been mentioned. Do you understand them? Carbon dioxide and oxygen have been mentioned. Do you know what they are? Light and heat were called forms of energy. Do you know what energy is? You were told that substances enter the plant. Have you any idea how they enter it? All these and some other matters need to be cleared up before we go on. They all have a great deal to do with plant life.

29. Solution. — Let us begin with solution. With it you have had experience. You have stirred sugar in tea or coffee or lemonade. You know that gradually the sugar disappears and gradually the tea or whatever it is becomes more or less sweet all through. How sweet it is depends on how much sugar you have put in, but no matter how much you put in it is as sweet in one place as it is in another. The sugar has become equally dissolved all through it. *Solution* has occurred.

Solution is the name of a process by which one substance becomes dissolved in another. A substance which will dissolve in another substance is said to be *soluble* in that substance. Thus salt and sugar are soluble in water. Many more substances are soluble in water than are soluble in any other substance. This fact that it is the greatest of dissolvers has much to do with the very important relation of water to all living things, as you shall see.

Solution, besides being the name of a process, is also used to name that which results from the process. Thus, solution occurs when sugar is put in tea, and that which results is also called a solution. It is a solution of sugar in tea. A solution is composed of two parts, the *solvent* and the *solute.* In this case the tea is the solvent and the sugar is the solute. The solute is the dissolved substance. The solvent is the substance in which the solute is dissolved. There may be more than one solute in the same solvent. Thus in the tea we have particles from the tea leaves dissolved as well as particles of sugar.

The law of solution is that particles of solutes tend to become equally distributed throughout the solvent. The process is not complete until such equal distribution has been attained, — until the particles of the solute or solutes have become equally distributed through all parts of the solvent which they can reach. So, until they have become equally distributed, there is a force which seems to pull them farther and farther apart, just as there is a force which seems to pull a falling ball to the ground.

Solution is of the utmost importance in life, both in plant and animal life. Though living things appear to be solid bodies, they are composed of cells whose contents are more or less fluid, and *practically all the processes of life occur in that physical condition of matter which is called solution.*

30. **Molecules.** — We have been speaking of particles. We have spoken of particles of sugar which become equally distributed throughout tea, and of the particles in the water of the soil which plants use in making food. A particle of a substance is, of course, any very small part of it. You may have wondered how small these particles of sugar are in a solution. We cannot tell you just *how* small they are, but we can tell you that they are *just as small as sugar can be and yet remain sugar.* They are *molecules.*

Molecules are the smallest particles into which any substance can be divided and yet remain the same substance. There are processes by which a molecule of sugar can be reduced, but it then ceases to be a molecule of sugar. It becomes molecules of other things.

This is another place where you must call on your imagination to help you out. We cannot actually show molecules to you. It requires thousands of them to make a particle large enough to be visible to the naked eye. Yet we know that they exist. We know that all substances are composed of them. This has been proved.

31. **States of Matter.** — All substances are examples of what is called matter. Anything that can be weighed and measured is matter. Matter occurs in three different conditions or states. There is the *solid* state, the *liquid* state, and the *gaseous* state. That is, matter may be a solid, a liquid, or a gas. And it may pass from one of these states into another. In passing from one of these states into another it may change the nature of its molecules or it may not; that is, it may become a different kind of substance or it may remain the same substance, different only in state.

Thus in the case of sugar in tea, the sugar changes from a solid state to a liquid state and yet remains sugar. But the molecules of sugar cannot be changed into gas without ceasing to be sugar; they become other substances.

There are some substances, however, which can exist in all three states of matter. Water is an example. Water when it evaporates changes from a liquid to a gas, but its molecules still remain molecules of water.

Thus we see that forms of matter can exist in one or in more than one state of matter. Water may exist in all three states; it may be ice, it may be liquid water, or it may be water vapor, which is a gas. Sugar may exist in two states; it may be a solid or a part of a liquid, but it cannot be changed to a gas. Gunpowder is an example of the substances which can exist in only one state; it is a solid which cannot be dissolved nor can it be changed to a gas without changing its nature. When gunpowder explodes it changes into gases which immediately expand with much force, and it has ceased to be gunpowder.

We found that water dissolves more substances than anything else. We now find that nothing exceeds it in ability to exist in different states without changing into other substances.

32. **Elements and Compounds.** — Men found long ago that they could change some substances into other substances. They found that when wood or anything else burns it changes into substances quite different from what it was before. The burning of wood leaves a solid called the ash; it also leaves certain gases which are *diffused* in the air. (By diffusion is meant a spreading abroad. The spreading of the molecules of a solute in a solvent is an-

other example of diffusion.) In burning, the molecules of wood have ceased to be wood. They have changed into molecules of other substances. Any molecules when they burn change into molecules of other substances.

The early scientists studied just such common things as the burning of wood, and scientists still study such things, for there is much yet to be found out about them. Since wood in burning produces two or more other substances, it seemed reasonable to infer that the molecules of these other substances are not so large as the molecules of wood. Does this seem reasonable to you?

Remember that to the early scientists molecules were a theory. That means that they had good reason to believe in them, but they could not absolutely prove their existence. But now the existence of molecules has been proved. Similarly, the early scientists formed the theory that in burning larger molecules break up into smaller molecules. This, too, has been proved to be a fact.

It is thus that science advances. A scientist studying a certain problem considers all the facts about it that he knows. He then forms an explanation of these facts. If this explanation cannot be proved it is called a theory. As more facts are learned, theories are either proved or disproved. All depends upon what the facts are, not upon what any one's theory is. So scientists constantly seek for undiscovered facts, and are always ready to change their theories in the light of facts newly learned.

As they tested various substances and noted what happened, the early scientists observed that some substances apparently could not be changed into other substances. These they called elementary substances or *elements*. They formed the theory that all other substances are composed of various combinations of these elements; that all matter, therefore, is composed either of elements or of *compounds* of elements. This theory, too, has proved to be a sound one. It is accepted to-day as fact. A compound, therefore, is any substance whose molecules are composed of two or more elements. If the molecule of a substance cannot be reduced into two or more elements, then the substance itself is an element.

As compared with the immense number of compounds which exist, the number of elements which have been discovered seems quite small. There are only about eighty. Probably some have not yet been discovered, yet there is good reason to believe that the list will never be very much lengthened. A few of the known elements have been discovered only within the last ten years. Thousands upon thousands of compounds have been examined and all but about eighty have been reduced to simpler substances. All but about eighty are composed of molecules in which two or more elements are united together. These eighty, then, are the elements out of which all subtances, so far as we know, are composed.

Some elements are quite familiar and common. Coal is one of the forms of a common element. This element, *carbon*, assumes other forms than coal. Diamonds are a form of carbon. Diamonds and coal are both solids. Thus we note that the same substance, even in the same state of matter, may have very different aspects. This is because the molecules which compose them may be differently related to each other. By very intense heat and pressure pure coal has actually been transformed into pure diamonds. The expense of the process is, however, very great, and the diamonds produced are very small.

Carbon is one of the few elements of which living bodies are composed. Four elements alone compose nearly all that there is to living bodies. The others besides carbon are *hydrogen*, *oxygen*, and *nitrogen*. Water is composed of hydrogen and oxygen, while nitrogen forms about four fifths of the air. It would be rash to conclude, however, that living bodies get the hydrogen, oxygen, and nitrogen that they need directly from water and from air. You will find that such is not the case.

Iron, *sulphur*, *gold*, *silver*, *lead*, and *copper* are also familiar elements. Of these iron and sulphur are found in living things.

33. **Atoms.** — We have spoken of water as an example of a compound. We have also said that the elements hydrogen and oxygen are found in water. The molecule of water is composed of these two elements. Now the question arises, how are the elements arranged in the molecules of a compound? If a molecule is the least particle of a compound, what are these least particles of elements of which molecules are composed?

To understand this we must understand something of what happens when molecules are broken up. A molecule of sugar is much larger than a molecule of water. It can be broken up into molecules of water and of other substances. Then the molecules of water can be broken up into molecules of hydrogen and oxygen. This can be done by electricity. But now the reducing process has come to the elements themselves. They cannot be reduced into other substances. Their least particles seem to be reduced as far as reduction can go. To the particles which could not be reduced at all the early scientists gave the name *atom*. The word atom means *uncuttable* or *indivisible*. It has turned out to be a poor name, for recent scientists have found that atoms are divisible. But these divisions of atoms need not concern us now. Just how the subdivisions of atoms affect life we do not know, and in this study we are concerned with these matters only as they affect life.

From what has just been said you may form the idea that an atom is just the same thing as the molecule of an element. This is not the case, for atoms do not exist independently if they have a chance to unite with other atoms. Thus elements are composed of molecules and their molecules are composed of atoms, just as is the case with compounds. The difference is that the atoms in the molecule of an element are all of the same kind, while the atoms in the molecule of a compound are never all the same. They are always of two kinds or more.

Atoms may be defined as *the units of which molecules are composed.*

Now, going back to our molecule of water, you should have some idea of what is meant when we say that a molecule of water is composed of two atoms of hydrogen and one atom of oxygen. Perhaps you have heard of H_2O. H_2O is a symbol which stands for the molecule of water. It means a molecule which is composed of two atoms of hydrogen and one atom of oxygen. Hydrogen and oxygen may form molecules of different proportions than this two-to-one proportion. But such molecules are not molecules of water. In molecules of water the two elements can occur in this proportion and in no other.

Another substance of which we have spoken and which is very important in plant life is carbon dioxide. The symbol or formula for carbon dioxide is CO_2, which means that its molecule is composed of one atom of carbon and two of oxygen.

34. **Compounds and Mixtures.** — Water is a compound, but air is not. Air is a mixture. It is composed of four or more kinds of gases. Nitrogen, oxygen, carbon dioxide, and water all exist in air as separate gases. There is no such thing as a molecule of air. Air is made up of entirely different kinds of molecules. It can be separated into the different kinds of gases which compose it. The molecules of these various gases tend to become equally spread through the air, somewhat as the molecules of a solute tend to become equally spread or diffused through a solvent. There is a law known as the law of diffusion of gases which is much like the law of solution.

A solution is also a mixture. It is not a compound. Thus in a cup of sweetened tea there are various kinds of distinct molecules. There are molecules of sugar, of water, and of the substance from the tea leaves which gives the flavor. These molecules are evenly distributed in this mixture in fulfillment of the law of solution, but they do not unite.

Another kind of mixture is called a mechanical mixture. In it no law of solution or diffusion makes the molecules tend to become equally distributed. You may have made such a mixture. Did you ever make a salad dressing? That is an excellent example of a liquid mechanical mixture. In making it oil and vinegar must be vigorously stirred together, for there is no tendency of their molecules to become equally distributed, as is the case with sugar and water. Oil is not soluble in vinegar, or even in water.

Mechanical mixtures of solids are quite familiar. Some grocers have been said to make them carefully out of sand and sugar. Before we had a pure food law, careful mechanical mixtures were also made of brick dust and red pepper, and of coal dust and black pepper.

34 A. **Soil.** — A certain mechanical mixture made by nature is of extreme importance to plant life and to our life. It is the soil. Our lives depend upon the lives of plants and, to an equal extent, the lives of plants depend upon the soil. To understand the nature of plant life we must understand the nature of the soil in which plants grow and from which they draw much of their sustenance.

In dry soil the molecules of many different substances lie mixed mechanically together. No law makes them tend to separate. Of course, when the soil is moist, as it usually is, many kinds of molecules dissolve in the water. Some dissolve readily, others dissolve quite slowly. Each kind has its own dissolving rate, but all these dissolved molecules, as you have seen, tend to become evenly distributed as far as they can travel through the water. Then, if the soil dries out, these molecules are deposited as solids again wherever they happen to be after their journey in solution. So after all, though it is proper to speak of dry soil as a mechanical mixture, we must remember that solution has had much to do with the way in which the molecules are distributed in it.

A. Origin and Kinds of Soil. — In order to understand how soil has been formed, we must think of the earth as it was before there was any soil. It is believed that at one time water and rock formed the entire surface of the earth. There was no soil and there was no vegetation. But this condition could not remain long. There are forces at work in nature which cause even the hardest rock to crumble at the surface, and thus the formation of soil begins. Soil has been aptly called " rock meal " or " rock flour," for it is principally composed of substances which once were hard rock. Another important part of soil is the material in it which has come from the decay of vegetation. This material is called *humus* or *mold*. It is dark colored. Rich surface soil, especially in shady forests, is largely composed of humus.

It is evident that the ingredients of soil may be classed as organic and inorganic. The organic ingredients result from the decay of plants and animals. The inorganic ingredients result from the decay of rock.

The inorganic ingredients of soil may be classed as to their physical condition, and upon this basis the two principal classes are *clay* and *sand*. Clay, when dry, is a fine powder. When wet, it forms a solid, plastic mass. It is derived from those parts of rock which, principally by the action of water, become reduced to fine particles. Sand, whose particles are much larger than those of clay, comes from those parts of rock which are insoluble, but which may, chiefly by grinding, become reduced to fine grains. Such grinding occurs on rocky beaches where the waves rub the rocks together. The sand thus produced may be carried away by waves and currents and form beaches which are composed of little else than sand. The sand bars in rivers show that river water may likewise produce sand.

We have referred to the " parts of rock " which form clay and the " parts of rock " which form sand. This indicates that rocks and, consequently, soils are made up of different kinds of substances. Thus sand, though composed of grains all about the same size, may contain many different substances. These substances of which rock and the soil made from rock are composed are called *minerals*. Thus clay is largely composed of the mineral called *feldspar*, while sand is chiefly made up of *quartz*. There are many hundreds of different kinds of minerals. When you hear of the *mineral constituents* of soil you will understand that the words refer to those substances in the soil which come from rock. These substances are very important to plant life. They furnish many elements which are essential to it. From what you have learned about molecules, you should understand that each kind of mineral is composed of the same kind of molecules.

Different kinds of rock do not produce soil of equal fertility. Limestone and sandstone are two of the common forms of rock. Soils derived from limestone " are famous for fertility, as in the blue-grass region of Kentucky and the prairies of Texas " (Dryer's *High School Geography*). Very much of the soil of the North-Central United States is largely composed of material deposited by the melting of the great ice-sheet which once covered these regions. Regions once covered by ice-sheets are said to have been *glaciated*, *i.e.* covered by a great *glacier*. *Glacial* soils contain a great variety of minerals, and are usually fertile. The glaciated region of the United States is one of the greatest food-producing areas in the world. The most fertile of all soils are the kinds which are *alluvial*. Alluvial soils are those which are deposited by streams, especially at flood time, upon the floors of broad valleys, *i.e.* upon the *flood-plains*. You have heard of " rich bottom lands."

These are alluvial. It is the alluvial soil of the Mississippi and Missouri valleys which have made their farms famous the world around. Another thing which adds to the agricultural value of such land is its levelness.

For agricultural purposes a mixture of sand, clay, and humus is desirable. Pure sand dries too quickly, while pure clay does not dry quickly enough. When wet, clay becomes compact, difficult to plow, and difficult for roots to penetrate. Clay mixed with sand makes a soil easy to work, easy for plants to grow in, and with about the right capacity for holding moisture. Alluvial soils usually contain sand and clay mixed in a proportion which is very favorable to plant growth. The humus part of soil is also very important. It adds to the mineral constituents material which is rich in substances used by plants, and it also contains that hidden, microscopic life of the soil which is very important to the life of green plants. Soil composed of a mixture of sand, clay, and humus is called *loam;* in some parts of the country this term is also used to describe mixtures of sand and clay which contain little if any humus. The proportions of sand and clay in common agricultural soils are given in Dryer as follows: —

Sandy soil	80% sand	10% clay
Sandy loam	60–70% sand	10–25% clay
Loam	40–60% sand	15–30% clay
Clay loam	10–35% sand	30–50% clay
Clay	10% sand	60–90% clay

B. Soil and Food. — We have said that the soil is as necessary to plants as plants are necessary to us. We get substances from plants upon which our life depends, and the plants get substances from the soil upon which their life depends. It is evident that if the supply of any one of the soil-substances necessary to plant life should give out, our own lives would very promptly be affected.

Now it is a fact that old farms whose soil has not been improved by the addition of *fertilizers* appear to lose their fertility. As the years go by, the amount of crop per acre becomes reduced. In recent years the question has been raised as to whether this is due to the gradual exhaustion in the soil of certain substances necessary to plant life. It is evident that this soil problem is of vital importance to the human race. Each year the population of the world increases. Each year there is more hunger to be satisfied. Are we sure that each year there will be more food? Most of the fertile land in the world is already under cultivation, and if fertility is decreasing while food-consumption is increasing, the outcome is evidently going to be very bad for the human race.

To solve the problem of maintaining the food supply of the human race it is evidently necessary to know just what are the properties of soil which cause it to be fertile in the production of food plants. It formerly was thought that soil fertility was due to the presence of certain substances, and soil sterility due to the absence of these substances. But now it is known that the relation of soil to plant growth is not so simple a matter as that. In fact it is a very complex matter, concerning which scientists are more inclined to speak of what is yet to be discovered than of what is already known.

It is only in cultivated soil that we note decrease of fertility. Farmers speak of the " exhaustion " of soil, but the exhaustion which they observe in their fields is not observable in nature. The plants which grow upon uncultivated soil die upon it and return to it that which they took from it. But from cultivated soil, crops are removed year after year, and the materials which the crop-plants took from the soil are not returned to it as they are in nature.

The elements essential to plant life which are obtained from the soil are *phosphorus*, *potassium*, *magnesium*, *calcium*, *iron*, *sulphur*, and *nitrogen*. Iron, magnesium, and sulphur are present in such great amounts beyond what the plant needs that there is no problem so far as these are concerned. It is about the supply of phosphorus, calcium, potassium, and nitrogen that the farmer has to concern himself.

C. Fertilizers and Cultivation. — Very early in the history of agriculture it was found that the crop-producing power of soil is increased by the addition of various kinds of substances. These substances, when applied to soil, are called fertilizers. Manure is the most widely used of artificial fertilizers. Humus may be regarded as a sort of natural fertilizer. Bone-dust, rock-phosphate, lime, and various specially prepared fertilizers are also widely used. The original idea in the use of these fertilizers was, of course, to enrich the soil in those substances which had been removed from it in the form of crops. It was believed that plants used the fertilizers as "food." It has been found, however, that fertilizers may produce their good effects without acting as sources of food-material. Thus they may cause the destruction of poisonous substances given off by the root-hairs of previous generations of plants. (See page 79.) Or they may stimulate the growth in the soil of certain microscopic forms of life which are essential to its fertility. Or they may change the general chemical condition of the soil in a way which increases its fertility without adding anything to the food-materials which are in it. Thus fertilizers benefit the soil in various ways, and will probably be even more used in the future than they have been in the past. In the future, however, the exact effects which they produce will be better understood, and this will lead to a more scientific and economical use of them.

As in the use of fertilizers, so also in the plowing and harrowing of soil, farmers have increased fertility without realizing just how this increase was caused. It used to be thought that the chief reason for plowing after the planting was done to keep down the weeds. However, it is now known that plowing increases the contact of air with the soil grains, and that to keep the surface of the soil broken up retards greatly the loss of water from it by evaporation. Both of these things are even more important to good crop production than is the keeping down of weeds.

35. **Osmosis.** — You should be able now to understand something of how substances enter the body of plants. You have been told that water and substances in solution in it enter the roots. You have learned that the principal work of roots is to secure this entrance of water and other substances. You know that the life of the plant depends absolutely upon this entrance. Yet have you any picture in your mind of how this entrance occurs? It is important to your understanding of plants that you have such a picture, and you should now be able to get it.

Let your thought go back to the root-hair. There it lies in the soil. Moisture and the soil grains surround it. The great task of its life is to take into itself that moisture, rich in the dissolved substances which the green leaves above it will build into food. How is this task to be performed? The root-hair has no mouth-like opening, nor can it, like a bit of blotting paper or of sponge, absorb all liquids that it touches. It is a living thing and it is surrounded by a continuous wall made of a substance called *cellulose.* Cellulose is the substance of which all cell walls are composed when young, and of which many cell walls are composed throughout their existence.

Even with a microscope we can see no openings in that cellulose wall. Yet evidently if the water is to get in at all, it must get through that wall. Are there invisible spaces between the particles which compose that wall? Although no one has ever seen them, we believe that there are such spaces. We are able to explain the entrance of the water in no other way. We believe that the spaces are large enough to admit molecules of water and other molecules, still larger, which are in solution in the water.

The root-hair has water in it from the first. As you already know, the young plant cannot grow unless water is present. Seeds will not sprout unless water is present. Root-hairs cannot be formed unless there is water present to fill them. This water inside the plant body has various other substances in it. It is called sap, but sap is always principally composed of water.

Now the moisture just outside the root-hair is a solution. In it as solutes are those various soil substances which the plant needs. As you know, the molecules of these solutes tend to become equally distributed as far as it is possible for them to travel through the water. Now if the pores of the root-hair are large enough to admit them, and the water inside is continuous with the water outside, these molecules are surely going to move through the spaces in the walls. They will do it in fulfillment of the law of solution. They will do it unless other molecules of their own kind are already as abundant in the sap inside the plant as they are in the water just outside the plant. This *is* the thing which they actually do, and it is called *osmosis*. The solutes pass through the spaces of the root-hair wall in accordance with the law of solution, and this is an example of osmosis. Osmosis appears to be simply the process of solution occurring through a *membrane*. (A membrane is a sheet of organic substance; the wall of the root-hair is an example of it.) However, a simple experiment shows that in osmosis we get evidence of a certain force which is not evident in the process of solution alone. This force is what is called *osmotic pressure*.

You can show the way osmosis works by a simple experiment. Parchment or the wall of an animal's intestine is a kind of membrane. Tie an unbroken piece of such membrane firmly over one end of an open glass tube. Partly fill the tube with a sugar solution and then place it so that the covered end stands in pure water. After you have let such an arrangement stand for some time, you will find that the liquid in the tube has risen, while the water surrounding the tube has sunk. Water seems to have entered the tube. This appears to indicate a power of osmosis which we have not yet mentioned. It appears that the abundance of sugar molecules inside the tube draws molecules of water from outside the tube into it; it appears that, to satisfy the law of osmosis, the molecules of the solvent will move as well as the molecules of the solute. If the sugar cannot easily get out through the membrane, the water comes in. Whatever it is that has caused the result, one thing is evident: the law of solution is more nearly satisfied than it was before, for the molecules of the solute are farther apart than they were before, and this has resulted from a movement of the solvent rather than from a movement of the solute. Some sugar, however, has found its way through the membrane and is in the water outside. Evidently it is easier for water molecules to pass through a membrane than it is for sugar molecules.

But to go back to the root-hair. It may be that some molecules in the soil water are too big to enter it, but many there are which do enter it, and which are sure to continue to enter it as long as the abundance of them is less within than it is without. The molecules of a solute appear to move from where that solute is more dense to where it is less dense as easily and as surely as a ball rolls down a sloping roof. But suppose that these molecules become equally abundant on both sides of the root-hair wall. That will establish a sort of equilibrium; the osmotic pressure will be the same on both sides of the wall so far as that particular kind of solute is concerned, and the movement will stop. How, then, are we going to account for the fact that the movement does not stop? Of the molecules which the plant needs there is a steady procession, constantly entering as long as the plant is at work. How are we going to explain this fact?

Not only the solutes keep entering the plant. The water in which they move also keeps entering. Both must keep entering, or the plant could not grow; it could not live. How is it that the procession of water is kept up as well as the procession of solutes? Here are two things to be explained.

36. **Causes of Continued Entrance.** — The water around a root-hair is a busy place. Molecules of solutes are entering and other molecules are coming out. It is like a hurrying crowd, and yet all are moving in an orderly manner, in accordance with law. More go in than come out, but the movement is in both directions. (See page 79.) Besides these movements of the solutes, there is the movement of the water itself. It is entering the plant. Day after day it keeps on entering and it does not come back. What becomes of it?

There are three things which permit the entering movement of water and of solutes to go on without ceasing. One of these relates to the water. The others relate to the solutes. You shall see that the movement of the water into the plant and the movement of the solutes into the plant are different things.

That which permits water to keep entering at the roots is the fact that water keeps going out at the leaves. It does not go out as a liquid. It goes out as a gas. It goes out by the process of evaporation and we do not see it. It goes out through the stomates. More water replaces that which evaporates. More comes in at the roots. Thus, as long as water enters the roots and passes out from the leaves, there is a stream which moves constantly through the body of the plant. The evaporation of water from plants is called *transpiration;* the ascending stream is called the *transpiration stream.* Water sometimes escapes from leaves in liquid form, but this is not usual.

That which permits solutes to keep on entering is not so simple as evaporation. The solutes do not evaporate. They have nothing to do with the water which becomes gas. Their movement is limited to the water which remains liquid.

Two things permit the solutes to keep on entering. The picture on this page (*Figure 37*) represents the region in which these things occur. In the center of this picture are the wood cells of the vascular system; these cells are dead. When the water and the solutes enter these cells, they begin to move together. The solutes move along

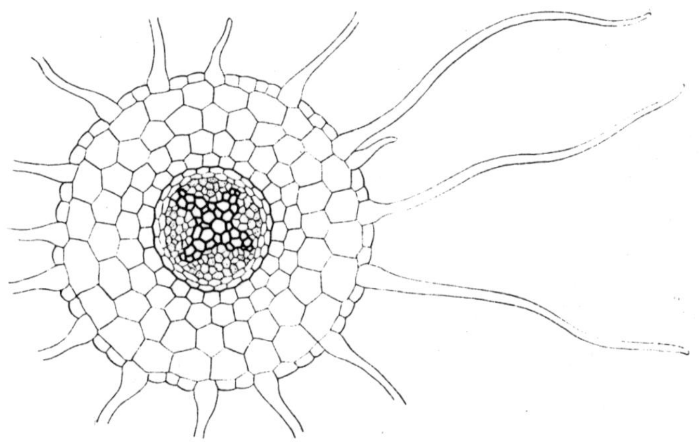

FIG. 37. — The region in which water and solutes enter the plant body. They move through the cells of the cortex by osmosis. These cells are alive. The cells of the xylem are dead. When the water and solutes enter the xylem they move in a mass together, not by osmosis.

with the water just as sugar dissolved in tea would move with the tea if you poured it from the cup. But before reaching the wood, that is between the root-hairs and the wood cells, the water and the solutes have to move through the living cells of the cortex, and through these they move by osmosis, and not together as one mass.

Now it appears that the sap current, ascending through the wood cells, carries away solutes from the cells of the cortex which border it. Then, to satisfy the laws of osmosis, more molecules of solutes move into these inner cells of the cortex and replace those which have been carried away. Thus we see cause for constant osmotic movement of solutes inward toward the wood cells and also cause for their continued entrance into the root-hairs.

The other thing which helps in allowing solutes to continue to enter is a thing which is important wherever osmosis occurs. *Many of the molecules which enter the protoplasts of the cortex are there changed to other substances.* So far as osmosis is concerned, this is about the same as if they had disappeared entirely. Fresh molecules come in to take their place, for they have ceased to be the molecules which they were when they entered the plant; they have become molecules of something else. This change of the molecules permits osmosis to continue and thus permits the plant to keep on getting the materials it needs. This second cause of the continued entrance of solutes probably does not have so much effect as the first, but it is a thing which is important in keeping up osmotic movements in other parts of the plant as well as in the roots.

What is it that thus changes the nature of these molecules? Think of the protoplasm in all the living cells. You remember that it is the living substance. You remember that it is constantly changing itself and constantly changing other substances which enter it. (See page 72.) Along comes a molecule of one of the solutes moving through the sap. If it happens to be a kind which the protoplasm uses, it very quickly ceases to be a molecule of that particular kind of solute. It becomes something entirely different. It has been lost; lost by transformation. It has made place for another molecule of the kind which it used to be.

You will remember that we spoke of certain solutes which tend to move out of the roots and into the soil water. Evidently such solutes are more abundant in the plant than they are outside of it. Evidently also they are the products of such changes, caused by the protoplasm, as we have just been considering. They are substances manufactured inside the plant. These outward-moving solutes do not amount to much in volume, but they may amount to a great deal in affecting the fertility of the soil.

Think, then, of the plant at work. All its cells contain water. A plant is composed of water far more than of any other substance; much over half the weight of common plants is the weight of the water they contain. The movement of substances from one part of the plant to another is absolutely necessary to plant life, and it is the water in plants which permits this movement. There are two kinds of movements in the plant body: the mass movement of water and solutes together which occurs in the dead wood cells, and the osmotic movements which occur in living cells. So we may picture to ourselves the solutes constantly moving through the tissues in fulfillment of osmotic laws, but always we must have in mind that the protoplasts change, and as they change they affect the osmotic movements. A kind of molecule which may be able to enter a protoplast at one time may not be able to enter it at another.

In the preceding section reference was made to osmotic pressure. It was said that when molecules of a solute become equally abundant on both sides of a cell wall, a sort of osmotic equilibrium is established. That is, the osmotic pressure of that particular solute is equalized; the pressure is no longer " high " on one side of the wall and "low" on the other side. A region of the solvent in which the solute is less abundant than in another region has a lower osmotic pressure than that other region, and osmotic movement sets up from the region of higher pressure (more of the solute) to the region of lower pressure (less of the solute). (This is suggestive of the way in which movements of the air are from regions of high pressure to regions of low pressure, a matter with which you are probably familiar if you have studied physiography.)

Having these facts in mind, you can probably understand the statement that the movement of solutes in plants "is from regions of high to regions of low pressure and may be downward toward the roots, outward toward the branches, or upward toward the flowers and fruits." (Cowles's *Ecology*.) It is important that you should have a true picture of this matter in mind, for osmotic movements are of primary importance in the life of plants. In fact, it is fair to say that *what is accomplished by osmosis in plants is similar and of equal importance to what is accomplished by the circulation of the blood in us.* Though the results of these two processes are similar, the processes themselves are, of course, by no means alike; they are very different indeed.

37. **Forms of Energy.** — Energy is the power to do work. You have energy. So has the falling stream of water. So has wind. Wind mills show the power of moving air to do work. Water wheels show the power of moving water to do work. Motion is one of the forms of energy. As matter exists in various states, so energy exists in various forms.

Energy may also exist without being in the act of doing work. Such energy is called *latent*, which means that power is present, but that it is not at work. It is energy of this sort that is stored in seeds. In fact, all food contains energy of this sort. It is not at work, but it may be set at work. After food has been digested, its energy is soon set at work. All food when it is dry will burn. Everything which will burn contains energy. The act of burning releases this latent energy. Heat is a form of energy. We see its power to work exhibited in the steam engine. It is heat which expands the water into steam; it is this tendency to expand, directly caused by heat, which enables the engine to drive machinery.

Heat and motion, then, are two forms of energy, and one may be transformed into the other. When a steam engine pulls a train, heat is transformed into motion. If the train gets a "hot box," we have a case of motion being transformed into heat.

Light is another form of energy. When things burn latent energy is transformed into both light and heat; the energy which has been stored up is released in both of these forms.

Man by means of his inventions has "harnessed" to do his work heat and motion and that other form of energy we call electricity. But he has not harnessed light to do his work. He produces artificial light by transforming into light the energy which is stored in kerosene, or gas, or in an electric current, but he has no invention for storing up again the energy of light or for transforming light into heat or motion.

Though man has no such invention, nature has. She long ago established factories in which the energy of light is stored up to serve her many purposes. She proved a far more ingenious inventor than any man. The heat and motion which he uses in his factories are expensive to produce, but the light which she uses falls everywhere and has no cost. Yet man has never been able to imitate the invention of nature; he has no such device by means of which he can use sunlight to do his work. But he does make use of the light-using device of nature. It furnishes him fuel for his engines as well as food for his body.

Already you know something of these factories of nature which use light for their power. You know that their product is more important than the product of any factories of man. You know that these factories of nature are the green leaves and that their product is food. You know that man uses the products of plants in his factories as well as at his meals, and you know that there would be no products of plants if it were not for the green leaves.

For fuel man first used wood. Now he depends more upon coal. Coal as well as wood owes its existence to the work of green leaves. Coal is the transformed bodies of plants which lived thousands of years ago. Those plants fell in the shallow water in which they lived and their bodies were covered. Other deposits covered them, and movements of the earth's crust, as well as the weight of the deposits above, caused much pressure upon them. At last they became transformed into layers of coal. It is common to find in coal outlines of the stems and leaves from which it came. When we burn it, we are releasing the energy of sunlight which shone on the green leaves of long-gone ages.

38. **Evaporation.** — Evaporation is the name of a process. It is the changing of a substance from a liquid or a solid into a gas. Usually we think only of liquids as

evaporating, but there are a few solids which change directly into gas. Camphor is an example.

Evaporation is a process with which all living things are much concerned. It occurs at the surface of your body as well as at the surface of a plant's body.

You have noticed on hot summer days that you are much more uncomfortable when perspiration sticks to you than you are when it quickly evaporates. People speak of "sticky heat" and of "dry heat." The former is much more uncomfortable than the latter. This illustrates a matter which affects plant life as well as our own.

On those "sticky hot" days of August you may have heard the expression that the "humidity is great." That expression refers to the air. It refers to the fact that in the air there is an unusual amount of water. This water is, of course, in the gaseous state. The humidity of air is the quality it obtains from the presence of water particles in it. The more water particles there are in the air, the less rapidly will evaporation take place.

This reminds you of solution and osmosis. There is the same principle involved. The rate of diffusion of particles in both solution and evaporation depends on the difference in their abundance. Solution is more rapid when the particles of the solute are less abundant in the solvent; diffusion of sugar in tea is most rapid when it is first put in; it gradually slows down. Evaporation is more rapid when the air has in it few particles of the substances evaporating than it is when the air has in it many of such particles; water evaporates more rapidly when the humidity of the air is slight than when it is great. This is a matter which affects plant life as well as our own, though in different ways.

Evaporation is a cooling process. The more it is checked at the surface of our bodies, the more we feel the heat. Why is it that evaporation cools the surface at which it occurs? Remember the molecules. Molecules are always in motion. Though the mass which they compose may be stationary, the individual molecules are always dancing about as much as their surroundings will let them. Now these dancing molecules at the surface of a liquid tend to dance off into the air, especially if it is dry air. This is the cause of evaporation. The warmest molecules are the most vigorous dancers. They are the most likely to fly off into the air, and, as they go, they take their heat with them. This lowers the temperature of the whole surface. If, on the contrary, the warmest particles do not fly off the surface on account of the abundance of them already in the air, and if the surface concerned is a human skin, the person inside of it is sure to be uncomfortable. This fact that evaporation causes loss of heat has been proved to be of advantage to plants that are exposed to high temperatures. It appears in some cases to prevent the internal temperature from rising to a point which seriously impairs the life processes.

With plants it is rapid evaporation rather than slow evaporation which is likely to cause trouble. In fact, plants thrive in greatest abundance in tropical forests where the humidity is always great and where evaporation is much less than in drier climates. On bright dry days of summer, especially during a drought, when the soil as well as the air has become dry, many plants are likely to suffer. Water evaporates from them more rapidly than the roots are able to supply it. Cells lose their turgidity, and this causes wilting. (See page 75.) The grass turns brown, and the corn in the fields " fires," as the farmers say.

So far as plants are concerned, evaporation appears to be a hindrance as well as a help. Plants can stand heat better then we can, but they cannot stand loss of water so well. It is easier for us to take a drink than it is for them to get more water. Plants which live in dry places, as in deserts, are mostly plants whose construction protects them against evaporation. Their power to retain water is much greater than that of plants which live in regions where water is more abundant. (See page 51.)

39. **Physics and Chemistry and Life.** — The matters we have been considering in this chapter belong in those divisions of science called physics and chemistry rather than in that division of science called botany. Such matters are studied by botanists only as they relate to plant life, and it is in this way that you have been studying them. In physics and chemistry they are studied for their own sake; not in order to understand life, but in order to understand the behavior of matter and energy whether in living things or in non-living things. Botany is science as related to plants. Physics and chemistry are science as related to matter and energy.

Nature, however, does not recognize these divisions of science which men have made. All her forces work together. You cannot study the life forces without also studying the other forces. You cannot tell where the " other " forces stop and the life forces begin. They are a part of the living world as well as a part of the non-living world. Life obeys them, and they serve life. This you have already seen.

Physics deals with the states of matter. That water exists in a liquid state is a fact which physics considers. Chemistry deals with the composition of matter. That water is composed of hydrogen and oxygen is a fact which chemistry considers. Since plant life, or any life, is based on states and composition of matter, you evidently need to know a little physics and chemistry; at least as much as we have given in this chapter.

Life a Procession of Changes. — Physics and chemistry also deal with changes in the states and composition of matter. Life, too, deals with changes. It is based on changes. It has no more fundamental characteristic than changes, the changes which constantly occur in protoplasm, the living matter. It cannot exist without these constant changes. When they stop, life is gone. It is like a whirlpool. When the water stops moving, the whirlpool is gone.

Here arises one of the great questions of science. Are these life changes simply physical and chemical, operating under the laws which are familiar to physicists and to chemists? Or are they outside the laws of physics and chemistry, operating under some mysterious law of life? The change of water into steam is a physical change. The change of wood in burning is both a physical and a chemical change. Apparently the life changes are different from the boiling of water and the burning of wood only in complexity.

Though we are yet very ignorant of the precise character of life changes, we have learned a great deal more about them in late years than was formerly known. Each year this knowledge increases. Yet all that science has revealed indicates the existence in living bodies of no laws of matter different from those which operate in non-living bodies, nor of any elementary substance different from the elementary substances which exist in non-living bodies. Seeming differences have disappeared as knowledge increased. Just why certain conditions do cause certain life changes we cannot say; just why certain conditions, for example, cause a seed to sprout we do not know; our knowledge may never reach that point. But the changes themselves, once started, operate under the same laws as all other changes. The powers of living things are different from the powers of non-living things, but all these powers are subject to the same laws of matter.

So plant life, as far as we understand it, is simply a great complex of physical and chemical changes. Evidently you can understand very little about plants without understanding something of the nature of such changes and of the laws under which they operate.

QUESTIONS AND SUGGESTIONS

SECTION 28. Make a list of questions whose answers would explain things recently mentioned in the book or in the class which you do not understand.

SECTION 29. 1. Describe what happens when one substance is dissolved in another. 2. Make a list of the substances which you know will dissolve in water. 3. What is the difference between "soft" water and "hard" water? 4. Why is soft water better than hard water for washing? 5. Why is it easier to wash in soapy water than in clear water? 6. Explain the relation of the temperature of water to the rate at which substances dissolve in it. 7. Define solute and solvent. 8. Why is it necessary to understand solution in order to study the processes of life? 9. Explain in what way the water which comes from springs is different from the water which falls as rain. 10. Explain why distilled water is pure water. 11. State the law of solution. 12. Having dissolved a solid in a liquid, how can you get it back in solid form again? Give an example.

SECTION 30. 1. What are molecules?

SECTION 31. 1. What are the three states of matter? 2. What is it that generally causes the change of matter from one state into another? 3. Of what advantage is it to man that water expands when it freezes? 4. Of what advantage is it that water also expands when it becomes a gas? 5. Watch water boil in a glass vessel and describe what you see. 6. Give examples of substances which may exist in three states; in two; in only one state. 7. Are liquids generally heavier than solids, or the reverse? Is wood always lighter than water? 8. Are some gases lighter than others? How would you prove this? 9. Why does a steel ship float?

SECTION 32. 1. What happens to molecules when they burn? 2. Is water produced when wood burns? How would you arrange an experiment to find out? 3. What is meant by diffusion? 4. What is an element? 5. What is a compound? 6. Give examples of elements. 7. What are the elements which principally compose living bodies? 8. What are the elements which compose water? 9. Of what elements and compounds is air principally composed?

SECTION 33. 1. What are atoms? 2. Is the molecule of an element the same thing as an atom? Explain your answer. 3. Explain the symbols H_2O and CO_2.

SECTION 34. 1. Give examples of things which seem to be of one substance but which are really mixtures of different substances. 2. What is the difference between a "mechanical mixture" and a solution? 3. Test a number of substances which are soluble in water, and rank them in the order of their solubility. The substances being in equal amounts by weight, the one which dissolves most quickly ranks first in order of solubility, and so forth.

SECTION 34 A. 1. Explain how the law of solution acts in distributing material in the soil.

A. 2. Describe the origin of soil, defining humus. 3. Define minerals, giving examples. 4. What is meant by glaciation? 5. Describe alluvial soils. 6. What physical kind of soil is best for agricultural purposes? Define loam.

B. 7. What is the "soil problem" in relation to the human race? 8. What elements does the plant get from the soil?

C. 9. Describe fertilizers and their use. 10. What are the various ways in which they may affect fertility? 11. The fertility of soil depends a good deal upon its ability to hold moisture; *i.e.*, other things being equal, soil which remains moist (if not *too* moist) will be more fertile than that which dries out quickly. Why is this so? Devise an experiment for testing samples of soil in order to determine their relative capacity for retaining moisture. 12. How does the farmer increase the capacity of his soil to retain moisture? 13. Devise an experiment to demonstrate how, by different treatments, the capacity of soil to retain moisture may be increased.

SECTION 35. 1. Describe the structure of a root-hair and the conditions which surround it. What is its chief function? 2. What is cellulose? 3. What is osmosis? 4. Describe a method of demonstrating osmosis. 5. Explain what is meant by osmotic pressure.

SECTION 36. 1. Define transpiration and the transpiration stream. 2. What is the principal cause of the continuous entrance of water into the plant? 3. What are the principal causes of the continuous entrance of solutes into the plant? 4. What is it that transforms the solutes which enter the plant? 5. Why do some solutes pass out of the plant into the soil? 6. Why is it that animals are not so dependent upon osmotic movements as plants are? 7. What are the two kinds of movements which occur within the plant body? 8. Give the description of the movement of solutes which is quoted in the text, and explain what is meant by high and low osmotic pressures. 9. How much of the weight of common plants is water? Why do they need so much water? 10. What are some plants which grow well in dry places? 11. What crops are most injured by drought? 12. What crops are least injured by drought? 13. What crops do well on sandy soil? 14. What plants do you know which grow in the water?

SECTION 37. 1. Define energy and give examples of it. 2. Define latent energy and give examples of it. 3. Give examples of various "forms of energy," and of the transformation of one form into another. 4. What are some of the principal practical uses which man makes of the fact that one form of energy can be transformed into other forms? 5. Describe "nature's invention" for storing up the energy of light. 6. Describe the formation of coal. 7. Does it seem possible for animals to have existed before plants? Explain your answer.

SECTION 38. 1. What is evaporation? 2. What is transpiration? 3. More plants die as a result of evaporation than from all other causes. Why is this so? 4. Evaporation is the greatest danger to plant life. Is it a necessary danger? Explain your answer. 5. Explain why hot weather is much more uncomfortable in a moist climate than in a dry one. 6. In Arizona water is kept cool by placing it in porous vessels and hanging them up out of doors. Can you explain this? Why is it better to hang the vessel up in the air? 7. Devise an experiment for proving that a plant gives off water by evaporation. 8. Devise an experiment for showing just how much water a plant gives off by evaporation. 9. Devise an experiment for showing the difference in rate of evaporation between a leaf surface and an equal area of water surface. 10. What are some of the ways in which you would expect the structure of plants which live in deserts to be different from the structure of plants which live in moist and shady forests?

SECTION 39. 1. Define physics and chemistry, and give examples of the kind of facts which each considers. 2. Why is it necessary to understand something of physics and chemistry in order to understand the life of plants? 3. In what respect is life like a whirlpool?

CHAPTER IV

ROOTS

40. Topics. — We must begin to go into details. We have considered the principal parts of a seed plant in a very general way. We have considered its work as a whole. Now we must consider each part in more detail.

To consider details we must divide our subject up into topics. Plant life is, however, a difficult subject to divide into topics. Life is not simply one thing plus a number of other things. It is a great number of things all taken together, and unless you can think of them all together instead of all apart you can have no true picture of life. You have already seen the difficulty in dividing the work of a plant into the work of its parts. The work of roots, stems, and leaves is so closely related that to understand the work of one of these parts you must understand the others as well.

Evidently plants were not made as schoolbooks are made. Schoolbooks have subdivisions and topics so that one thing may be studied at a time. But plants are doing everything at once and each thing as a part of the rest. It is hard to make a good schoolbook about them. As soon as you begin to divide their work up into parts so that it may be simple for beginners to understand, you find you are in danger of telling something that isn't so. Yet to tell the whole truth makes the subject too difficult for beginners.

The work of plants is subdivided, but it is also united. So our study must be subdivided, but also united. You will read chapters on roots, stems, and leaves separately, but you will think of them together. You will have your facts classified into topics. You will also understand their relationship to each other. You will perceive their unity as well as their separateness.

41. **Introductory as to Roots.** — You already know a good many important things about roots. You know that they are the burrowing parts of plants; that they seem to be seeking in the soil for things the plant needs. You know that they anchor the plant; that their extent may be even greater than the extent of the stem and its branches. You know that they seem to respond to the nature of the soil in which they grow, suiting their structure to the conditions they meet. You know that there is a dense growth of delicate hairs just back of the growing tips, and that these hairs are organs of absorption. You know that a great deal of water is absorbed by roots, though it is only through the tips of young roots and through root-hairs that this water enters. You know that, in the water of the soil, substances which the plant uses in food manufacture exist as solutes, and that by osmosis these solutes enter the roots. You know that the internal structure of all roots is quite similar, and have noted the relationship and appearance of epidermis, cortex, and stele. You know that the absorbed substances pass through the cortex, reach the stele, and move through the xylem up toward the stem, while through the phloëm food descends from above. You know that food is often stored up in the roots, and that the roots of some plants live through the winter, while the

upper parts of these plants die. You know that roots of some plants can directly produce new plants.

All these facts are important, but they form no more than a starting point in the study of roots.

42. **Roots as to Origin.** — Roots come from the embryo, or from other roots, or from the shoot. Those which come from the embryo or branch off from other roots are called *primary;* those which arise from the shoot are called *secondary.* Leaves as well as stems sometimes give off roots as branches. The leaf of begonia is an example. This plant may be propagated by means of cuttings from the leaves.

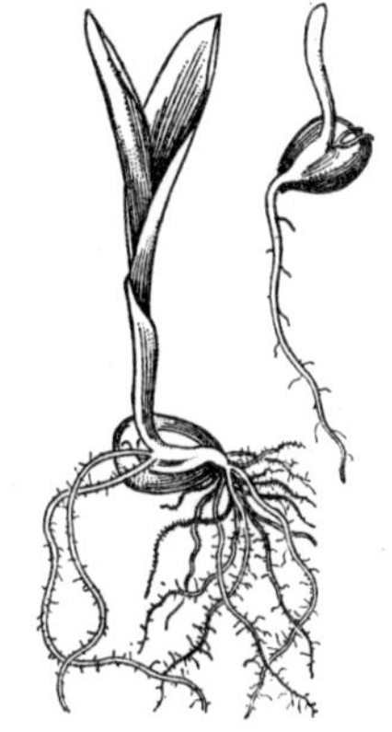

FIG. 38. — Seedling of corn showing the development of primary roots.

A. Primary Roots. — You remember that the embryo comes from a single cell which is produced by the union of two other cells. (See page 58.) This new cell by repeated divisions and by growth forms the embryo. Certain cells of the embryo develop into the original primary root. (See *Figure 38.*)

Sometimes this original primary root grows straight down and becomes much larger than any of its branches. Such a root is called a *tap-root.* Dandelions, radishes, turnips, carrots, and parsnips have tap-roots. (See *Figure 39.*) Tap-roots are more commonly used for storage than are other forms of roots.

More often, however, a tap-root is not formed. The primary root divides into many branches of about equal size. Such roots are called *fibrous.* Corn has fibrous roots.

B. Secondary Roots. — These come from the stems or leaves. Creeping or subterranean stems usually give off roots at their nodes. Strawberry and grass stems are examples. The creeping stems of the strawberry are called *runners.* At their nodes the runners may develop both roots and shoots and thus establish new independent plants. (See *Figure 40.*)

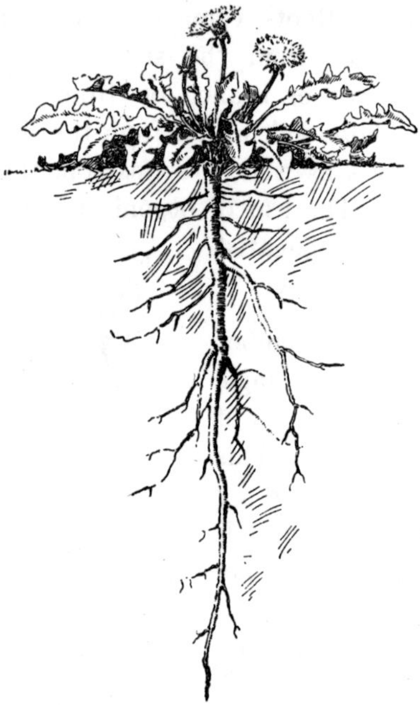

FIG. 39. — Dandelion showing the tap-root. — *Redrawn from Bailey.*

Berry bushes are sometimes reproduced by a process called *layering.* Layering consists in bending down the yielding outer stems of the berry bushes and covering them with soil. From the buried nodes roots and shoots arise. Presently a new little plant is thus established. It is then cut away from the parent stem and transplanted. The black raspberry spreads by means of new plants which arise naturally from the ends of its trailing branches.

Some secondary roots have special functions; that is, their work is quite different from the work usually done by roots. The prop roots of corn are examples of this. (See *Figure 4,* page 48.) They rise from the first node above the soil and sometimes from the second node as well.

Some tropical trees have unusual roots which come from the stem. The banyan is a tree from whose lower horizontal branches roots descend. At first they are like swaying ropes, free at one end. When a free end reaches the soil, it begins to burrow like any root. These prop roots may develop in thickness until they seem like second-

FIG. 40. — A strawberry plant showing the way in which the creeping stems put off secondary roots and establish new plants. — *Redrawn from Bergen.*

ary trunks of the parent tree. The largest banyans have hundreds of these secondary trunks. They cover several acres. At a distance one tree may look like a forest in itself.

The mangrove swamps along tropical shores are named from plants which gradually grow out into the sea. Like the banyan, they have the habit of dropping new roots from their branches. Among the new roots on the seaward side drifting material becomes entangled. This helps in the formation of soil about them. Thus, where the water is shallow, the mangroves "march out to sea."

Stems which are in contact with water are apt to develop what are called *water roots*. These differ from soil roots in that they occur in dense clusters. Such roots may also develop from the primary roots, as when a root finds its way into a tile drain. A mass of roots may result. Drains are frequently choked up by these water roots.

Secondary roots may act as *tendrils*. Tendrils are curving, slender organs which hold climbing plants to the things on which they climb. Probably you have noticed the tendrils of vines. Tendrils are conspicuous on the grapevine and on the climbing cucumber. The poison ivy, the English ivy, and the trumpet creeper are vines whose tendrils are secondary roots. The tendrils of other plants may be modified leaves or may be modified stems.

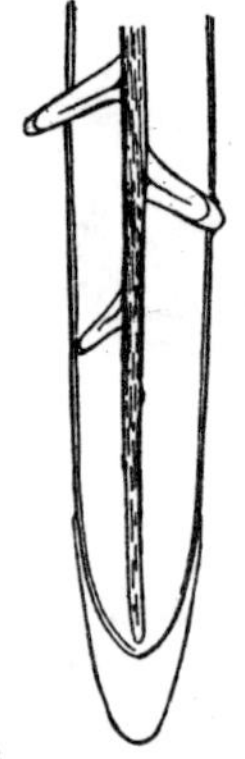

Fig. 41. — Diagram showing the origin of the branches of roots.

C. Branches of Roots. — The branches of roots do not arise from the epidermis of the parent shoot. They arise from the outermost layer of cells of the stele. (See *Figure 41.*) Thus they are compelled to begin their burrowing habit even before they enter the soil. The first burrowing they do is through the cortex to the epidermis. With a knife you can easily strip the soft cortex of a young root away from the hard stele. You will then see holes in the cortex which were made by the branches.

In this matter of the origin of branches, roots are different from stems. As you already know, the branches of stems first appear as buds in the axils of the leaves. They arise near the surface.

43. **Structure and Growth of Roots.** — Plants grow because their cells grow. New cells are formed, but the formation of new cells does not alone cause growth. One cell may divide into two, but the two together are at first no larger than their parent was. Gradually they enlarge. It is this cell enlargement rather than cell division which causes growth.

All living cells may enlarge, but not all living cells appear to have the power to divide. Comparatively few of the cells of a plant appear to have this power. Tissue composed of cells having this power is called *meristem*.

So in considering the growth of roots we are interested to find two things: (1) Where are those cells of the root which have the power to divide? (2) In what part of the root does cell enlargement principally occur?

(1) The meristem of roots we find in two places. We find it in the center of the young root just behind the tip. (See *Figure 42*.) We also find it in the older parts of the roots near the edge of the stele. The meristem at the tip forms those cells whose enlargement causes growth in length and some growth in thickness. The meristem in older parts forms cells whose enlargement causes growth in thickness only. The growth resulting from the activity of meristem in the older parts of plants is called *secondary growth;* that resulting from the activity of meristem at growing tips is called *primary growth.* In long-lived plants secondary growth goes on year after year. As a result, the wood of long-lived roots appears to be formed in layers, like the wood of long-lived stems. A meristem which causes secondary growth is called *cambium.* Cambium occurs in stems more prominently than in roots, and in connection with stems we shall study it in more detail.

(2) That part of the root in which cell enlargement principally occurs lies directly behind the *apical* meristem; that is, directly behind the meristem which is at the apex or tip of the root. This part is called the *region of elongation.* In this region the cells grow in length more than they do in width.

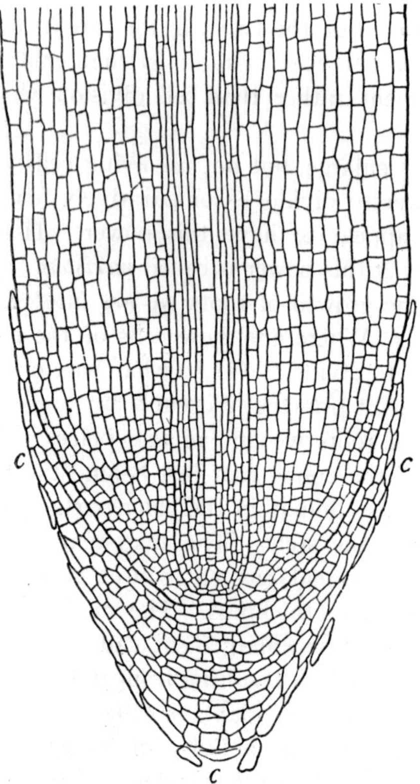

FIG. 42. — A lengthwise section of a root tip. The part marked *c* is the root-cap. Just above this is the *meristem,* the region of cell division.

In stems also the region of elongation lies directly behind the growing tip. It is usually several inches long; often more than a foot long. In roots, however, this region is much shorter than in stems. It is hardly ever more than two fifths of an inch long, and often not more than half of that. Evidently root cells finish their growth in length more quickly than stem cells do; they mature more quickly than stem cells.

It is simple to show by experiment what part of the root

elongates most rapidly. You can do it readily with peas, beans, or corn. You sprout the seeds and when the young roots have become about an inch long, you rule the surface off into small equal spaces. The ruling should be done with India ink and a fine-pointed brush. The lines should be as close together as you can get them and still keep them distinct. After one or two days, you will find the spaces between the lines are no longer equal. That part of the root in which these spaces have become widest is the region of elongation. (See *Figure 43*.) The cells elongate slowly at first, then rapidly, and then slowly again until they cease to elongate. What would happen if the cells on one side elongated more rapidly than those on the side directly opposite?

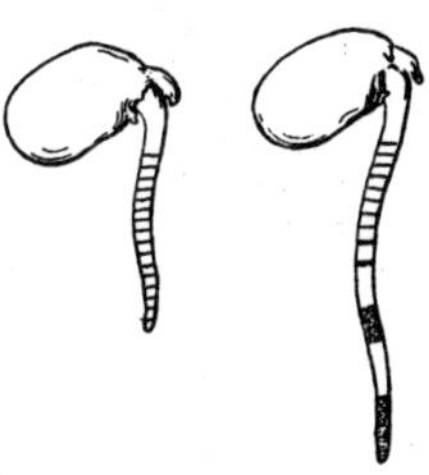

Fig. 43. — Seedlings of bean which have been marked so as to show the "region of elongation."

A. Root Curvature. — As the root grows it is almost sure to curve. The young tips curve around the soil grains. They curve toward moisture. It is evident that this curvature can be more easily accomplished by the region of elongation than by any other part. If the cells elongate more rapidly on one side than upon the opposite side, a curve is sure to result, whereas if the cells have finished their growth in length, and have become somewhat more rigid, curvature evidently would not be so easy to accomplish. So it does not surprise us to find that the region of elongation is also the region of curvature.

Suppose that the growing tip meets an obstacle. It is always meeting obstacles. The tip itself cannot curve around an obstacle, but it does seem to be able to cause a curve. It seems to be able to cause the region of elongation to curve, and, as a result, the tip itself is turned away from the obstacle.

Think of your fingers feeling in the dark. They may touch something which gives you an unpleasant sensation. You snatch them away. Now the very ends of your fingers which touched the object cannot move. They can be moved, of course, but they cannot move themselves. What they do is to start a message to your muscles which, by their action, produce the movement. The fingers are organs of the sense of touch. Similarly, root tips seem to be organs of a sense of touch. They seem to be able to start a message to which the region of curvature responds. This sense of touch of plants is very delicate. It has been measured by experiments upon the tips of tendrils. They seem to have a sense of touch far more delicate than our own. A gentle stroke of a very fine silk fiber, so gentle that the end of your finger or your tongue would not perceive it at all, is enough to make a tendril curve.

B. Tropisms of Roots. — We have just been considering the way in which contact with an obstacle may affect the direction of a root's growth. Other things also have an effect upon the direction of its growth. Water and gravity are two of these things.

Tendencies of living things to turn in response to the influence of things outside of them are called *tropisms.* (The word tropism is derived from a Greek verb meaning *to turn.*) You know, for example, that leaves turn to the light. Tropism caused by the light is called *phototropism.* (Recall the meaning of photosynthesis.)

When the tendency is to turn toward that which exerts the influence causing the turn, the tropism is said to be positive. When the tendency is to turn away from that which exerts the influence, the tropism is said to be negative. Thus a tropism toward light is positive phototropism; a tropism away from light is negative phototropism. Leaves, unless the light be too intense, are positively phototropic, while roots are negatively phototropic. Roots tend to turn away from light.

Early in this book it was said that "stems grow up and roots grow down." We may now express this by saying that roots are positively *geotropic*, while stems are negatively geotropic. (*Geo* comes from a word meaning the earth. Recall *geography*.)

You remember that plant organs, especially roots, appear to be modified to suit their surroundings. You remember that we called these modifications responses. Tropisms evidently are one kind of responses. Thus the geotropisms of which we have just been speaking are the responses which the plant shows to gravity, to the pull of the earth's mass.

That which causes a response is called a *stimulus*. The tendency of the root to grow down appears to be a response to the stimulus of gravity, for, if we remove the effects of gravity, roots continue to grow in whatever direction they may have started. To observe the result of removing the effect of gravity, sprouting seeds may be pinned to an upright revolving disk, being kept moist by dripping water. As the disk rotates, each seedling constantly changes its position with reference to gravity. It occupies no one position long enough for an effect to be produced. Thus the effect of gravity is practically removed, and the result is that both roots and stems continue to grow in whatever way they began, at least so far as gravity is concerned.

The tropism induced by water is called *hydrotropism.* Water seems to produce even more effect upon the direction of root growth than gravity does. If a seedling is so arranged that gravity attracts it in one direction while water attracts it in another, the water seems to have a stronger influence. (See *Figure 44.*) If root tips come through the bottom of a wire cage in which seeds have been sprouted in damp moss, they turn back toward the water against the influence of gravity.

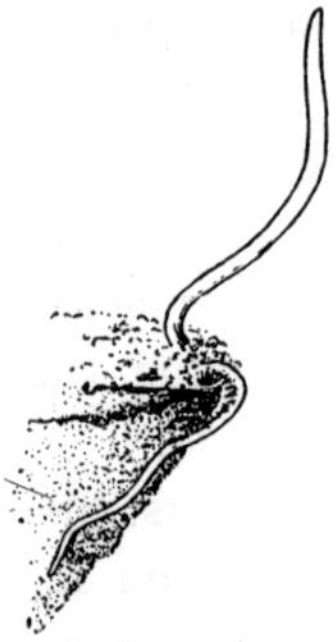

FIG. 44. — Seedling of corn grown at the edge of a funnel whose surface has been kept moist. Note that the root follows the moist surface instead of growing straight down.

C. Root-cap. — It is evident that the tender growing tip of a root would be injured if it kept rubbing against the soil grains as it presses forward. It needs protection. It is protected by what is called the root-cap. (See *Figure 42.*) This cap is composed of several layers of cells. As the root pushes and turns through the soil, the outermost layers of the root-cap are gradually worn away. They are replaced by new cells which are constantly being formed beneath. That same apical meristem which forms the cells of the root proper also forms cells which become a part of the root-cap. It produces cells forward as well as back.

Root-caps are hardly ever to be found on roots which you pull up from the soil. The tips of the roots and the caps are almost sure to be broken off. The screw-pine, a plant which is common in conservatories, shows the root-cap very well. This plant has prop roots which develop prominent root-caps even while they are growing through the air. They fit over the end of the root as a glove fits over the end of your finger. Their development while the root is in the air indicates that such structures may develop even when there is no need for them. Evidently there are forces within the plant as well as forces without which affect its growth. The forces within are called the forces of *heredity;* the plant inherits them. The forces without are called the forces of *environment;* these the plant does not inherit. All the conditions which surround a plant form its environment.

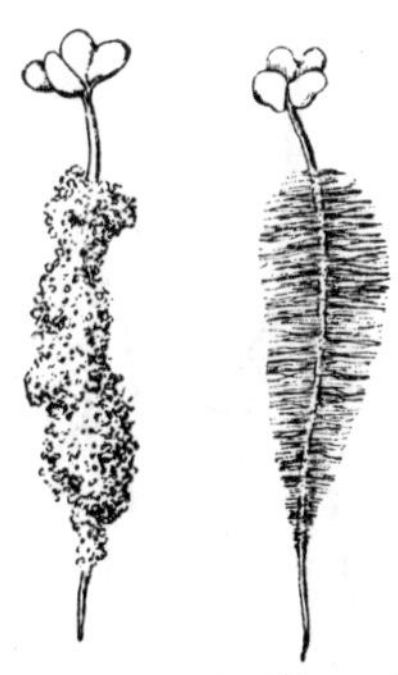

FIG. 45. — Seedlings of mustard. The one at the left shows soil sticking to the root-hairs. The one at the right has grown in moist air. It shows root-hairs free from soil.

D. Root-hairs. — Just behind the root-cap the root-hairs appear. (See *Figures 45* and *46*.) These hairs are really cells of the epidermis whose outer walls are enormously extended. They grow nearly as thickly together as the little hairs which compose the surface of velvet. Suppose there were no root-hairs, as is the case with water roots. Absorption could still go on, but not nearly so rapidly, for the absorbing surface would be very much reduced. The hairs greatly increase the "coast line of absorption." A surface through which liquids can pass is called a *permeable* surface. A root with hairs has five or ten times as much permeable surface as the same root would have if it had no hairs. Water roots make up for the absence of root-hairs by having their whole surface permeable. They can afford to have their whole surface soft better than soil roots can. Why?

As you may already know, the absorption of food in your own body occurs in the small intestine. Here, too, we find an arrangement of absorptive hairs. The walls of the small intestine are densely covered with hairs called *villuses*. Their structure is not at all like that of root-hairs, but they are like root-hairs in that they permit absorption to go on much more rapidly than if they were not present.

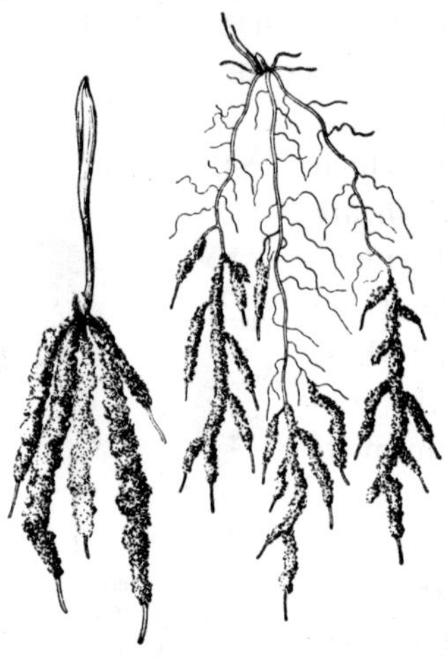

Fig. 46. — Seedlings of wheat, the one at the right being four weeks older than the one at the left. Note the regions of the root-hairs as indicated by the adherent soil.

Root-hairs are very short lived. They last only a few days or weeks. Indeed the whole of the outermost layer of cells of the youngest roots becomes worn off soon after it is formed. The layer under it becomes the outermost layer, and in this layer a substance called *cutin* is formed. It is this cutin which makes the older surfaces of roots impermeable.

As rapidly as the old hairs die, new hairs are formed. All the while the root is growing forward through the soil. Thus the root-hair region is constantly being moved into new and fresher soil regions. These fresher regions may be better for the work of the root-hair. This is not only because they may be richer in the substances which the plant needs, but also they may be poorer in some of the excretions of the plant which appear to be harmful to the plant itself.

The root-hairs adhere very closely to the soil grains. It is almost as though they were glued to the grains. This is shown by the fact that the hairs all break off when you pull the root up, or else they come up, carrying the grains sticking to them. This close relation to the grains appears to be an advantage to the plant, for the water that is closest to the grains, the water that forms a sort of invisible film around them, is that which is richest in the solutes used by the plant as food materials.

You remember that substances pass out from the roots as well as enter them. One of the substances which passes out from the root-hairs is carbon dioxide. This substance may exist as a solute in water as well as a gas. You have already heard about it as a gas. You know that leaves absorb it as a gas and use it in photosynthesis. You now find that roots excrete it dissolved in water. It is one of the wastes resulting from the changes which go on inside the plant, as well as one of the substances used in building up food. This carbon dioxide excreted by the roots as waste is of aid in the work of absorption. In the presence of carbon dioxide some of the substances which the plant needs will dissolve in the soil water much more readily than if carbon dioxide is not present.

E. Cortex and Stele. — The internal structure of a root is better explained by a picture than by words. (See *Figure 47*.)

The cortex is composed of thin-walled cells. From cell to cell of the cortex osmosis readily occurs; the walls are easily permeable. The cells are much alike in shape and size. It is in these cells of the cortex that food is principally stored. It is the cortex which is principally enlarged in tap-roots.

The stele of roots is also called the *central axis*, or the *vascular cylinder*. It is a part of that *vascular system*

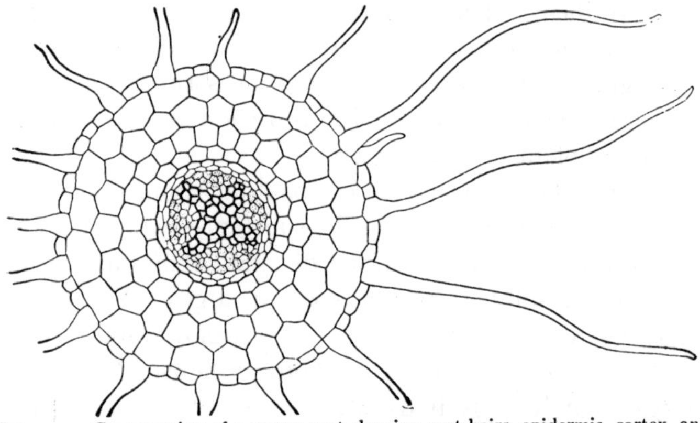

FIG. 47. — Cross section of a young root showing root-hairs, epidermis, cortex, and stele. The stele or central region is chiefly composed of xylem and phloëm. The cells of the xylem have heavy walls; the cells of the phloëm are shaded.

which extends throughout the whole body of the plant. The veins of the leaves and the vascular bundles of the stem are also parts of this system. You are already familiar with xylem and phloëm. The vascular system is principally composed of these tissues. Xylem is also called *wood*, and phloëm is also called *bast*.

You note that in the stele of the root the tissue xylem (heavy-walled) occupies the center. It also radiates from the center. Between the radiating arms of the xylem are groups of cells which are phloëm. The xylem or wood and the phloëm or bast lie along alternating radii. This arrangement of phloëm and xylem is called the *radial arrangement*. The radial arrangement is peculiar to roots. It does not occur in the stems.

44. **Duration of Roots. Annuals, Biennials, and Perennials.** — The length of the life of a plant depends principally upon the length of the life of its roots. Upon the basis of their length of life plants are classified as *annual*, *biennial*, and *perennial*. An annual is a plant which completes its life in one growing season. Thus a plant which begins its growth from the seed in spring and dies, root and all, when winter comes, is an annual. A biennial is a plant which lives in two growing seasons, while perennials are plants which live on from season to season indefinitely.

Trees and woody shrubs are perennials, and there are many other perennials which are alive in the winter though we do not see them. They are plants whose parts aboveground die down just as annuals die down. But their underground parts continue to live, even in frozen soil. Dandelion and grass are very common examples of this sort of perennial, and most of the flowers which you find in the woods in early spring are the flowers of perennials of this sort.

Some plants are either annual or biennial. Wheat is an example. Winter wheat is wheat which is sowed in the fall. It sprouts and the whole field may be green with it before the snow falls. But it is a hardy plant. The snow blanket protects it and it lives on through freezing weather. In the spring it starts to grow again, and the crop is ready for harvest in early summer. Since this winter wheat lives in two growing seasons, it is a biennial. Spring wheat, however, is an annual. It is sowed in the spring and harvested in the same growing season. Shepherd's-purse and pepper-grass are common weeds which show this same habit of being sometimes annual and sometimes biennial. Apparently, hardy plants which are usually annuals are capable of living as biennials if their seeds sprout in late summer. Similarly there are plants, usually biennials, which behave as annuals if they get their start very early in the spring; that is, they will complete their life history and die before winter comes. Also some plants, like the castor bean, are annual in temperate regions and perennial in tropical regions.

Biennials usually devote their energy in the first season to the manufacture of food and its storage; they usually give no attention to reproduction. Food is stored in the underground parts. The next season, using this surplus of food, the plant produces flowers, seeds, and fruit, and so completes its life history. The burdock is a common biennial weed which has the habit just described. In trying to get rid of weeds, it is evidently important to know which weeds are annual and which are not. Only annual weeds are destroyed by destroying their tops. Carrots, beets, and parsnips are biennials. In the first season of growth they develop a large bunch of leaves, but not much stem. The food is stored in the tap-root. If this is left in the ground, next season the growth of the stem will be much greater. Flowers and fruit will be produced, and the fleshy root will wither. Its food will have been used up.

45. Contraction of Roots. — Some plants have roots which seem constantly to be pulling the stem down into the soil. If you have ever dug dandelions, you have noticed how tightly the center of the cluster or *rosette* of leaves seems to be pulled down into the soil. (See *Figure 39.*) Rosette is the name given to such a cluster of leaves as the dandelion has. The dandelion rosette not only *seems* to be pulled down into the soil; it is actually pulled down. It is pulled down by the contraction of its tap-root. How this contraction occurs is not fully understood, but it is easy to understand that it is an advantage to the dandelion to keep its rosette close to the ground. For one thing, it thus escapes the lawn mower. More important, however, is the fact that in this lowly position evaporation is less than in a more exposed position. Plants with this habit are quite common in dry places. They grow in the crevices of rock.

QUESTIONS AND SUGGESTIONS

SECTION 40. 1. What is the principal difficulty in making a school-book about plant life? 2. What are the topics which form the titles of the chapters of this book? 3. Does your knowledge about plants seem to exist in your mind as separate facts or as facts which are related to each other? Explain your answer. 4. Is it easier for you to remember facts separately or to remember them as related one to another? Make a list of ten ideas which the word *leaf* brings to your mind.

SECTION 41. 1. State ten of what seem to you the principal facts about roots.

SECTION 42. 1. Define primary and secondary roots. 2. Describe the secondary roots of corn. 3. Define tap-root and give examples. 4. Of what advantage is a tap-root to the plant? 5. Describe what is meant by fibrous roots. What advantages have fibrous roots over tap-roots? 6. Describe the process of layering. 7. Describe the growth of the strawberry plant. 8. Pull up a piece of blue-grass sod about as large as a saucer, shake and wash away the soil, and then examine the structure. Can you be sure where one plant ends and another begins? Are the roots primary or secondary? 9. Describe the secondary roots of the banyan. 10. Describe the growth of mangroves. 11. How do "water roots" differ in appearance from ordinary roots? 12. What climbing plants use secondary roots as tendrils? 13. Describe the way in which the branches of roots arise.

SECTION 43. 1. By what means do plants grow? 2. What is meristem? 3. Locate the meristem of roots. 4. Distinguish between primary and secondary growth. 5. Define cambium. 6. Locate the region of elongation of roots. Compare it with the region of elongation of stems. 7. Describe a method of demonstrating the region of elongation in roots.

A. 8. Describe the curvature of roots. 9. Do plants have a sense of touch? Compare it with your own.

B. 10. What are the external factors which affect the direction of root growth? 11. What are tropisms? 12. Define phototropism, geotropism, and hydrotropism. 13. State the differences between roots, stems, and leaves as to their tropisms. 14. What is a stimulus? 15. Describe the growth of roots and shoots with the effect of gravity removed. 16. Does water or gravity produce a greater effect on the growth of roots? How may your answer be proved?

C. 17. Describe the root-cap. 18. Describe the root-cap of a screw-pine. What does its development indicate? 19. Define the terms heredity and environment.

D. 20. Describe root-hairs. 21. What is meant by the word permeable? 22. What is cutin? 23. Is there any advantage to the plant in the fact that it must constantly produce new root-hairs? 24. What is the advantage in the close adherence of the hairs to the soil grains? 25. What advantage in the excretion of carbon dioxide by the root-hairs?

E. 26. Describe, making a sketch on the board, the internal structure of roots. 27. What advantage is there in having the xylem and phloëm radially arranged in roots?

SECTION 44. 1. Define annual, biennial, and perennial. 2. Give examples of annuals, biennials, and perennials. 3. To which of these classes do the plants most valuable to man belong? Explain your answer by giving examples. 4. Explain the fact that the same plant may be sometimes an annual and sometimes a biennial. Give an example. 5. Does the annual or the perennial habit appear to be more successful in plant life? State the advantages of each.

SECTION 45. 1. Describe the contraction of roots in the dandelion. 2. What advantages are there in this habit?

CHAPTER V

STEMS

46. Introductory. The great object in life for most stems appears to be to get the leaves which they bear in good position with reference to light. The roots seem to strive to place the root-hairs advantageously as to moisture and the stems seem to strive to place the leaves advantageously as to light. Many things other than leaf position affect the character of stems, and many stems have nothing to do with getting leaves into the light, yet, speaking of stems in general, it is fair to call them the *leaf-related* organs. Principally, they serve the leaves.

To have leaves high in the air and root-hairs well-placed in the soil seem to be the ends for which many plants are striving. In the forests high leaves and deep roots are secrets of plant success, for leaves high in air are almost sure to get light enough and roots deep in soil are almost sure to get water enough. Shade and drought are enemies. To overcome them is the first great task of the plant. In forests, where hosts of plants compete, the struggle for light and water goes on all the time. The cultivation of plants by man, the whole art of agriculture, has consisted chiefly in making this struggle easier for the plants useful to man.

It is with the struggle for light that stems are principally concerned. The history of forests is the history of an endless struggle for light. In that struggle it is the stems which play the principal part, and those forms of stems which exist to-day are those which have been successful in that struggle. Old competitors of theirs which were not successful have disappeared.

In Chapter I you learned of stems as helping organs. You learned that they place flowers and fruit as well as leaves in advantageous positions. You learned that through them run paths of movement from roots to leaves and from leaves to all other living parts of the plant. You learned, in Chapter II, that the root-to-leaf movement is through the xylem or wood and the leaf-to-root movement largely through the phloëm or bast.

You have learned something of the structure of stems; you know of the nodes and internodes. You know that green stems make food just as leaves make it, and that in some plants all the food-making is done by stems. You have learned something of the tissues of stems. You learned that their arrangement is quite different from the arrangement of the tissues of the roots. While the tissues are arranged in practically the same way in all roots, there are two quite distinct types of tissue arrangement in stems, and neither of these is like the tissue arrangement of roots. In one of the stem arrangements the vascular bundles form a cylinder which in cross section appears as a ring (see *Figure 48*); in the other arrangement the vascular bundles are scattered (see *Figure 49*).

All stems may be divided into those which are aboveground and those which are underground. Since the stems which are above the ground are more familiar, we shall consider them first. We may call them *aërial stems*.

47. Aërial Stems. — We may classify aërial stems as to their general character or as to their internal structure. As to internal structure they fall into two great classes based on the arrangements of vascular bundles which have just been mentioned. These will be considered in the section on the structure and growth of stems (*Section 50*). As to their general character, there are various ways in which stems may be classified; as to position they may be classified as *erect*, *prostrate*, or *climbing;* as to length of life they may be classified as *annual* or *perennial;* as to hardness and strength they may be classified as *woody* or *herbaceous*.

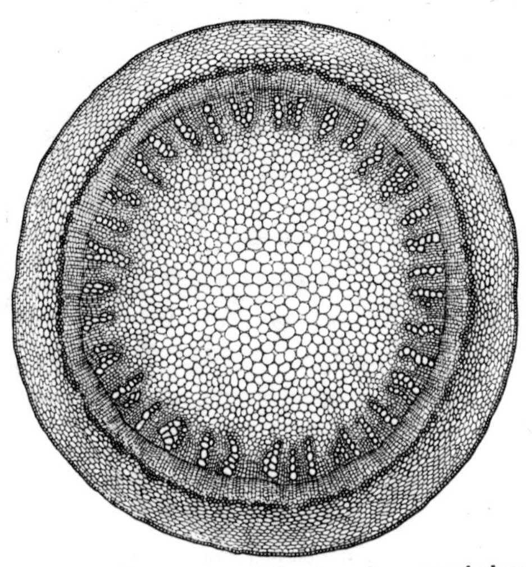

FIG. 48. — Cross section of the stem of an annual plant showing the arrangement of the vascular bundles in a cylinder which here appears as a ring. The center is occupied by pith. Outside the vascular bundles is the cortex.

Herbaceous means having to do with herbs. Herbaceous stems are the stems of herbs; they are generally soft and easily broken. Though they have xylem, they are not woody in the ordinary sense of the term. Herbaceous stems are generally annual, though the plant to which they belong may be a perennial on account of its perennial underground parts. Woody stems are perennial.

Probably the oldest classification of plants is the classification into *herbs*, *shrubs*, and *trees*. This classification is based upon qualities which pertain to the stems. Herbs are smaller and weaker than shrubs and trees, the wood or xylem in them is not largely developed, and they die down, at least to the ground, when winter comes. Shrubs are intermediate in size between herbs and trees, and their woody stems are usually arranged in a cluster of approximately equal shoots which arise at or near the ground, thus forming what are called bushes. Trees have tall, woody, perennial stems, and usually one main stem, called the trunk or *bole*.

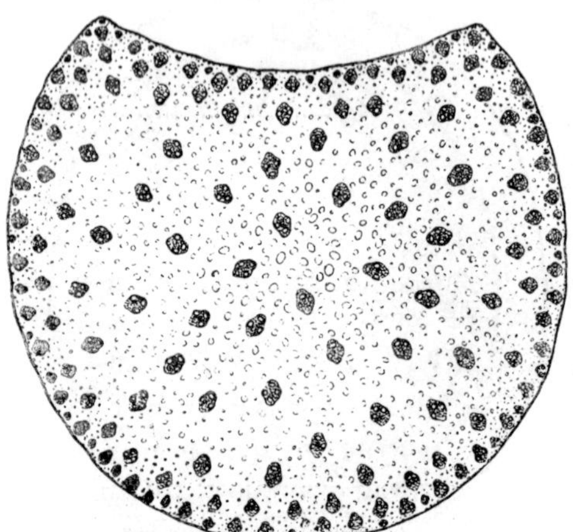

FIG. 49. — Cross section of a corn stem showing the way in which the vascular bundles are scattered through the pith.

This classification into herbs, shrubs, and trees is often convenient in general descriptions of vegetation, but it is not satisfactory for other purposes, for the reason that there are so many gradations between trees and shrubs and between shrubs and herbs.

A. Erect Stems. — Stems are usually erect. Under the ordinary conditions of plant growth an erect stem is the best device for properly relating the leaves to light. Erect stems usually have more or less horizontal branches. In the arrangement of its leaf-bearing branches the plant appears to avoid as much as possible the shading of one branch by another; the branches are usually arranged so that they secure for the leaves the *maximum illumination*, that is, as much light as possible.

The building of strong, erect stems is a problem in construction which plants solve in two distinct ways: (1) Some erect stems are hollow; the hard tissues are arranged in the form of a hollow cylinder; stalks of wheat and the fruit-bearing stems of dandelion are good examples of such stems. (2) More commonly, however, erect stems are solid; the hard tissues are arranged in the form of a solid cylinder; the trunk of a tree is an example of this arrangement. Erect stems which are annual are either hollow, or else their centers are occupied by that soft tissue called pith; erect stems which are perennial are usually solid.

You have already noted that the tissue called xylem or wood is composed of cells with thick walls. It is this thickness of the walls of the xylem which principally gives to stems their strength and stability. The strength of the xylem is frequently supplemented, however, by the strength of what is called *mechanical tissue*. Mechanical tissue is tissue which is of service to the plant simply by its presence; it does not participate directly in the life processes of the plant; its cells may be dead. Similarly, in your own body, the hard parts of your bones are a sort of mechanical tissue. Xylem, as it grows older, may become purely mechanical tissue, as, for example, the older wood of trees.

The stems of trees are the most conspicuous of erect stems; the most conspicuous, in fact, of all aërial stems.

Their beauty impresses us as well as their utility. Have you ever seen an ugly tree? Tree stems show great variety in form. This variety suggests that there is no one form that has great advantages over all other forms. If this were so, the trees having that one best form probably would have crowded out all the others. Probably each common form has certain advantages of its own.

The two principal types of tree form are illustrated by the pine and the oak. Trees like the pine in form have one great central trunk running clear to the top and from it the branches spread out horizontally. The *crown* of such trees is cone-shaped; by crown is meant the branch-bearing part of a tree. Trees like the oak in form have spreading crowns, more rounded than cone-shaped in outline. Their main stem does not continue to the top; it soon divides into large branches.

The elm is one of our most beautiful trees, and its form is quite different from that of the pine or the oak. It has a form which is very characteristic; that is, it is quite different from the forms of other trees. You can tell an elm by its graceful and characteristic form almost as far away as you can see it. Instead of becoming rounded or conical in outline, the elm tends to assume the form of a huge vase with broad, flaring top. There are many trees besides the elm which by their form you may recognize at some distance if you study the matter a little and make observations with care. Usually we give trees a hasty glance and are satisfied with a general impression of their beauty or of their novelty to us. But it will be found far more satisfying to give them enough study to be able to recognize them. Different kinds of trees are quite as distinctive as are the faces of the people we meet. We notice faces and remember them. Similarly we might notice trees and remember them.

Hollow stems have an advantage over solid stems in that they attain erectness and height at much less expense; that is to say, the amount of material which the plant must manufacture in building them is much less than if they were solid. To be sure, they are not always erect. They sway in the wind. Perhaps you have noticed the beauty of a field of grain or tall grass bowing as the breezes blow over it. But though they bend, the hollow stems of grass and grain do not break easily; they bend without breaking, and a storm may do more damage to the high stiff stems of a forest than to the swaying stems of a field of wheat. Hollow stems, however, generally endure for but one season and never attain the great heights of solid stems.

B. Prostrate Stems. — If to secure light for leaves is the main work of stems, why should any stems which bear green leaves be prostrate? Surely this is a position unfavorable for leaf display; it seems about the most unfavorable position possible. Yet there must be some advantage in it, since prostrate stems are very common. They lie sprawling on the ground and run the risk of being shaded by even the lowest of plants with erect stems.

An answer to our question may be found in the expression, " Many kinds of places, many kinds of plants." In other words, the principal reason why we find so many kinds of plant forms appears to be that there are so many kinds of places in which plants grow. A plant form which is sure to fail in one place may be just the thing to succeed in another.

Plants with prostrate stems are especially abundant on sterile soil. You may have noticed them on sandy beaches where plants with erect stems are very few and far between. In such places prostrate plants have certain advantages. There is the advantage that they are not in danger of being shaded by other plants; there is the advantage over plants with erect stems that they are not so much exposed to the danger of loss of water by evaporation more rapidly than the roots can supply it; there is the advantage that they are not called upon to devote nearly so much work and material to the making of their stems as are erect plants. Prostrate stems have another advantage over erect stems in their chance to reproduce at every node. So, all in all, there is a good deal to be said in favor of prostrate stems, in spite of the fact that they are poor devices for getting leaves up into the light.

Some plants with erect stems also produce prostrate stems which appear to be for the express purpose of reproduction. You recall this in the case of the strawberry plant whose prostrate stems or runners have been described. (See *Figure 40.*) The common white clover also reproduces by runners; this matter is also illustrated in the process of layering. (See *Section 41.*) Since all the nodes of prostrate stems are in contact with the soil, there is the chance for them to develop root branches as well as stem branches and so establish new plants.

Prostrate stems send up leaves from their upper side only. Frequently you find prostrate stems twisted between one node and the next, so that side of the stem which was under at one node is above at the next node. Thus, though the leaves all arise from the side which is upper at their point of origin, yet those at one node, on account of this twisting of the stem, usually arise from a different side from those at the neighboring nodes; it is a different side which is upper. (See *Figure 50.*)

Evidently it is a disadvantage to prostrate stems to have leaves arising first from one side of the stem and then from the other; the stem must twist to overcome this disadvantage. Erect stems, however, have the same habit and the advantage to them is apparent. Erect stems are

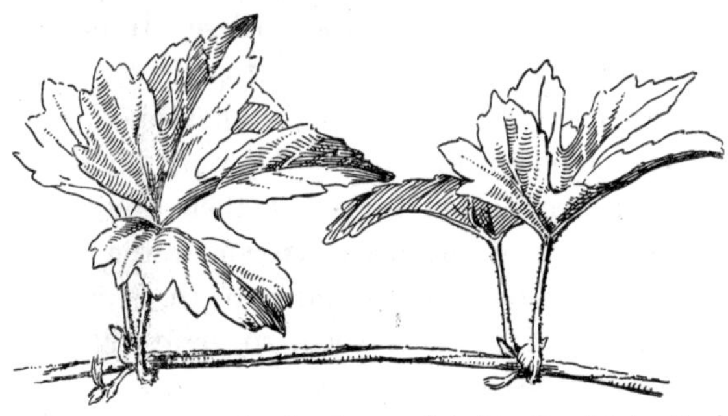

FIG. 50. — A prostrate stem showing the way it twists, thus helping the leaves in attaining a favorable relation to light.

not twisted, but adjoining nodes give rise to leaves on different sides, and this is evidently a good device for the avoidance of the shading of one set of leaves by the set next above it. (See *Figures 1* and *5.*) In prostrate stems we seem to have in this matter a case of an inherited habit which is of no advantage. It suggests that the ancestors of prostrate plants with twisted stems may themselves have been erect.

C. Climbing Stems. — In the introductory part of this chapter you read of the struggle of plants for light, especially in forests. It is in forests, and apparently it is on account of this struggle for light, that climbing stems show their greatest display.

In the dense forests of the tropics great woody vines abound. They hang from the trees and grow up from the ground, for they come from seeds which sprout on the branches of the trees as well as in the soil. Those which are growing down in the air as well as up in the air have free strands which swing like ropes from lofty branches, but when they have taken root, these strands become stretched taut. The whole forest top seems sometimes interlaced with these woody vines. They make easy highways for the troops of monkeys. They mount to the tops of the tallest trees where sprays of their flowers may be seen in the midst of the dense crown of the tree which supports them. Sometimes their growth is so vigorous and the burden of carrying them so great that the supporting tree perishes under its load.

Such woody, rope-like vines are called *lianas* (pronounced *lē-ä′nȧ*). (See *Figure 51.*) Our own wild grape often grows as a liana. In river bottoms it is common to find trees which have been broken down by the weight of grapevines upon them.

In the intense struggle for light in tropical forests some plants are found climbing which under other conditions stand erect. This is evidence of a rather unusual capacity to respond to environment.

There are three quite distinct ways in which climbing plants attach themselves to their supports: (1) by twining their stems about the support, (2) by tendrils which twine about the support, and (3) by tendrils which end in sucking disks or holdfasts. The poison ivy is a common climbing plant whose climbing organs seem more like roots than tendrils. They are roots as to origin. They penetrate the tissue of the plant on which the ivy climbs.

Beans, peas, and the morning glory are examples of plants which climb by means of twining stems. In these plants the young elongating part of the stem swings free. It does not stand erect. Its end seems to be describing an arc of a horizontal circle. It moves slowly, following a horizontal curve, until it comes in contact with something. Around this something it swings itself if it can, growing as it swings.

FIG. 51.—A tropical forest showing many lianas.—*Photo. by Whitford in the Philippine Islands.*

You have already read a little about tendrils. You have learned of their delicate response to contact. (See *Sections 41* and *42*.) The climbing or star cucumber is a plant whose tendrils have been carefully studied as to the delicacy of their response. Its tendril tips have been found to respond to the touch of a swinging object which weighs only one eighth as much as the lightest swinging object which our skin is able to perceive. Such tendrils as those of the star cucumber respond to contact in the following manner. First the touched side of the tendril grows less rapidly than before, and the untouched side more rapidly than before. This results in a curve about the object touched, just as you noted in the case of root tips. After this, the rest of the tendril begins to curve about the support, until at last the plant is fastened as though by spiral springs. (See *Figure 52.*)

FIG. 52. — The curving tendrils of the star cucumber vine.

Woodbine and some kinds of ivy are common vines, the ends of whose tendrils, instead of twining, develop sucker-like disks. These sucker-like disks attach themselves firmly to walls or to the trunks of trees; so firmly, in fact, that when you try to pull them off the tendril breaks and leaves the suckers attached to the surface of the support. It is evident that this manner of climbing is particularly well suited for vines which grow on walls.

D. Exceptional Forms of Stems. — The branches of stems usually resemble in structure the stems from which they arise, but there are some striking exceptions.

The plant commonly called smilax, or wedding smilax,

has branches which look exactly like leaves. So far as function is concerned they are leaves, but so far as origin is concerned they are stems. If you examine this plant closely, you will notice that the apparent leaves grow from the axils of small, scale-like structures. These small, scale-like structures are the true leaves, while the leaf-like branches which arise from them are exceptional forms of stems. Such exceptional forms of stem branches are called *cladophylls*. (The word means *branch-leaf*.) Asparagus has already been given as an example of a plant whose stems do work ordinarily done by leaves. Its needle-like branches may be also properly called cladophylls, though their resemblance to leaves is not so striking as in smilax.

Thorns, as to origin, are sometimes stems, sometimes leaves. The same is true of tendrils. Thus we note that stems and leaves appear to be equally capable of becoming the same organ as to function if not as to origin. The honey-locust and the hawthorn are plants whose thorns are stems as to origin.

48. Underground Stems. — Horizontally elongated underground stems are called *rhizomes* or *rootstocks*. If a portion at the end is much more swollen than the rest, it is called a *tuber*. Underground stems sometimes bear leaves. If the leaves are fleshy and arise from a short stem, the structure thus formed is called a *bulb*. A bulb, therefore, is an underground shoot, the word shoot, as you know, denoting stem and leaves together. A *corm* is another kind of underground shoot. In it the stem is more prominent than the leaves, most of the food is stored in the stem part, and the leaves may be simply protecting scales.

A. Rhizomes. — The word means *root-like.* Rhizomes sometimes form the entire stem system of the plant. Some violets and nearly all ferns have no true stems above ground. The leaves arise directly from the stems and are lifted into the air by their petioles. It is more common, however, for rhizomes to give rise to true stem branches. These, of course, push up into the sunlight instead of remaining in the ground. Such branches are negatively geotropic instead of transversely geotropic like their parents. They bear leaves and flowers and manufacture food. Much of this food is stored up in the rhizome to be used in the formation of young shoots in the following season. The rhizome

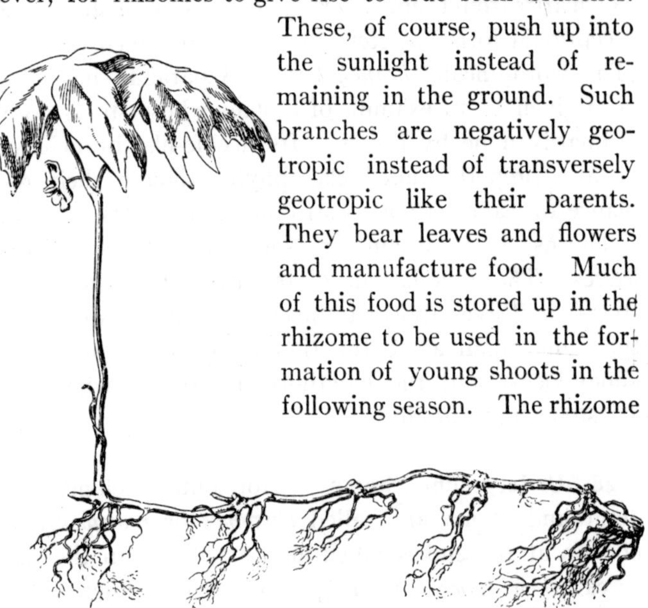

Fig. 53. — A single plant and the perennial rhizome of the May apple or mandrake. The rhizome is several years old, the growth of each year being marked by a cluster of roots. — *Redrawn from Bergen.*

itself is perennial. The ferns and many of the early spring flowering plants arise from perennial rhizomes. The May apple or mandrake, the bloodroot, the violet, the trillium, and the Solomon's seal all have this habit. (See *Figure 53.*) Iris and water lily also grow from rhizomes. It is the rhizome habit that makes some weeds so very hard to kill out. The rhizomes of quack-grass, a common weed, may be hoed to pieces, and still new plants will spring up from the nodes. Among garden plants, asparagus and rhubarb grow from rhizomes. It is evident that a rhizome is both a reproductive and a nutritive organ.

Rhizomes possess power to assume certain rather definite positions in the soil. Each kind appears to do well at a certain depth, and, if changed from that depth, will grow up or down until the proper depth is attained again. Thus if a trillium rhizome is transplanted and not put at the right depth, it will grow up or down until its particular soil level is reached. It will then begin its horizontal growth again. The causes of this kind of response are not well understood.

By growth forward and death behind rhizomes gradually progress through the soil, putting up new shoots from season to season as they go. Occasionally they branch. When death reaches the point of branching, the branch is severed from its parent and thus separate rhizomes are produced. It is evident that this rhizome habit of reproduction has some advantages over the seed habit. Indeed some plants are reproduced by their rhizomes far more than they are by seeds. The bamboo, for example, produces flowers and seeds only about once in twenty years, apparently having given up the seed habit almost entirely in favor of the rhizome habit. Taking plants in general, hundreds of seeds die for one that develops. This great loss is chiefly due to the fact that on ground already occupied by other plants it is next to impossible for the seedling to get its root into the ground and at work before its food reserve is consumed. This difficulty, however, is no difficulty at all for rhizomes. They can burrow along among other plants just about as well as where there are no plants. Indeed they seem to be most successful in soil which is already well occupied by other plants.

On freshly cleared land, or on fields left fallow after a crop, annual plants spring up in great abundance in the first season. This is because the light, wind-blown seeds of the annuals are the first to occupy the new territory, and because their seedlings have an unusual chance to take root. But gradually rhizome plants and other perennials encroach on the annuals, and, if undisturbed, will finally drive them out.

B. Tubers. — It is the potato which has made tubers famous. If it were not for this celebrated plant, the word tuber would be no more familiar than the word corm. But the potato has proved one of the best friends which man has in the plant world, and every one should know something of its culture.

Probably you have noticed the eyes on the skin of the potato. The eyes are at the nodes of this swollen, underground stem. They are composed of small scale-like leaves and buds. The bud is at the axil of the leaf. (See *Figure 54.*)

FIG. 54. — A potato showing the eyes. These are composed of scale leaves with buds in their axils. One of these buds has begun to sprout.

When potatoes are to be planted, selected tubers are cut into pieces, each piece bearing one or two eyes. These pieces are planted a few inches below the surface of the soil and about a foot apart. From the eyes new plants develop. Potatoes do well in soil which may not be good for grain crops. A loose soil is better for them than a hard, compact one; it offers less resistance to the enlargement of the tubers. After the shoots have come up, the soil is heaped up into hills about the base of the plant. This stimulates the formation of tuber-bearing branches.

Very likely you have noticed potatoes sprouting in the cellar. Long, pale shoots grow from the eyes. They seem to be seeking the light. If they find it, they soon turn green. This is on account of the manufacture of chlorophyll. Evidently the plant does not manufacture chlorophyll in the dark.

Another tuber important for food is the Jerusalem artichoke. The plant is a sort of sunflower. In this tuber the nodes and the buds are much more prominently developed than in the potato.

C. Bulbs. — An onion is the most familiar kind of bulb. Lily, tulip, and hyacinth bulbs are also well known. A bulb is a modified shoot in which food is stored, the leaves being more prominently developed than the stem. In onion and tulip bulbs the leaves are so broad that they completely cover those within. In lily bulbs the leaves overlap, but do not completely cover each other. (See *Figure 55.*)

Tulip beds are usually made by planting the bulbs rather late in the fall. The cold of winter does not injure them, and they prepare for growth in early spring. It is well to cover the beds with straw and manure. Such a covering is called a *mulch.* The mulch prevents sudden changes in temperature in the ground beneath it. It also enriches the soil. In the spring, when the ground is not likely to freeze again, the mulch may be raked off. Soon the young shoots of the tulip appear. At first they draw nourishment from the bulbs, but after they are fully developed, they will store nourishment back in the bulbs again. New bulbs may develop as branches from the old ones, and new and old ones may both be used in making beds for the following season.

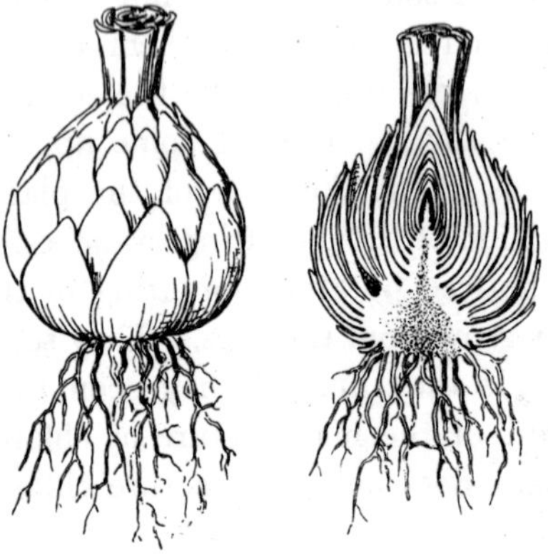

Fig. 55. — A lily bulb. In the view at the right the short stem and the overlapping scales are evident.

D. Corms. — Most boys and girls who live in the country have tasted the Indian turnip. It has a taste which no one forgets, and no one cares to taste it more than once. After that, you are willing to let others taste it for you. It does not poison you unless you eat too much of it, but it does make you wish for quite a while that you could get away from your mouth. By cooking, however, the Indian turnip may be made palatable and the Indians did use it for food.

The Indian turnip is a corm. It is the base of the stem of that interesting plant, Jack-in-the-pulpit (see page 279). The flowers are quite common in the woods in May. The corm is a fleshy, rounded, underground part of the stem with scale leaves on it. It grows from year to year and sometimes forms small corms as branches. These separate from the parent corm and form new plants. The spring-beauty (*Claytonia*) and the pepper-and-salt (*Erigea*) are other common spring flowers which arise from corms.

E. Stools. — Some garden plants, like the canna, form what are called stools by which they are propagated. These stools are composed of both the bases of the stems and the tops of the roots, in both of which food is stored. The clustered, fleshy roots of the dahlia are used in its propagation, and they also are called stools.

49. **The Vernal Habit.** — Underground stems have much to do with what is called the *vernal habit*. Vernal means having to do with the spring of the year. It is in spring that flowers bloom in abundance in the woods. These are not the flowers of summer or fall. They are flowers which are gone before summer comes, flowers which are peculiar to spring, flowers which reach their greatest abundance in late April or early May, flowers which do much to make spring the most attractive season of the year and botany an interesting study. Soon after the flowers, or with them, the leaves of these plants appear, and before the leaves of the trees are full grown, before the shade they make has become dense, the principal work of these lowlier plants is done. Their season of growth and reproduction is finished even before summer comes, and they are ready to rest until spring comes again. It is this habit of growth and reproduction before the leaves of the trees are out that is called the vernal habit.

The vernal habit illustrates well the apparent capacity of plants to take advantage of all their opportunities. In what you read about climbing stems you learned something of the intense struggle for light in tropical forests. Not only are myriads of lianas to be seen there, but orchids and other *epiphytes* grow upon the branches of the trees. Sometimes the trees support dozens of different kinds of stems and leaves besides their own; advantage seems to be taken of every inch of space available for leaf work. (Epiphytes are plants which grow, root and all, upon other plants; *epi* means *upon; phyte* means *plant.*) Similarly in our own forests there is competition for light, and the vernal habit appears to be a result of this competition. In our own forests the struggle for light is not so continuous as in the tropics; there the struggle goes on month after month throughout the year, and at no season do low plants have much chance. With us, however, there is a growing season of about thirty days before the trees are in full foliage, and at this season the low plants of the forest have their best chance; they take advantage of this brief period when sunshine strikes down to the forest floor.

Most of the plants usually studied by high school classes on field trips in the spring are plants which arise from bulbs or rhizomes or corms and have the vernal habit.

50. **The Structure and Growth of Stems.** — This division of the study of stems is a little more difficult to understand than what has preceded. We have been discussing stems as to their outside, and with the outside of stems you were already somewhat familiar. We now must discuss stems as to their inside, and with the inside of stems you are not familiar. We have been discussing things which you can see easily with the naked eye. We now must discuss things which it takes a microscope to see. No matter how clear the description, facts which it takes a microscope

to see are always more difficult to understand than facts which may be seen with the naked eye.

It will not be so easy to understand the structure of stems as it was to understand the structure of roots. The structure of roots is of just one general kind; the structure of stems is of two general kinds. The arrangement of tissues in the root is simple; the arrangement of tissues in the stem is complex.

So you are warned to give this section particularly careful attention in order to get it straight; the structure of stems is a matter which high school students are in the habit of getting mixed. Before reading ahead, you should review the section on the structure and growth of roots. (*Section 43.*)

A. Contrast with Roots. — Epidermis, cortex, and stele are the three great divisions of the tissues of roots. They are also the three great divisions of the tissues of stems. The stele of stems often includes a central tissue called pith. Roots do not have pith.

Meristem is tissue whose cells retain the power to divide into new cells. You learned of it in connection with roots. In stems it is found principally in two places: in the center just behind the growing tip, and in older parts near the edge of the stele. The meristem near the growing tip is called primary meristem; that in older parts is secondary meristem. Primary growth is growth due to the enlargement of cells formed by the primary meristem; secondary growth is due to the enlargement of cells formed by secondary meristem. Secondary meristem in stems is also called cambium.

Roots have the radial arrangement of xylem and phloëm.

(See *Figure 30.*) Stems do not have this arrangement. The growing tip of a root is covered by a protecting cap. The growing tip of a stem has no such protecting cap. Stems are jointed; roots are not. Stems have outgrowths called leaves; roots have no such outgrowths. The branches of roots arise at the edge of the stele and burrow through the cortex to the epidermis. The branches of stems arise from the outer part of the cortex, at the axils of the leaves. Roots are positively geotropic, positively hydrotropic, and negatively phototropic. Stems are usually negatively geotropic, negatively hydrotropic, and positively phototropic.

B. Epidermis, Cortex, and Stele. — It is the arrangement of the tissues of the stele which is the most complicated feature of stem structure, and to that feature we must give especial attention. As compared with the stele, the epidermis and the cortex are simple and show little variation. Most of the cells of the cortex, as you remember, have thin walls and nearly equal diameters. Tissue composed of cells like this, whether it be in the cortex or elsewhere, is called *parenchyma.* (The word means *parent-tissue.*) All undifferentiated young cells are parenchyma; some remain parenchyma, others change to tissues of other kinds. Parenchyma is the name of a tissue as judged by its appearance rather than by its position; thus most of the cells of the cortex are *cortical parenchyma;* the loose cells inside the leaf are *spongy parenchyma*, etc.

You will recall that in the cells of the older epidermis of the root a substance called cutin is manufactured and that this substance renders the cells which contain it impermeable to water. This and other substances also appear in the cells of the epidermis of stems and they, too, are thereby rendered impermeable to water.

C. Arrangement of the Vascular Bundles. — In stems, which are unlike roots in this respect, the tissues of the stele are organized into separate strands or bundles called vascular bundles. Each bundle is composed of xylem and of phloëm. The xylem furnishes the ascending or outward path of movement, while the phloëm furnishes the descending or inward path of movement. Note, however, that the movement through the phloëm to flowers and fruits is usually ascending and outward.

There are two distinct ways in which the vascular bundles are arranged in stems, and the two distinct kinds of stem structure are based upon these two ways of arrangement of the vascular bundles. In one of these ways the bundles are arranged in a cylinder; in the other they are scattered. We will call one the *cylindrical arrangement* and the other the *scattered arrangement.* (Compare *Figures 48* and *49*, pages 145 and 146.)

D. The Cylindrical Arrangement. — Nearly all perennial stems and many annual stems have the cylindrical arrangement of vascular bundles. We will consider it first in perennial stems. A good way to understand the cylindrical arrangement and the growth which occurs in connection with this arrangement is to study the twigs of trees and what happens to them.

The cut end of a twig looks very different from the cut end of the trunk that bears it. (Compare *Figures 56*, *57*, and *58*.) Yet the trunk once had exactly the same structure as the twig has now. How have all these changes come to pass? If we are to understand them, we had better study the twig first and then see what happens to it as it grows older.

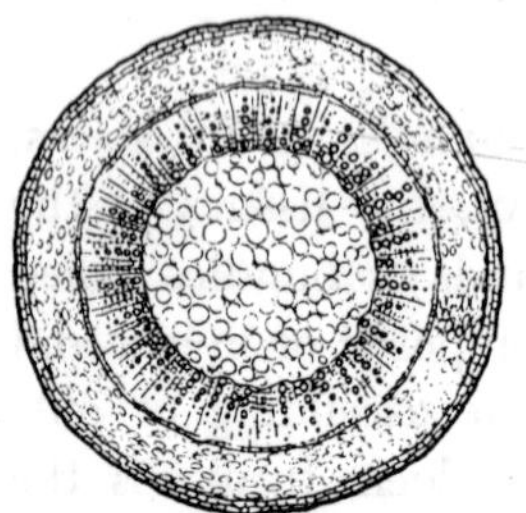

Fig. 56. — Section of a one-year-old twig of box elder.

A one-year-old twig of any common tree shows four distinct tissue regions. (See *Figure 56.*) Outermost there is the epidermis; it is a thin skin that often may be easily stripped off. Under the epidermis lies soft tissue usually containing chlorophyll; this is the cortex. Next comes a zone of hard tissue; this is the vascular cylinder; it is chiefly composed of wood. Finally, in the center, there is the pith, a white, spongy tissue most of which soon dies.

If we examine a three- or four-year-old twig of the same tree, we find that a great change has occurred. (See *Figure 57.*) The tender epidermis has disappeared and a new and darker protective covering has been developed by the outer part of the cortex. This new covering is called *bark.* The cortex has lost much if not all of its chlorophyll; it is ceasing to be an organ of photosynthesis; the bark does not let light or gases through as the epidermis did. The wood cylinder has become

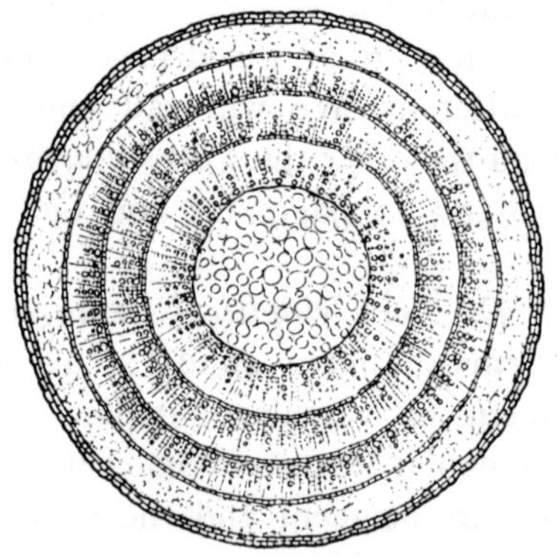

Fig. 57. — Section of a three-year-old twig of box elder.

greatly enlarged; it now forms the principal part of the stem; the new wood has been added in the form of rings. The pith is much reduced in area; it seems to be disappearing.

Now if we turn to the trunk itself, we find that the same kind of change has continued. (See *Figure 58.*) The pith is gone entirely. Where once there was but a cylinder of wood, there is now a practically solid column which makes up nearly all of the stem. From the center clear out to the bark, ring after ring of wood shows the growth of many seasons. The cortex is now nothing more than a sap-filled inner part of the bark. The bark itself is thick, and furrowed, and corky.

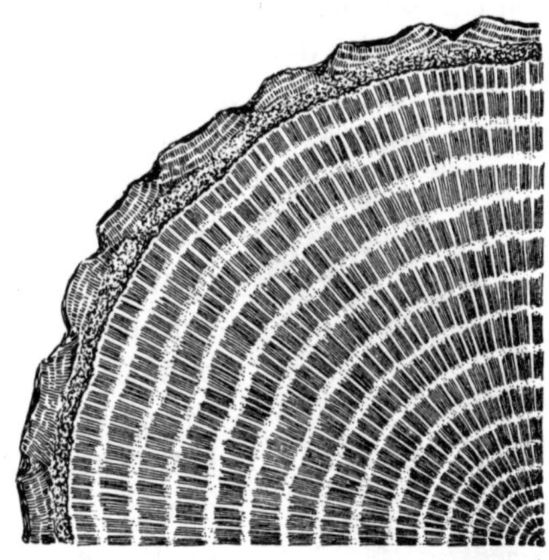

FIG. 58. — Section of the trunk of a young oak.

How have all these changes come to pass? By what method has all this secondary growth been accomplished, all this wood been formed? What is the meaning of the rings in the wood? What can we learn of the process which thus produces timber for the world?

By the secondary growth of perennial stems the timber supply of the world is maintained. There is no other way. If the timber is cut more rapidly than this growth can renew it, the supply is sure to become exhausted.

In order to understand this formation of wood, it is necessary to understand the structure of an individual vascular bundle. That cylinder of wood that you saw in the one-year-old twig is made up of vascular bundles, but they lie so very close together that the individual bundles are not distinct. They are separated only by very thin plates of pith-like cells which run between them out to the cortex. In some stems, especially the stems of pine, these *pith-rays* are quite prominent; they form radiating lines which run through the wood from the center to the cortex; you can see them easily. (See *Figure 58.*) In other stems the pith-rays disappear as the stem grows older. Pith-rays are also called *medullary rays* (*mĕd'ŭ-la-rĭ*).

The structure of an individual bundle may be best understood by study of a stem whose bundles are distinct, being separated by wide pith-rays. (Study *Figure 59.*) You note that the xylem part of the bundles (the inner part) is larger than the phloëm. Do you see any relation between this difference in the extent of these two tissues and the difference in the work which they do? The phloëm or bast lies outside the xylem and is separated from it by a layer of cells which are narrowly oblong in cross section. This layer of narrowly oblong cells is cambium (see page 129). This cambium has a very important part to play in the subsequent growth of the stem.

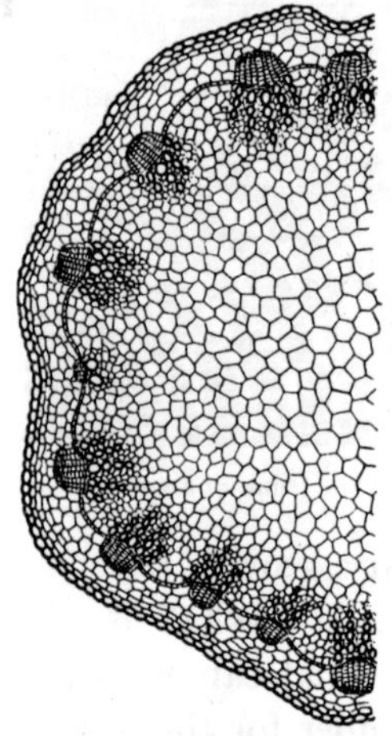

FIG. 59. — Section of a stem whose vascular bundles are distinctly separate. From bundle to bundle runs a thin belt of tissue, the *cambium*. Within the bundles the cambium passes between the xylem and the phloëm.

As the young bundles lie side by side, touching or nearly touching, the layer of cambium in one bundle soon becomes connected with the cambium of its neighbors. Thus there is formed an unbroken cylinder of cambium, which is referred to as the *cambium ring*. In this way the bundles become connected; in some stems they are closely side by side from the first, in others they are at first bridged by the extensions of the cambium and gradually grow together, or, as in many stems of annuals, they may remain distinct. (Compare *Figures 24* and *59*.) In any case, a complete cylinder of cambium is soon formed. Remembering this cylinder of cambium and its power to produce new cells, it is not difficult to see where the secondary wood comes from. In each growing season the cambium forms new cells. This is done by lengthwise divisions of its old cells. Of the new cells thus formed, those on the inner side of the cylinder enlarge and become wood cells (xylem); those on the outer side become cells of the phloëm. The new layers of wood, thus formed each season, increase the diameter of the stem, and the cambium which formed them is compelled to adjust itself to this constant increase within it. The cambium does this by growth of its cells in length and by some crosswise divisions. Growth on the outside of the cambium is not so great as the growth within it. The cells of the phloëm are not so large as those of the xylem. With these facts in mind we can see how it happens that trunks of trees are composed almost wholly of wood; we can understand how it happens that the cambium lies quite near the surface, just under what is called the *inner bark*. This inner bark is composed of cortex and phloëm.

The outer bark is composed of tissue called *cork*, often with dead cortex in addition. This cork is produced by what is called the *cork cambium*, a layer of cells which forms bark only. The cork cambium has no connection with the inner cambium we have just been describing. It produces cork cells and produces them only on its outer side. It is the outermost layer of the cortex, or sometimes a deeper one, which is transformed into this cork cambium. The cork from which corks for bottles are made comes from the cork-oak, a tree which is abundant in Spain. The outer bark of the cork-oak is very thick and is formed in layers like the layers of wood. It is stripped off in sheets, which are sold in the market as cork. Cork is very impermeable to water. It is this quality, along with its pliability and lightness, which gives it commercial value. Its cell walls have become thickened with *suberin* (*sū' bĕr-ĭn*), a substance somewhat similar to the cutin of which you learned in connection with your study of roots.

In view of the continued growth of the wood, it is not difficult to understand the cracks and furrows which appear in the outer bark. It is constantly being ruptured by pressure resulting from the growth of the wood within.

The cylindrical arrangement in annual stems differs from the cylindrical arrangement in perennial stems chiefly in the fact that in the former the vascular bundles are more widely separated than in the latter. Sometimes they are not even connected by extensions of the cambium. Sometimes, in stems of this kind, the cambium does not appear at all. Since such stems die at the end of their first growing season, evidently they have little need for cambium and the secondary growth which cambium causes. For their purposes primary growth is often sufficient. Stems of this kind are usually either pithy or hollow. The stem of the sunflower is a good example of a stem of this type which is pithy; asters and golden rod and many road-side weeds also have pithy stems of this type. Of hollow stems of this type that of the jewel weed or balsam (*Impatiens*) is an excellent one for observation. It is abundant in damp woods after the middle of May. The stem of this plant is translucent; that is, when you hold it between you and the light, the light shines through. This permits you to see the vascular bundles separately, somewhat as you see the bones in an X-ray picture of the hand.

E. The Scattered Arrangement. — Reference was made to the scattered arrangement of vascular bundles on page 82. It was said of the corn stem that the vascular bundles lie, in a cross section of it, "like islands in a sea of softer tissue." That softer tissue is a sort of combination of cortex and pith. In the cylindrical arrangement we have the bundles forming a hollow cylinder; pith is on the inside of it and cortex on the outside. But in the scattered arrangement we have no such dividing cylinder; the soft cells form a continuous mass, the bundles being scattered through it. There is no such thing as dividing this mass of soft cells into cortex and pith; they are all about alike; cortex and pith are undifferentiated.

In scattered bundles *there is no cambium.* Since there is no cambium, there is no forming of new cells in the bundles. With respect to cambium, bundles which lack it are called *closed;* bundles which possess it are called *open.* Closed bundles have no growth save primary growth; open bundles are "open" for secondary growth. This possession or lack of cambium in the vascular bundles greatly affects the general form and habit of a plant. The lack of it generally means that there is no gradual enlargement of the plant from year to year, no increase in the amount of foliage display, no gradual spread of branches. In temperate regions this lack of cambium seems to compel the annual habit, at least so far as the aërial parts are concerned; very few stems with closed bundles live through severe winters. Evidently the annual habit and the lack of facilities for secondary growth have something to do with each other.

The stems of some tropical and semi-tropical plants which have scattered bundles, especially those of palms and yuccas, do increase in diameter from year to year. This is due, however, to the activity of a cortical cambium. There is no cambium in the vascular bundles.

Many familiar plants besides corn have the scattered arrangement of vascular bundles. One of the greatest of the *families*[1] of plants is the grass family. To it belong all the cereals except buckwheat, as well as all the kinds of grass. Another large plant family is the lily family, of which asparagus and onion and tulip are familiar examples. All the members of the grass and of the lily families, and many other plants which are related to them, have this scattered arrangement of their vascular bundles.

You will recall that the tall grass stems you have pulled in summer are hollow, and a stalk of wheat has already been given as an example of a hollow stem. Evidently hollowness in stems is not limited to those with the cylindrical arrangement; in fact, it is more common among stems with the scattered arrangement. As hollowness

[1] Frequently from this time on you will meet the word *family* as applied to a group of more or less similar plants. Botanists have arranged plants into groups which they call families for the reason that such groups are determined by what are believed to be natural relationships or kinships.

develops in stems with scattered bundles, the bundles gradually become limited, of course, to that hollow cylinder which forms all that there is of the stem. This is not to be confused, however, with a cylinder which is composed exclusively of the bundles. All stems are solid at their growing tips; that is, at the point where the primary meristem lies. It is in the region of elongation that hollowness first appears. An old-fashioned bamboo fishing pole is an excellent example of a hollow stem whose bundles have the scattered arrangement. As you may have noted, however, and as the bamboo shows very well, hollow stems are solid at their nodes.

F. Cells of the Bundles. — It is not important that you should know many of the very many kinds of cells which may be found in the vascular bundles of different plants; special cells which appear in connection with the bundles of one plant may not appear at all in connection with the bundles of another, and with such special cells we are not now concerned. It is important, however, that you know something of the few principal kinds of cells found in wood and bast. Already you know something of the difference in appearance between phloëm cells and xylem cells, at least as they are seen in cross section. When viewed lengthwise, however, cells show the real nature of their structure far better than when viewed crosswise. (See *Figure 60.*)

The cells of the wood that conduct water are called *tracheary vessels.* Wood also includes cells of smaller diameter called *fibers;* the fibers have pointed ends and are usually very strong and tough; they, more than the vessels, give toughness to timber. The tracheary vessels are not simple cells. They might be regarded as compound cells. They have been formed by rows of cells whose end walls disappear and which thus become united into a single vessel or canal. As you have noted in the picture (*Figure 60*), the walls of the tracheary vessels are peculiar. They are thick, and the thickenings are not evenly arranged over the surface. Thus at *A* in *Figure 60* we have what are called spiral vessels; the thickened portion of the wall takes the form of a spiral. The advantage of this arrangement is not clear, but it has been found that the first tracheary vessels to be formed are of this type. At *B* in *Figure 60* we have what are called pitted vessels. In these the thickening covers nearly all of the surface, but it is marked by frequent pits. This is the type of vessel of which the *secondary* wood is chiefly composed. (By secondary wood is meant, as you will recall, wood which is produced by the cambium and not by the original growth of the stem.)

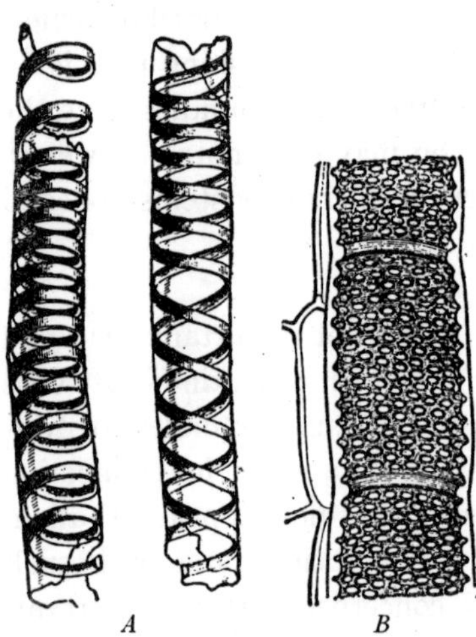

FIG. 60. — Tracheary vessels of the xylem viewed lengthwise, very highly magnified; *A*, spiral vessels; *B*, pitted vessels.

The tracheary vessels are of more service to the plant when they are dead than when they are alive, and they die soon after they are formed. This death, however, is not accompanied by decay. Completely surrounded as they are by live tissue and by bark outside of that, the wood cells may be preserved indefinitely. Their death means only that their protoplasts cease to exist.

The reason why dead tracheary vessels are of more service to the plant than live ones is not difficult to understand. The principal use of these cells to the plant is that they furnish the path for the ascending water, the *transpiration stream.* This water, and the solutes which are in it, can move much more rapidly through dead cells than it can through living ones. You recall that when water and solutes pass through living cells, they do it by osmosis, being independent of one another in rate and direction of movement. You remember that the osmotic movement of any solute depends upon whether the cell walls and the protoplasmic membranes are permeable for that particular solute, and upon differences in its abundance; that is, upon differences in osmotic pressure. Movement of this character, taking place through living cells, is slow. But movement through dead cells may be relatively fast. Through dead tracheary vessels, water and solutes may move together for considerable distances without having to pass the barrier of a single protoplasmic membrane, and they move much more rapidly than when moving by osmotic diffusion through living cells. "Except for a few cells in the epidermis and cortex of the root and in the mesophyll of the leaf, the entire course of water through a plant is in dead tissues." (Cowles's *Ecology.*)

The most important conducting cells of the phloëm or bast are the *sieve vessels.* (See *Figure 61.*) It is through these sieve vessels that food principally moves. They are called sieve vessels or sieve tubes because their end walls are full of holes like a sieve. Side by side with the sieve vessels are found what are called *companion cells*. The function of these companion cells is not understood. Phloëm also contains mechanical tissue in the form of fibers. These *bast fibers* resemble the wood fibers in appearance and in function.

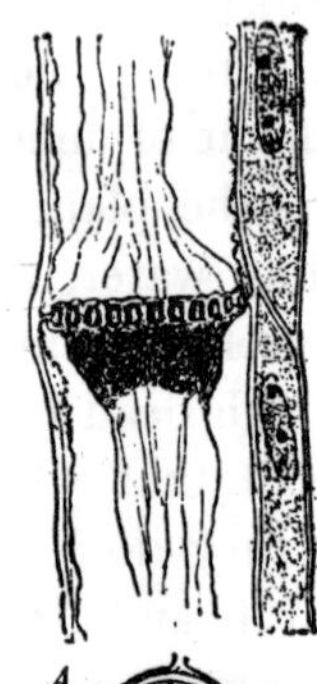

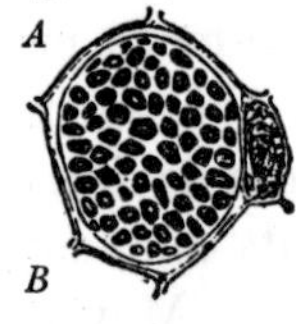

Fig. 61. — Cells of the phloëm. *A*, the adjoining ends of two *sieve vessels*. The shaded cells which accompany them on the right are *companion cells*. *B*, the end partition between two sieve vessels showing by shaded dots the pores which form the sieve.

Reference has already been made to mechanical tissue (page 147). This expression refers to all tissues which serve the plant merely for mechanical support. It is now evident that the cells of a vascular bundle are, as to function, of two principal kinds — *mechanical* and *conductive*. The fibers, both of wood and bast, are principally mechanical in function, while the sieve and tracheary vessels are chiefly conductive in function. Two useful words in connection with this topic are *stereome* and *mestome*. Stereome means the mechanical tissues of a plant taken as a whole; mestome means the conductive tissues of a plant taken as a whole.

G. Rings in the Wood. Age of Trees. — (See *Figure 58*.) Have you ever noticed the cut end of a freshly felled tree? There before you lies the record of a history which has run through many years, perhaps through hundreds of years. To look at the rings in the wood, to notice carefully the beauty of the layering, to remember the generations upon generations of green leaves whose work in each season's sunlight went into the making of this wood, to think of all the intricate processes and forces of which we know so little except that they have been concerned in this matter — this stirs our wonder and our admiration.

If you were there when the tree fell, you smelled its fragrance and perhaps saw the sap oozing from that wood which lies just beneath the bark. You noted that the wood is divided into two quite distinct zones. The outer and usually lighter zone is the *sapwood;* the rest is *heartwood.* It is common to speak of the sapwood as alive, and of the heartwood as dead, but this is not exactly accurate. Many of the cells through which sap runs are already dead. So, though the trunk as a whole may have endured for centuries, only a comparatively small part of it has been alive at any one time. This live zone lies between the outer bark and the heartwood, and the most active part of it is, as you know, the cambium.

By understanding the way in which the cambium works we can understand how the rings in the wood are formed. Perhaps you have heard that each one of these rings indicates a year of growth, and that by counting the rings you can tell the age of the tree. This is not strictly true. These rings are often called *annual rings*, which implies, of course, that each one of them represents the growth of one year. It is better to call them *growth rings*, for more than one may be formed in a year.

It is true that a ring usually represents a season's growth, but that does not explain why the ring is there. The fact that a bricklayer stops work at night does not mean that the wall of bricks he is building will show by lines where he left off at night and where he began in the morning. So with the rings. They are not caused by the fact that the tree stops work over winter. They are caused by the fact that the last cells formed in the fall are much smaller than the first cells formed in the spring. Just so the brick wall would be marked off into distinct layers if the mason laid small bricks the last thing every night and large ones the first thing every morning.

The rings are caused, then, by the difference in size between adjacent cells; layers of small cells lie against layers of larger cells. During freezing weather the cambium is not active. When in spring the sap ascends from the roots and conditions for growth are right again, the cambium begins to form what is called spring wood, and the cells of this spring wood are much larger than the last cells which were formed the fall before.

A long period of drought may cause the cambium to become inactive just as winter causes it to become inactive; in other words, very dry weather has the same effect upon growth that freezing has. So, if a long, summer drought is followed by a warm, wet fall, two growth rings, as you can readily see, are likely to be formed. In some tropical regions it has been noted that alternating wet and dry seasons appear to have about the same effect in producing rings that winter and summer have.

H. The Bark and Things related to It. Lenticels.—With bark most boys are familiar. If you live in the country, you have almost surely cut your initials in it. As you cut them you noticed the difference between the outer, dry, and corky bark, and the inner, moist, and fibrous bark. If you were doing a good job of initial cutting, you cut clear through to the wood itself. That is, you cut through cork, cortex, phloëm, and, last of all, the cambium.

The outer bark of a tree may be cut through without seriously injuring the tree. It is usually able to heal wounds of that kind by the growth of tissue from all sides of the wound; wounds appear to stimulate to division adjacent cells which otherwise would never divide. Sometimes, however, especially if the wounded tree is not vigorous, decay will start at such wounds and gradually cause the death of the entire tree. The bark of a tree protects it from decay somewhat as our skin protects us from disease. When our skin is broken, we do what we can to prevent the wound from becoming infected. Similarly, shade trees often have their wounds treated to prevent or to remove disease and decay. Boring insects are one cause of the decay of wood. Probably more deaths of trees are caused, however, by insects like the gypsy moth which eat the leaves.

To cut through the inner bark of a tree and completely around the trunk will cause its death. To cut through to the heartwood of trees is a method often used to deaden a forest where land is to be cleared; this process is called *girdling.* Girdling stops the descent of food into the roots and most of the ascent of sap to the leaves. Sometimes there appears just above the girdle a considerable swelling of the trunk. This swelling is due to the accumulation of food which has been interrupted in its downward movement. This accumulation of food induces an abnormal growth, hence the swelling. Sometimes in young plants, just above a girdle, roots also will appear; the plant seems to be trying to save its life by developing an entirely new root system.

If you have been where there are birch trees, you have noticed their papery, white bark. You have been interested in the use of it in making canoes, a use which shows very well the impermeability of bark to water. Birch trees also show very well those structures of the bark called *lenticels*. In birch bark the lenticels form conspicuous markings which run crosswise of the trunks. (See *Figure 62*.) On other trees than birch, especially upon the younger bark, lenticels are frequent, but they are not so conspicuous as upon the birch. Lenticels, though quite different in structure, are similar in function to the stomates of the leaves; they permit the passage of gases. In them the cork cells, such as are present in the rest of the outer bark, are replaced by loose cells through whose intercellular spaces it is possible for gases to pass. You will recall that elsewhere in this section it was mentioned that the cortex, as the stem grows older, usually is compelled to give up its earlier work of photosynthesis. If it were not for the lenticels, the cortex would have to give up photosynthesis much more quickly than it does, for if the covering of cork were continuous, the exchange of gases which is necessary to photosynthesis could not occur. In

FIG. 62. — Trunks of birch trees showing many transverse lenticels.

some trees with thin bark, like birch, lenticels permit the cortex to manufacture food throughout the life of the plant.

I. Grafting and Budding. — In the culture of fruit trees it was long ago found to be very desirable to join different parts of different plants together and thus form one individual. Some apple trees have strong and hardy roots and stems and leaves; their nutritive organs are excellent for the purposes of apple culture, but the fruit which they bear is small and sour. Other apple trees, which bear delicious fruit, are rather weak and not hardy as to their vegetative organs. So, long ago, fruit growers experimented, trying to get twigs which bore good fruit to grow upon stems which were hardy. They were successful in their experiments. They perfected a process by which many kinds of fruit besides apple are now cultivated. The name of that process is *grafting*. It is a process which is based upon the activity of the cambium layer.

The hardy plant whose roots are in the ground, and upon which the graft is to be made, is called the *stock*. The twig which is to be made to grow upon the stock is called the *scion*. The scions are usually cut in the fall, kept in moist soil or sand, and grafted in the spring. Thus they are cut when their resting period has begun, and grafted just when their activity is about to begin again; these times of cutting and grafting have been found to be most favorable for securing successful results from the operation. When the graft is made, the cambium of the scion and the cambium of the stock are brought into close contact, and, if both are active, they grow together. The stock and scion soon become as closely related as the stock and its natural branches. Just before grafting, the ends of the scions are cut on a long diagonal so that a good deal of cambium is exposed at the surface. The stock is cut so that its cambium and the cambium of the scion may be brought together, and, after this adjustment has been made, the wound is covered with clay or with grafting wax. (See *Figure 63.*)

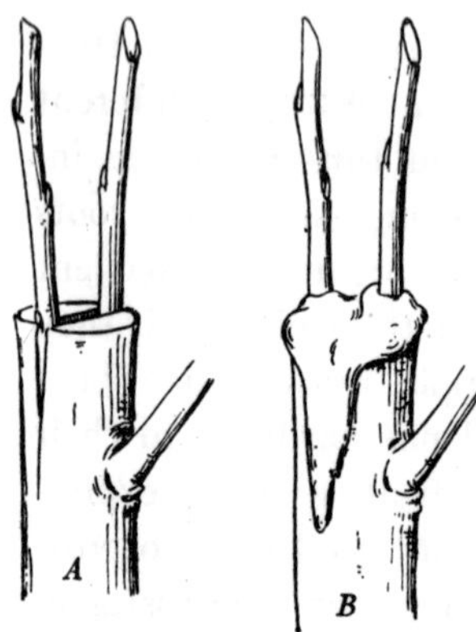

Fig. 63. — Grafting: *A* shows two cuttings from the scion. These are wedge-shaped below. In this process they are thrust into a cleft of the stock as shown. *B*, the same, the wound being covered with grafting wax.

In the cultivation of peaches a process called *budding* is used. This process is much the same as grafting, the principal difference being in the character of the scion. Instead of a twig, only a strip of bark bearing a bud is used. This strip of bark is cut so that cambium is exposed on its inner face. A T-shaped cut is made in the young bark of the stock, and the bark is turned back a little on each side of this cut. The scion is then slipped into the cut, the lifted bark on each side holding it in place. To hold it more firmly, strings are tied around the stem just above and just below the inserted bud. (See *Figure 64.*) Budding, like grafting, is done in early spring.

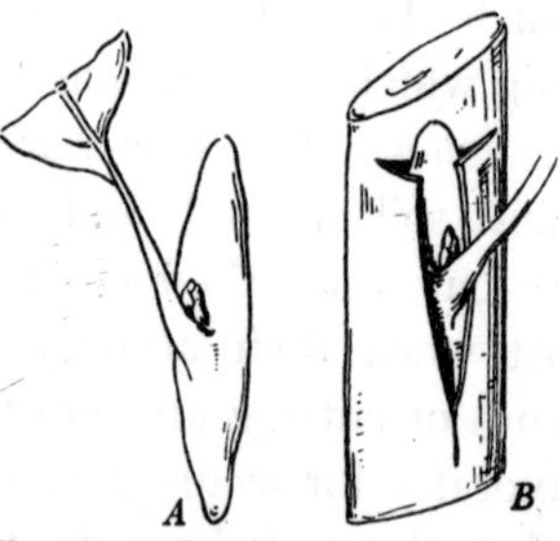

Fig. 64. — Budding: *A*, a bud and a bit of the bark as it is cut from the scion in performing this process. *B*, the scion placed in position on the stock; after being so placed it is securely tied. — *Redrawn from Bailey.*

51. Stems in the Classification of Seed Plants. — The scientific name for all plants which produce seeds is *spermatophytes*, a word whose first part means seed and whose last syllable means plants.

The spermatophytes are composed of two great groups, the *gymnosperms* and the *angiosperms*. The word *gymnosperm* means naked-seed; the word *angiosperm* means seeds-in-a-box. Pines and other common northern evergreens are gymnosperms; all other common plants which produce seeds are angiosperms. From a ripe pine-cone you can shake out the naked seeds; all other seeds are inclosed within the structure called ovary (see page 60). The ovary usually develops into what we call fruit. In some angiosperms, like corn, the seed case or ovary seems like the outer part of the seed itself, but in origin it is perfectly distinct from the seed.

The angiosperms are composed of two great divisions of plants known as *monocotyledons* and *dicotyledons*. *Monocotyledon* means one cotyledon; *dicotyledon* means two cotyledons. Cotyledons, as you know, are parts of the embryo. Your attention has already been called to the fact that plants like corn have but one cotyledon, and that the bean is an example of plants with two (see page 66).

It is evident that this classification of seed plants is not based upon differences in their stem structures. It is based rather upon the way the seeds are borne and upon the number of cotyledons; the meanings of the names used indicate that such is the basis. As a matter of fact, in deciding what the great subdivisions of plants are, botanists do not depend upon single characters to guide them. The great groups of plants are based upon similarities or dissimilarities as to a number of characters. By using a number of characters rather than just one, botanists have been able to determine what are the *natural* great groups of plants; that is, those groups which have resulted from the processes of nature. Similarly, the processes of nature, working through long ages, have formed the natural groups of mankind, and we do not distinguish a white man from an Indian by one character, but by a number of characters.

It took botanists a long time to find out the real kinships of the great groups of plants, and even now the system of plant classification is revised from time to time in the light of newly discovered facts. Thus, for example, a plant which was thought for a long time to be a sort of cousin of another kind of plant may be found to have certain characters which prove it to be no such relation at all, but the kin of quite another family of plants.

You know already that there are two great types of stem structure among seed plants. Not many years ago botanists divided all seed plants on the basis of these two types of stem structure. Plants with scattered vascular bundles were called *endogens;* those with the cylindrical arrangement were called *exogens.* It was found, however, that this was not a natural division and the system had to be readjusted. Yet stem structure is still used as *one* of the great characters upon which classification is based. You should know which of the great groups have the cylindrical arrangement and which have the scattered arrangement.

It is the monocotyledons which have the scattered arrangement. All the members of the grass family are monocotyledons. This great family includes all the cereals. Bamboo is the largest member of this family.

As a food producer, this is by far the most valuable family of plants to man. The lily, the palm, and the orchid families are also monocotyledons.

The dicotyledons, though not so close in kinship to gymnosperms as they are to the monocotyledons, are like the gymnosperms in that both possess the cylindrical arrangement of vascular bundles. Most of our common kinds of seed plants, except pines and other evergreens and members of the grass family, are dicotyledons.

52. **Buds and Branches.** — A bud may be defined as an undeveloped shoot or as an undeveloped branch. In a bud a branch lies in embryo much as the whole plant lies in embryo in a seed. All branches arise from buds.

You may be a little confused as to the use of the words branch and shoot. A branch is an example of a shoot, for shoot is a word descriptive of stems and leaves together; it may refer to a mere twig with one bud on it, or it may refer to the whole plant except the roots. Shoot is a more inclusive term than branch.

By means of its buds the plant is able to take quick advantage of growing conditions. The buds also enable it to replace lost or injured branches. Buds are evidences of what we might call the preparedness of the plant.

It is in spring that we get the most striking evidence of the advantages of buds to plants. Through the winter the trees and shrubs have been bare and apparently lifeless. Yet, if you examine a twig in winter, you find on it what are called *winter buds*. You find them just above the scars which mark the places where last summer's leaves were attached. These winter buds contain, under many covers, growing points which are dormant. Their formation was complete even before the frosts came and the leaves fell. Thus, even before winter comes, the plant appears to prepare for spring. Embryo shoots and flowers are formed in the buds and food accumulates there. With the coming of the first warm days of spring we note a change. We say "the buds are beginning to swell." Presently the brown, scale-like leaves which cover the buds are burst apart and the first soft green of the leaves appears. Or, if it be the bud of a flower-bearing branch, the tender flower buds begin to show. This is a danger period, especially for fruit trees. If a freeze comes now, these tender, inner buds are killed. For this reason we often hear of the loss of fruit crops in early spring. If, after a frost, you find that the fruit buds have turned black in the center, you may know that no fruit will come from those buds. Some fruit may be produced, however, by buds which were not yet open; trees bear many buds which seem to be held in reserve, and are of much advantage in just such an emergency.

A. Kinds of Buds. — Buds may be variously classified. As to the nature of the organs they produce, they are classified as *flower buds* or *leaf buds*. Flower buds are usually thicker than the leaf buds of the same plant. They often produce leaves as well as flowers, and buds of this kind are sometimes called *mixed buds*. The parts of flowers are themselves, as you shall see, the same as leaves in origin.

As to the position in which they occur, buds are classified as *terminal*, *axillary*, *accessory*, or *adventitious*. The expression terminal bud explains itself. It contains the terminal growing tip; all growing shoots end in a terminal bud.

Axillary buds are those which appear at the axils of leaves. Buds regularly arise in this position; in fact, the presence of buds in their axils is one of the ways in which leaves are recognized. *Compound leaves* are leaves which are subdivided into *leaflets;* these leaflets often have the appearance of true leaves, but their real character is revealed by the absence of buds at their axils. (See *Figure 65*.) Axillary buds may become terminal buds. Thus in the lilac the buds in the axils of the first pair of leaves below the apex develop more vigorously than the terminal bud; in fact, the terminal bud may not develop at all. The result is that with each new season a forking oc-

FIG. 65. — Twig of the ash, one compound leaf showing. Note the buds in the axils of the leaves and the lack of them in the axils of the leaflets. The enlarged bit of stem at the right shows lenticels, and also the scars left by the leaves of the preceding season. Note the alternating arrangement of the attachment of the leaves.

curs, each axis giving rise to a pair of branches which have developed from the last pair of axillary buds.

In the axils of the leaves of some plants more than one bud appears. In such cases there is usually one principal or central one. The others are called accessory buds. Thus in the axils of maple trees three buds are borne, though usually only one develops. The advantage of having accessory buds becomes evident when injury, like freezing, befalls the first branches which are put out. Usually it is only under such circumstances that accessory buds develop into branches.

Adventitious buds are those which occur elsewhere than at the tips or in the axils. They appear on the internodes or even on roots and leaves. Often they arise near where an injury has occurred. Young sprouts or *suckers* are often seen arising from near the base of the trunks of trees. These come from adventitious buds. Willow stems which are used in making baskets are obtained in abundance by injuring the trunks of willows in such a way as to stimulate growth from adventitious buds. Elms very often show short, leaf-bearing branches which arise from the main stem below the crown; such branches are adventitious in origin.

As to their protective covering, buds are either *scaly* or *naked*. Nearly all buds in our climate are protected by firm, dry, brown scales. Naked buds lack scales. Though sometimes occurring in temperate regions, naked buds are principally found in the tropics. Scaly buds are often further protected by an excretion which resembles varnish. This may be seen on buds of the horse-chestnut tree or of the balsam-poplar. Perhaps you have noticed the rich, spicy odor of the balsam-poplar when its buds open. This odor gives the name to the tree. It is the odor of the varnish-like, sticky secretion which is abundant on the scales of the buds.

It is common to think of cold as the principal thing from which the inside parts of buds need protection. However, loss of water by evaporation is an even greater menace to plant life than low temperatures. It is certain that these bud scales, and the hairs and coatings which they often have, are of great advantage in preventing water loss at a time when the roots are unable to make such loss good. This use of the bud scales is more important to the plant than their use in preventing freezing; in fact, it has been observed that ice forms in some winter buds without injuring them.

Buds which remain undeveloped are called *dormant*. Dormancy appears to be due chiefly to lack of opportunity to develop. The stronger and more favorably located buds "get a start" on the others. Buds may remain dormant for four or five years before they lose their power to develop. In case of serious loss of its other buds and of branches, it is evidently of much advantage to the plant to be well equipped with dormant buds. After emergencies, such as the loss of foliage by a heavy hail storm, dormant buds have their chance to develop.

B. Overproduction of Buds. — There is a great overproduction of buds. That is to say, very many more buds are produced than ever develop into branches. There is not room for the development of all the buds; those which expand first arrest the development of the others. Those first started use up all the space available for exposure to light, or, it may be, they tax to the utmost the capacity of the root system to supply water. Whatever the restraining cause, certainly many buds remain undeveloped.

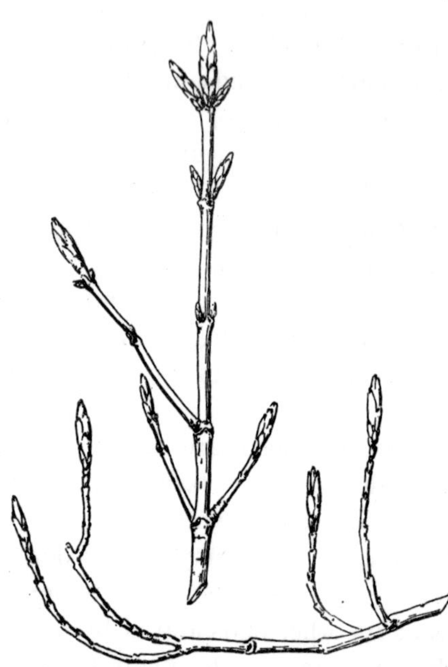

FIG. 66. — Twigs of maple. The upper twig is taken from the top of the tree; the lower one from the side of the tree. Note the differences in the direction of growth. Note that the terminal buds are larger than the lateral buds. Only a small proportion of such buds can develop into branches.

This overproduction of buds reminds us of the overproduction of seeds. Both kinds of overproduction are of advantage to the plant; there are always other seeds and other buds to take the places of those which fail. On the ground there is always a sort of struggle for existence among the great host of young seedlings. Similarly, in the tree tops, there is a sort of struggle for existence among the young branches. (See *Figure 66.*)

53. Accumulation in Stems. — This topic well illustrates a very common and very important misunderstanding of plants; namely, the misunderstanding that plants do things which involve forethought.

This misunderstanding is important because it causes an entirely wrong conception of plant life; it is the sort of misunderstanding which is worse, from the standpoint of science, than no understanding at all. It is not important for you to remember a great number of facts about plants, but it is important that you should have no such misunderstandings about them. Your knowledge, so far as it goes, should be based on sound conceptions, for, if so, it is possible to *understand* plants without *knowing* a great deal about them. The right understanding of a little knowledge is the principal aim of this course.

A. Why not "Storage"?—You note that the title of this section is *accumulation* in stems. Why not *storage* in stems? The latter expression would probably suggest to you about the same thing as the former, and yet there is a great difference involved. The use of one or the other of these words implies a correct or an incorrect conception of plant life.

Many people think of plants as "storing" food for later use somewhat as men and animals store it for later use. But the word storage implies forethought, and for that reason it is objectionable as applied to plants. We have no evidence at all that plants have anything like forethought or intention; indeed all our evidence is to the contrary. All our evidence goes to indicate that plants, in their growth and behavior, respond to various stimuli, such as have been noted. These stimuli may be inside the plant or they may be outside of it or they may be both, but, wherever they are, they are the causes of plant behavior, and the plant, we think, reacts to them with no more power to interfere than a falling ball has power to interfere with gravitation. There is no such thing as judgment in a plant. The plant does what it does simply because internal or external conditions or both cause such doing. So far as we know, its responses are automatic; certainly they are not guided by any intention of the plant's own.

Some expressions have been used in this book which may have misled you in this matter. Such, for example, is the expression in the preceding section, "the plant *appears* to prepare for spring," or, farther back, "the plant *seems* to be trying to save its life." The only thing which saves such expressions from being wrong is the use of words like *seems* or *appears*, and even with such qualification such expressions are apt to be misleading. Yet we find it difficult to speak or write about plants without using such expressions. The things we observe about plants look to us as though they were *intended* by the plant, but this is chiefly because we are in the habit of judging the behavior of plants by the behavior of animals, including ourselves. We do this unconsciously. We have our own relation to the world in mind instead of theirs, but the world is a very different place to a plant from what it is to us. In attempting to explain plants, we take our own point of view instead of theirs, yet even to suggest that plants *have* a point of view is to make a wrong implication. However, it is not important that our speech about plants should be absolutely free of expressions which imply wrong conceptions. *The thing which is important is that we should not have the wrong conceptions which our speech may imply.*

B. Substances Accumulated. — If a plant manufactures more food than it uses, or if it takes in more water than it uses or gives off, under such circumstances food or water or both necessarily accumulate. Sometimes such surplus food and water are subsequently used to the advantage of the plant; sometimes they are not used. Evidently, however, plants which have the habit of making such accumulations may live successfully in certain surroundings wherein plants without this habit may die out. Waste products as well as food are often accumulated by plants.

The accumulation of water in the stems of some desert plants is quite striking. During rainy periods such plants absorb much water which is retained and used during periods of drought. The nail-keg cactus, so named from its shape and size, has been used by travelers in the deserts as a source of water. When split open, the "keg" may yield a pint or more of free water.

On bright days the accumulation of food occurs in nearly all tissues in which photosynthesis takes place; this applies to some stems as well as to leaves. In bright sunlight food is usually manufactured more rapidly than it moves away. During the night this food is transformed into soluble forms and moves to other parts of the plant.

Rhizomes and tubers are the forms of stems in which the accumulation of food principally occurs; these, along with tap-roots and seeds, form the principal organs of food accumulation. The food in seeds is, of course, used by the progeny of the plant. The food in underground structures may be used by the progeny of the plant or it may be used by the plant itself or much of it may not be used at all. Thus, for example, potatoes contain much more food than is used by the plant.

C. Latex. — Reference has already been made to the milky juice which many plants contain (see pages 53 and

71). This milky juice is called *latex.* It is an accumulation in which water, food, and waste products are all present. Its function is not well understood. It is usually contained in tubes which result from the fusion of many cells, or, as in the milkweed, result from the growth of special cells which, as they grow, crowd their way through the softer cells of the cortex. The latex vessels run through the entire plant somewhat as the vascular tissues do. (See *Figure 67.*) A familiar example of latex, besides that of the milkweed, is the reddish juice of the bloodroot. Members of the spurge family are common plants all of which have abundant milky latex.

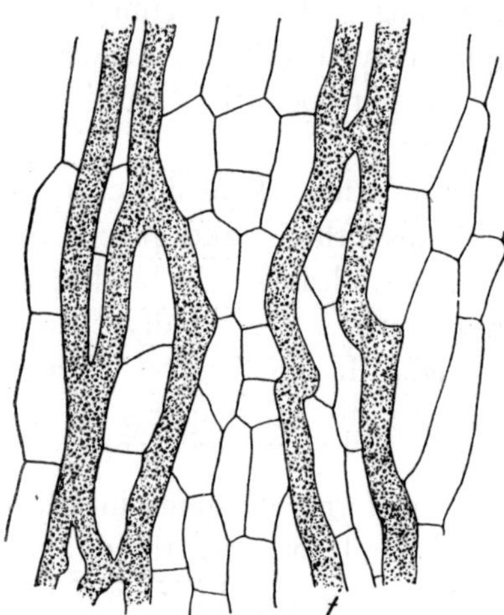

Fig. 67. — Tissue of the wild lettuce showing the way in which the latex tubes (*t*) branch among the other cells.

Latex is of much economic importance for the reason that it contains *caoutchouc* (*kou'-chŏŏk*), the substance from which rubber is made. Although more or less of caoutchouc seems to be present in all latex, that of only a few plants has been found profitable to use in the manufacture of rubber. Certain trees, either native to South America or thriving there under cultivation, produce the world's principal supply of caoutchouc. Other tropical trees, similar to these rubber trees, produce a latex from which gutta-percha is derived. Besides its use for golf balls, gutta-percha is the one substance which furnishes satisfactory insulation for submarine cables.

53 A. **Forestry.** — You have noted that the stems of trees, both on account of their usefulness and on account of their beauty, are of great importance to man. That division of science which concerns the culture of trees, and especially of their stems, is called forestry. (See page 52.)

A. Importance of Forests. — Primarily forests are important as sources of timber. Other things add to their value and to the great need for their preservation. One of these things is the fact that forest soil is very retentive of moisture; it acts as an absorbing reservoir in which the water after heavy rains is held, and from which the excess of water runs away slowly. This tends to prevent floods which wash away valuable soil and in other ways cause great damage. Man's enjoyment of forests also adds much to the desire and need of all of us that they be not ruthlessly destroyed.

B. Aims of Forestry. — It is the aim of forestry not merely to preserve forests, but so to manage them that they will be continuously productive. Foresters learn by careful study the habits of growth and reproduction of trees, and make plans for forest management which will insure the usefulness of the forests to future generations as well as to this one. The Bureau of Forestry of the United States Government manages large tracts of public land which have been set aside as forest reservations. The lumbermen who cut timber on these reservations do so under government direction, and their permanent productivity is assured. *Thus forestry aims to do for timber land just what scientific agriculture aims to do for farm land.*

By reckless and wasteful lumbering the United States has lost very much of one of its most valuable resources. Governmental action was necessary to prevent the coming of a lumber famine, and, even as it is, the price of timber is many times what it would have been if the forest resources of the United States had been scientifically managed.

Foresters are concerned not only with the management of forests now existing; they are also concerned with the *reforestation* of lands from which the timber has been cut or burned. Forest fires have caused losses of millions of dollars annually, and the forest service of the government fights and seeks to prevent these fires. It has also replanted with new trees many thousands of acres which have been cleared by fire or ax.

C. Forestry outside of Forests. — Forestry is not restricted to forests. The term is also applied to the scientific culture of trees which do not grow in forests. Many cities employ foresters whose business is the planting and care of trees in parks and along streets. This is a division of the subject which directly concerns even more people than does forestry proper, for nearly every one is constantly interested in the presence and health of trees on lawns and along sidewalks, while they may visit forests but rarely.

Ignorance of how to set out trees often causes unnecessary loss. Trees should be set out only in early spring or in the fall; fall planting is preferable. The holes should be considerably larger than the root mass. After the roots have been placed naturally, the holes should be gradually filled with pulverized soil, gently tamped. The young trees should be pruned before they are set out. If the soil is quite sterile, as city soil commonly is, fertilizers should be used. While the tree is young, the soil around it should be kept broken and free of weeds. If branches are cut away, wounds larger than two inches in diameter should be covered with thick paint or tar, or by a plate of zinc. The

View in a beech forest in Indiana.

life of decaying trees is often saved by removing the decayed parts and filling the cavities with asphalt or concrete. Smoke or gas frequently causes the ill health of trees in cities. Plants are sensitive to quantities of gas far too small to be detected by the sense of smell; a gas leak may cause injury to plants many feet away on account of the gas-poisons which work through the soil.

Some insects cause great damage to trees, both by devouring leaves and boring into the stems. One of the most striking examples of this is the case of the gypsy moth. In recent years this insect has caused immense damage to trees in the eastern United States. Many thousands of dollars have been spent in efforts to exterminate it. In combating this and other insect enemies of plant life the best general method which has been found is to encourage in every way the abundance of birds and of other insects which prey upon those which do the principal damage.

QUESTIONS AND SUGGESTIONS

SECTION 46. 1. What appears to be the principal function of stems? Compare them with roots as to principal function. 2. Make a summary of your previous knowledge of the structure of stems.

SECTION 47. 1. What are various ways in which aërial stems may be classified? 2. What are herbaceous stems? Give examples. 3. Give examples of shrubs.

A. 4. Describe the arrangement of the branches of some common plant with an erect stem. 5. What are the two types of structure of erect stems? 6. What is mechanical tissue? Give an example of it. 7. What is the crown of a tree? What are the two principal types of crowns? 8. What advantages have hollow stems over solid stems? 9. What advantage have solid stems over hollow ones?

B. 10. What appears to be a principal reason for the existence of so many kinds of plants? 11. Discuss the advantages and disadvantages of prostrate stems.

C. 12. Describe the growth of vines in a tropical forest. 13. What is a liana? Give an example. 14. What are the three ways by means of which plants climb? Give examples of each.

D. 15. What are cladophylls? Give examples. 16. Name plants whose thorns are like stems in their origin.

SECTION 48. 1. Define rhizomes and give examples. 2. Define tuber, bulb, and corm.

A. 3. Describe the habit of growth of the May apple. 4. What garden plants have rhizomes? Do such plants require attention after their growth has once been established? 5. Will fertilizer placed on an asparagus bed in the fall cause it to produce better shoots the following spring? Explain your answer. 6. Describe the rhizome habit of reproduction and its advantages. 7. Describe the sequence of plants on freshly cleared land.

B. 8. Describe the structure of an Irish potato. 9. Describe potato culture. 10. Describe and explain the appearance of potatoes sprouting in the dark.

C. 11. Describe the structure of an onion bulb. 12. Describe the culture of tulips.

D. 13. Describe the Indian turnip. 14. Describe the "flower" of the Jack-in-the-pulpit. What cultivated plant has a similar "flower"?

E. 15. Describe the stools of canna or dahlia.

SECTION 49. 1. What is the vernal habit? 2. What plants do you know which have the vernal habit? 3. What are some of the results of the struggle for light in tropical forests? 4. What are epiphytes? 5. What plants do you know which are epiphytes but are not seed plants? 6. Assuming that both have the vernal habit, what advantages would a plant coming from a rhizome have over a plant coming from a seed?

SECTION 50. *A.* 1. Describe the structure of roots, defining epidermis, cortex, and stele. 2. Define meristem and cambium. 3. Contrast stems with roots as to their tropisms.

B. 4. Define parenchyma. 5. Compare epidermis, cortex, and stele as to complexity.

C. 6. Contrast xylem and phloëm as to functions. 7. Describe the two ways in which vascular bundles are arranged.

D. 8. Describe the appearance of the cut end of a one-year-old twig. 9. Discuss the changes which occur as the twig grows older. 10. Describe bark and its function. 11. Describe the appearance of the cut end of a log. 12. Describe the structure of a vascular bundle, defining pith-rays. 13. Describe the formation of the cambium ring. 14. Describe the formation of secondary wood. 15. Describe cork and its production, defining suberin. 16. Explain the furrows and cracks in old bark. Do you know any trees whose old bark is smooth? If so, how do you explain this fact? 17. Contrast the cylindrical arrangement in perennial stems with that in annual stems.

E. 18. Describe the scattered arrangement of vascular bundles, giving examples of plants which have it. 19. State the difference between open and closed bundles. 20. State the relation of vascular cambium to the duration of life of plants in temperate regions. 21. How is the secondary growth of plants with scattered bundles accomplished? Give examples. 22. What are some of the common kinds of plants which have scattered bundles?

F. 23. Describe tracheary vessels and fibers as to structure and function. 24. Explain why dead tracheary vessels are of more service to the plant than live ones. 25. Describe sieve vessels as to structure and function, defining companion cells. 26. Define mestome and stereome.

G. 27. Describe heartwood and sapwood. 28. Explain the formation of the rings in wood. To what extent are these rings annual?

H. 29. Describe the structure and functions of the bark of trees, defining girdling. 30. Describe lenticels and the advantages to the plants in having them.

I. 31. Describe grafting and budding, defining stock and scion.

SECTION 51. 1. Name the two great divisions of spermatophytes and describe a fundamental character in which they are different. 2. What is the ovary of plants? 3. Name the two great divisions of angiosperms and explain the meaning of these names. 4. What is meant by "natural" groups of plants, and how are they determined? 5. What two of the three great divisions of spermatophytes have the cylindrical arrangement of bundles? 6. Give examples of monocotyledons. 7. Explain the terms endogen and exogen and tell why these terms were abandoned.

SECTION 52. 1. Describe the structure of buds and their use to plants. 2. When did the buds which open in spring first appear? 3. Explain why a freeze while the fruit blossoms are open may not entirely prevent a crop. How can you tell whether the young fruit has been destroyed by a frost? 4. What is the difference between a frost and a freeze? 5. Describe two ways of classifying buds, defining the terms used. 6. Explain a method of distinguishing leaflets from simple leaves. 7. Describe the method of branching of the lilac. 8. Discuss the advantage of overproduction of buds, giving examples.

SECTION 53. 1. Explain the objection to the use of the word *storage* as applied to plants. 2. Give examples of the accumulation of substances in plants and state advantages which ensue therefrom. 3. Describe latex and give examples of it, defining caoutchouc.

CHAPTER VI

LEAVES

54. **Introductory.** — Leaves are usually the most expanded parts of plants. They are the principal organs wherewith the plant establishes those relations to light and to air which are essential to its life. They are the principal organs of photosynthesis. They have been called the *light-related* organs of the plant, for, although their relation to air is equally important, light appears to be that external factor which chiefly affects their behavior.

It is difficult to construct a perfect definition of leaf, for it is a word which came into general use long before any effort was made to give it an exact meaning. On the basis of origin leaves may be defined as *organs which arise at nodes and have buds in their axils.* According to this definition, a leaf is not a leaf by virtue of the work it does, but by virtue of its position. For example, the parts of flowers and many thorns and tendrils have the origin just described, and your attention has been called already to the fact that these parts are often *leaves as to origin.* So it appears that a definition of an organ based on its function may be far from agreeing with a definition of the same organ based on its origin. It appears that we can understand leaves best, and perhaps be nearer the truth, by not trying to crowd our idea of them into one precise group of words. Concerning leaves, as concerning many other things, we can form a right idea even though we find an exact definition of that idea difficult to make. Common green leaves are the kind this chapter principally considers.

A. Use and Beauty of Leaves. — No organs of living things exceed leaves in beauty, utility, and abundance. Concerning them it should be possible to say some compelling thing — something that compels you to give them that attention, interest, and appreciation which you have need to give them. Often leaves seem to say that compelling thing for themselves; they do not need to have it said for them in a book; they say it for themselves far better than any book can say it, but they can say it only to those who are ready to listen, only to those who have kept alive their natural interest in nature and are able to appreciate and enjoy keenly the constant beauty and usefulness of her ever changing forms. To most of us leaves seem commonplace; we take them for granted. We take pleasure in their reappearance in spring, their color and the shade they furnish are grateful to us, and we exclaim over the beauty of their autumnal colors. But our curiosity about them is soon satisfied and we dismiss them with little thought. Not one person in ten can tell what leaves do. Yet, with understanding, our enjoyment of leaves is sure to increase, and the nine out of ten people who do not understand them would be glad to do so if they knew what a difference such understanding makes.

What makes the beauty of a summer landscape? Nothing has a larger part in it than the leaves. Stems determine the rounding, graceful forms of plants, but the leaves clothe the stems and multiply their beauty. In spring leaves come like a green mist over the gaunt forms of winter.

Some morning there appears a film of green over fields that were brown the day before. It is the green of the first leaves, the promise of the harvest. Whether it be the green of the grass blades, or the green of the stirring leaves in the tree tops; whether it be the velvet green of lawns or the pale green of leaves that grow under the shadow of forests — always this green delights us. It rests our eyes. We respond to it with pleasure. Everywhere it is this green of plants which dominates the summer landscapes. It fills the fields of the farms and the parks of the cities. It covers the hills and the valleys; the whole world seems green. And this green is always the green of leaves; the green of numberless millions of leaves. In the sunlight of each growing season these myriads of leaves are at their work, and upon that work all familiar life depends.

B. Points already Noted. — Blade and petiole, vein and midrib, axil and stomate, are terms which have to do with leaves, and with these you are already familiar. You know in general what photosynthesis is, and that it occurs principally in leaves. You know that the loss of water by evaporation is a very important phenomenon in plant life, and that it, too, occurs principally in leaves; you know that this process is called transpiration (see page 108), and that the water which ascends from the roots to the leaves is called the transpiration stream. You know what the mesophyll is, and that it is usually composed of palisade tissue and spongy parenchyma. (See page 84.) You know that air which comes in through the stomates diffuses in the intercellular spaces of the spongy parenchyma, and that with this air in the intercellular spaces the plant makes various exchanges of gases — oxygen, carbon dioxide, and water vapor being the gases principally concerned in these exchanges. You know that the cells of the mesophyll contain chloroplasts, and that the chloroplasts contain chlorophyll, and that chlorophyll is the substance which makes leaves green. You know that the chloroplasts are the organs of photosynthesis. (See page 74.)

55. **External Characteristics.** — In this section we shall consider characteristics of leaves which are visible to the naked eye. Flatness and greenness appear to be the principal qualities which leaves have in common. In features of their shape other than flatness we find a great deal of variety among leaves, more than we find in the shapes of stems and roots. Since the shapes of leaves depend very largely upon the ways in which their veins are arranged, we shall consider vein arrangements first.

A. Venation. — Compare a blade of grass with the broad leaf of any tree and you will notice a great difference in the way the veins are arranged. The visible veins of the grass blade are almost parallel, while the veins of the broad leaf are arranged somewhat like the meshes of a fine-meshed net. The arrangement of the veins is called *venation*.

There are two principal types of venation and these are the two which have just been described. Leaves are either *net-veined* or *parallel-veined*. Nearly all dicotyledons have net-veined leaves and nearly all monocotyledons have parallel-veined leaves. (See *Figure 68.*) Thus, you observe, net veins are associated with the cylindrical arrangement of vascular bundles, and parallel veins with the scattered arrangement. The veins are, of course, the extensions of the vascular bundles into the leaves. It should be noted that parallel-veined leaves have net veins also; that is, the very small, invisible veins which branch off from the principal ones are of this character.

Much of the beauty of the individual leaf is due to its venation. Sometimes the veins have all the delicacy and symmetry of beautiful lace work, as may be seen by holding the leaf up to the light. The large veins send off smaller branches and these send off still smaller ones, until the smallest veinlets are not visible. With the utmost precision, each tiny portion of the blade is equally penetrated by the veins, and all together they form a perfect framework and conducting system.

FIG. 68.—The two principal types of leaf venation. The figure at the left represents a leaf of Solomon's seal, a monocotyledon; the one at the right represents a leaf of willow, a dicotyledon.

The principal vein is called, as you know, the midrib, but often there is no principal vein. There may be a number of principal veins of about equal strength. In such cases the principal veins are called *nerves*, the leaf being described as three-nerved or five-nerved or whatever the number may be. Such leaves occur among both the net- and the parallel-veined types of leaves.

One of the commonest types of net-veined leaves is the

pinnate type. (The word means feather-like.) Most common trees and shrubs have pinnately-veined leaves. In this type there is a central midrib from which branches are given off somewhat as *barbs* are given off from the *shaft* of a feather. (See *Figure 69*.)

Fig. 69. — Twig of elm showing *pinnate* venation of the *alternate* leaves and their toothed margins.

Another common type of the net-veined leaf is the *palmate* type. (The word means palm-like.) Palmate leaves have a number of principal nerves or veins which arise from the base of the blade somewhat as spread fingers arise from the palm of the hand. The common plantain has palmate leaves; so has the geranium. (See *Figure 70*.) A palm-leaf fan is an example of a palmate leaf of a monocotyledon.

B. Shape. — The shape of a leaf is a characteristic which is not very constant. That is, the shapes of leaves vary more or less as their surroundings vary. Thus, for example, some plants which grow part in and part out of water, such as the water hemlock, have submerged leaves which are very different in shape and structure from those which develop in the air.

This tendency to vary in shape is not a property equally possessed by all leaves. Leaves, like all other organs, *vary in their power to vary*. Many plants have leaves of distinct and characteristic shape which vary but little and furnish a sure means of identifying the plant. Many others, however, have leaves of such variable shape that they are of little value as a means of recognition.

The shape of a leaf depends a good deal, as has been noted, upon the character of its venation. Parallel-veined leaves tend to be long and narrow; net-veined leaves tend to be oval. Long and narrow leaves, like grass blades, are called *linear*. Leaves whose shape approaches that of an oval are called *ovate*.

As to the nature of their margins, leaves are called *entire*, or *toothed*, or *lobed*. Entire leaves are those whose margin is an even line, unbroken by teeth or lobes; all parallel-veined leaves are entire. Net-veined leaves quite usually have their margins broken into teeth or lobes, or even into deeper divisions such as are shown by maple leaves. The leaf of the white oak is a deeply lobed leaf. The leaf of the elm is a toothed leaf. (See *Figure 69*.) If the teeth resemble in their outline the teeth of a saw, the leaf is called *serrate*.

FIG. 70. — Leaf of geranium showing palmate type of venation. Note also the stipules at the base of the petiole.

As to the extent to which their blades are divided or branched, leaves are classed as *simple* or *compound*. Compound leaves are those whose expanded part is in the form, not of one, but of a number of separate blade-like parts which are called *leaflets*. Leaves may be very deeply divided, but still they are classed as simple unless the blade is completely divided into leaflets. Compound leaves are nearly always net-veined. A branching pinnate leaf is said to be *pinnately compound;* a branching palmate leaf is said to be *palmately compound.* Leaves of locust or ash (see page 187) are examples of the former; leaves of clover are examples of the latter. (See *Figure 71.*)

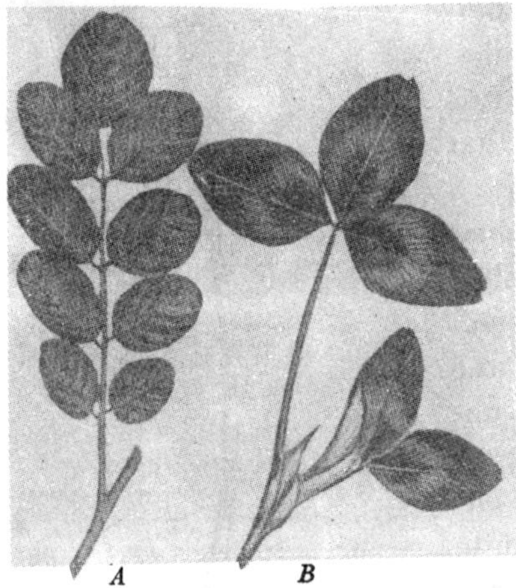

FIG. 71. — Compound leaves: *A*, pinnately compound leaf of black locust; *B*, palmately compound leaves of red clover. The clover leaves have sheathing stipules at the base of the petioles.

C. Attachment and Arrangement. — Leaves are commonly attached to stems by petioles. The end of the petiole which is attached to the stem is usually somewhat enlarged. When this enlargement appears to form a special organ it is called the *pulvinus.* Leaf movements are generally due to change in the water content of the cells of the pulvinus; such movements are characteristic of the clovers and locusts. In some groups of plants, especially in the great rose and pea families, a pair of structures like small leaf blades are borne at the base of the petiole; these are called *stipules.* (See *Figures 70* and *71.*)

Many leaves do not have petioles; their blades begin

right at the axils; such leaves are called *sessile.* (See *Figure 72.*) Sessile leaves often embrace the stems; that is, their blades extend somewhat below the point of attachment; such leaves are said to be *clasping.* This development of the bases of the leaf blades along the sides of the stem sometimes completely covers the stem. The leaves are then said to be *sheathing.* Corn is a plant whose leaf bases completely sheathe the stem.

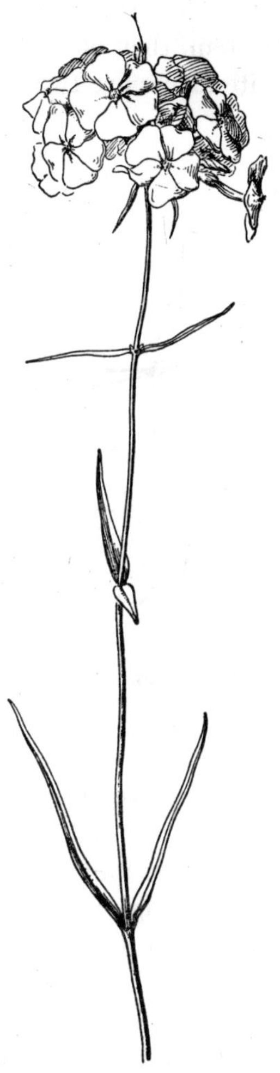

FIG. 72. — Phlox, showing opposite and sessile leaves.

The attachment of leaves and the length of their petioles often have a good deal to do with the relation of the leaves to light. Leaves are attached to stems singly, or in pairs, or in clusters, but, whatever the method, one principle appears to be observed by all. That principle is the avoidance of getting in each other's light. A leaf which is shaded by another leaf has its photosynthesis much reduced, and, if the light has to pass through two leaves first, it has been noted that no photosynthesis occurs at all.

The arrangement of leaves is described by the terms *alternate*, *opposite*, and *whorled*. Alternate leaves are one at a node; they appear to alternate, first on one side of the stem, then on a different side. (See *Figure 69*.) Opposite leaves are two at a node. (See *Figures 72* and *73*.)

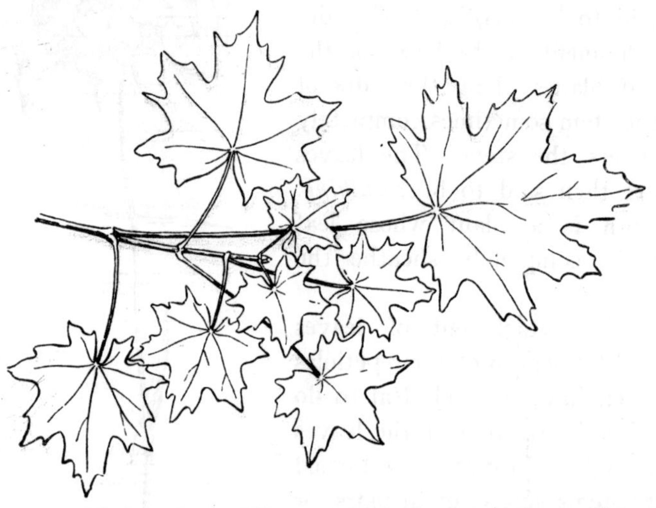

FIG. 73. — A horizontal twig of Norway maple having opposite leaves. Note also the differences in the lengths of the petioles whereby the shading of one leaf by another is avoided.

Whorled leaves are more than two at a node. The parts of a flower are usually arranged in whorls.

If you examine a stem with alternate leaves, you find that an imaginary line drawn through the points of attachment of the leaves forms a spiral around the stem. The advantage of this arrangement is evident. It avoids shading. (See *Figure 74*.) Even more striking in getting this result is the arrangement of opposite leaves. Often in the woods, or in greenhouses, you may find young plants with opposite leaves whose arrangement has the precision of a geometrical drawing. As you look down from above you see that each pair of leaves alternates in position with the pair above and with the pair below it; that is, neighboring pairs are set at right angles to each other and with almost perfect regularity. Here again the advantage is evident. Members of the mint family, which have square stems as well as opposite leaves, show this arrangement particularly well.

Fig. 74. — A plant (*Cineraria*) showing the way in which the points of attachment of the petioles form a spiral around the stem. Note also the differences of the petioles in length.

Petioles often vary in length on the same plant, and their growth in length appears to be stimulated by lack of light. The result is that they often keep on growing, and the blades as a consequence are kept in the light instead of being overshadowed by the leaves above them. (See *Figure 73*.) This habit of growth of the petioles is often found in plants with opposite leaves such as have been just described; the lower down on the stem the leaves are, the longer are their petioles. But it is in connection with whorled leaves that the advantage of petioles which are graded in length is most apparent. This is especially true of plants which, like the dandelion or the plantain, have very short stems and their leaves in whorls. (See *Figure 75.*) Such an arrangement of leaves as these plants have is called a *rosette*, a word which describes their form very well. The outer and under leaves of a rosette have the longest petioles, and they decrease in length toward the top and center. Such plants lack the advantage of long stems whereby to reach the light, but they certainly take advantage of all the light that reaches them. Also this form gives excellent protection against that deadliest of all plant enemies—the loss of too much water. Another arrangement of leaves which is favorable to light exposure is called a *leaf mosaic*, being so named from the fact that the edges of the leaves as viewed from above fit together like the little tiles of a real mosaic. Leaf mosaics are common among vines. (See *Figure 76.*)

FIG. 75.—A dandelion showing the rosette type of leaf arrangement.

D. Leaf Surfaces.—Leaf surfaces are usually either smooth or hairy. Smooth plant surfaces are sometimes covered by a film known as the *bloom;* such surfaces are said to be *glaucous*. The bloom is a thin coating of wax which is secreted by the epidermal cells. It retards evaporation. Surfaces from which it has been rubbed off have been observed to lose water much more rapidly than before.

Cabbage and tulip leaves show this bloom and you have doubtless noticed it on the skins of grapes.

A plant surface covered with soft hairs is said to be *pubescent;* the skin of a peach furnishes a familiar example of pubescence. Many leaves are slightly pubescent, usually more so on the under side than on the upper. Some leaves are densely pubescent, the leaves of the common mullein

FIG. 76. — A leaf mosaic formed by the leaves of the Japanese ivy.

of the pastures being an example. The dense hairs of mullein leaves may have something to do with the mistaken tradition of boyhood that when they are dried they are good to smoke. The stinging hairs of nettle, and the stiff, spine-like outgrowths of thistle leaves are also familiar.

These epidermal hairs of leaves when examined under the microscope often are found to have strange and symmetrical forms; some are *stellate* (star-shaped) and quite beautiful. (See *Figure 77.*)

The possession of hairs may have some advantages to the plant which are not understood; one advantage that has been proved, however, is that hairs when abundant retard evaporation. It is significant that hairs are usually more abundant on the under surfaces of leaves, and that it is on the under surfaces that the majority of stomates are usually located. Some leaves are densely hairy below and perfectly smooth above. Tender, young leaves often have hairs which disappear when these leaves have become older and their surfaces have become more waterproof.

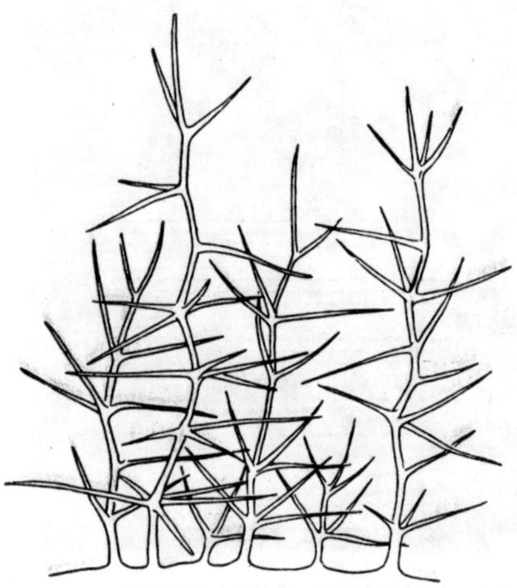

FIG. 77. — Microscopic view of the hairs of a leaf of mullein.

It is not surprising to find that hairiness and thickening of the epidermal walls by cutin are quite general among alpine and northern plants, for the danger of death by loss of water is even greater in cold regions than in warm regions. In cold regions the soil is often much colder than the air, and this makes it more difficult for absorption to keep pace with evaporation. By alpine plants we mean those that grow high on the mountains. The heath family is a large family of plants which is more abundant in cold and alpine regions than elsewhere, and the leaves of these plants are nearly always either hairy or coated with wax or cutinized. Bear in mind that we have no evidence that cold conditions *produce* hairs or leaf coatings; we simply note that plants *having* such structures appear to do better in such regions than plants which lack them.

Cutin, which is present in the older epidermal cells of roots, is also present in the epidermal cells of leaves. Just as it prevents absorption in roots, it prevents evaporation in leaves. Sometimes it is present in such abundance as to give a characteristic appearance and texture to leaves. Probably you have noticed how stiff the leaves of Christmas holly are; this stiffness is due to an abundance of cutin.

E. Leaf Size and Foliage Area. — Leaves are of all sizes; they range from the tiny leaves of a moss, which may be no larger than the head of a pin, to the huge, round leaves of the *Victoria regia*, a tropical water lily, whose leaves may be seven feet in diameter. (See *Figure 77 A.*) The largest

FIG. 77 *A*. — Leaves of a tropical water lily (*Victoria regia*), growing in Lincoln Park, Chicago.

leaves are found in the tropics. Banana leaves are often eight feet long, while palm leaves are just as large though of different shape. The more severe the climate, the smaller, in general, leaves tend to be. Thus it appears that small leaves, in general, are an advantage to plants which have to contend principally with the elements, while large leaves, in general, are an advantage to plants which have to contend principally with one another. In other words, small leaves go with severe climates and larger ones with mild climates. Can you think of any possible reasons for this?

In our own climate large leaves are more frequent in shady places than in sunny places. You find them in damp woods where there is little danger from drought, but a good deal of danger from lack of light. Large, thin leaves can make food in dim light better than small, thick ones; they take advantage of all the light there is. (See *Figure 78.*) Direct sunlight is not essential for photosynthesis; much of it is done in *diffused* light such as there is in shady forests. On cloudy days when the sun does not shine at all, photosynthesis is reduced, but it is not stopped.

You may recall some plants with large leaves which you have seen growing in the open. Do not think of these as exceptions to a rule we have laid down, for we have laid down no rule. Dim light is not the only thing which may make it an advantage to have large leaves.

In studying plant structures it is desirable to consider how this form or another form of the same organ may be related to the fulfillment of the laws of plant life; it is desirable to consider what the advantages or disadvantages of this form may be as compared with another form; it is desirable to consider how certain forms of structure aid the plant in living where we find it and necessarily restrict it to certain kinds of surroundings. In so considering plants, we are studying them as living things, and it is for this reason that we have been considering leaves as related to their environment; we have not been trying to explain

FIG. 78. — Wild spikenard (*Aralia racemosa*), a shade plant. This kind of leafage is common in the woods. Such large, thin leaves are able to do their work in dim light.

why some leaves are large and some are small. It would be quite wrong for you to get the impression that botanists know the *causes* which produce the structures whose advantages or disadvantages they discuss. What seems to be the cause may be only one cause among many, and this whole matter of *why* plant structures are as they are is not to be explained by simply noting a few external conditions to which they seem to be well suited. It is well to bear in mind the principle that structures must appear before their suitableness can be determined; it is no longer believed that the need for a certain structure is a cause of its appearance, though of course the advantage or disadvantage of a structure to a plant may determine whether it continues to exist or not. It is believed that some structures which plants possess are of no advantage to them at all. Similarly, in our own bodies, the vermiform appendix is a structure which seems now to be more of a disadvantage than an advantage.

By *foliage area* we mean the area of all the leaves of a plant. In connection with foliage area, root area should be in mind. Evidently a plant is limited as to the amount of its leaf area by the amount of its absorptive area, its root-hair area. It cannot maintain more leaves than it has root-hairs to supply them with water. Foliage area must be limited also by the amount of water in the soil. In dry soil more absorptive area is required to maintain a given amount of foliage area than in moist soil. Thus in desert plants the root area is usually much larger in proportion to the leaf area than in plants which grow where moisture is more abundant.

Large foliage area is evidently an advantage so far as photosynthesis is concerned, but a disadvantage so far as the danger of drying out is concerned; small leaf area is good for protection, but poor for food making. These two things are checks on each other, and you will find that in many other features than in leaf area the plant seems to preserve a balance between photosynthesis and protection. Too much of one is sure to result in too little of the other.

56. **Functions.** — It will be easier to understand the structure of leaves if we consider their functions first; it would be easier, also, to understand their functions if we were to consider their structure first. Something is to be said in favor of either way. You have already read something about the structure of leaves, however, and it may be better to consider some of the processes with which leaves are concerned before going on to consider in more detail the structures concerned with these processes.

A. Transpiration. — Transpiration is the evaporation of water from living plants. Strictly speaking, it is not a function at all. A function is something which an organ does, and transpiration, strictly speaking, is not a thing which leaves do. It is rather a thing which is done to them. If transpiration is a function of leaves, then we may speak of the evaporation of perspiration as a function of our skins. Transpiration from leaves, like evaporation from our skins, is a thing which happens to them rather than a thing which they cause to happen; with reference to evaporation the leaves (and our skin) are passive rather than active. Transpiration is a process of great importance in plant life, however, and, since it occurs principally in leaves, it is customary to consider it among their functions.

We have considered the large quantities of water absorbed by roots (see page 47) and have noted that the continuous entrance of this water is permitted by continuous transpiration (see page 108). We must now consider some of the relations of this phenomenon to plant life in general. We must bear in mind that the transpiration stream passes out into all the veins of the leaves, and from these into the mesophyll. Here the water is contained in cells whose walls are permeable and which are in contact with air. This air, having entered the stomates, diffuses freely among the intercellular spaces. Under such conditions, unless the air already contains all the moisture it can hold, evaporation is sure to occur. The area of the moist cell walls which are exposed to the air in the spongy parenchyma is many times as great as the total leaf surface. (See *Figure 79.*) With these facts in mind, it is not difficult to understand the source of the drops of water which collect on the inside of a glass bell jar which is placed over a healthy potted plant.

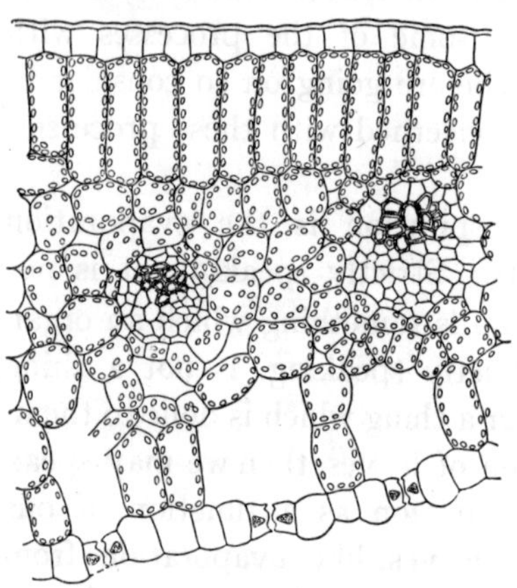

FIG. 79. — Cross section of a leaf of lily showing the internal structure.

Water has more influence in determining the way plants grow and where they grow than any other single factor; it is usually water which makes all the difference between fertile land and desert land; it is the supply of water which, more than anything else, determines the size of the crops of the world. Thus, for example, if varieties of wheat or corn could be made to grow with less water than is usually required for them, millions of acres which are now not used at all could then be used for growing these important crops. Varieties of wheat and corn hardy enough to grow under such conditions have been discovered, and experimenters are now at work trying to improve these varieties as to the weight of food which they produce, and as to other properties.

Thus the need of plants for water is a thing which concerns our lives as well as theirs. The reason that they need so much is that they transpire so much. Nearly all the water that goes into them evaporates out again. Evidently, from the standpoint of the farmer as well as from the standpoint of the plant, this loss by transpiration is a great waste of good water unless such loss is in some way necessary to the life of the plant. It is for us to consider in what ways this constant loss appears to be a good thing for the plant, or in what ways it appears to be a bad thing.

a. The *disadvantages* of transpiration are more evident than the advantages. More plants die on account of it than on account of anything else. That is, water evaporates from them more rapidly than they can replace it, and so they die. Much that we see in the structure of plants is explained by the need to protect against this loss of water. By bark and by cutin, by hairiness and by waxy coatings, by reduced foliage and by thick walls, and in many other ways plants are protected from water loss. To get necessary gases, however, air must be let in, and as surely as air gets in, water gets out. The carbon dioxide and the oxygen of the air are necessary to the plant, but any structure which permits them to enter will surely at the same time permit water to escape. The former is necessary to the life of the plant, but the latter may cause its death, and they must occur together. So we are not surprised to find leaves so constructed that, though the entrance of air is permitted, the exit of water is restricted. The plant appears to be trying to let in air and keep in water at the same time.

Under these conditions it is not surprising that botanists came to look upon transpiration as a sort of necessary evil in plant life, especially after it was found, as you have noted, that the entrance of solutes into the plant does not necessarily depend upon the entrance of water at the same time.

b. The *advantages* of transpiration have been emphasized by certain recent discoveries. It has been found that transpiration, though a menace to the life of the plant in one way, is a protection to its life in another way. Since evaporation is a cooling process, it often prevents leaves from becoming overheated; it may cause their death by drought, but it may also prevent their death by heat. That this death from heat is a real danger is shown by the *scalding* of leaves; this sometimes occurs when bright sunshine follows quickly after a summer shower which has filled the air with moisture. Under these conditions transpiration is checked. The internal temperature of the leaf increases, and, in the absence of transpiration enough to carry off this heat, the leaf is sometimes killed. The heat which may thus kill the leaf is derived from the sunlight. It has been found that only a very small part of the energy which the leaf absorbs from light is used in photosynthesis; it has been found that the unused portion of this energy, being transformed into heat, may be very injurious to the leaf if transpiration is checked too much. It is transpiration which rids the leaf of this dangerous excess of heat. It has been found in the case of some leaves that the excess of heat, if transpiration be stopped, may raise the internal temperature of the leaf to the death point in even less than ten minutes.

Succulent plants of the deserts, such as cactuses, have

been found to be able to endure a considerably higher internal temperature than ordinary plants. Such plants evidently cannot afford that loss of water which would be involved in their cooling off by evaporation. Their ability to endure high temperatures indicates that their protoplasm has power to withstand heat which is much greater than that usually possessed by protoplasm.

In this connection we see that transpiration may be a disadvantage on one day and an advantage on the next. Thus on a hot dry day after a long drought the leaves of corn may be in danger of drying up. They are losing far more water by evaporation than is needed to keep the inside of the leaf from overheating or for any other purpose. Under such conditions excessive transpiration certainly seems a disadvantage. Yet the very next day may be hot and moist and the leaves will be needing all the transpiration that is possible in order to keep them from overheating. Evidently we cannot judge fairly of the advantages or disadvantages of any feature of plant life without taking into consideration the changes in external conditions which are constantly occurring.

Another and a very decided "advantage" of transpiration is that it is a necessary part of the process whereby the ascent of water and solutes is accomplished. You know that these solutes are necessary for the work of the leaf. You have already learned that their entrance into the root from the soil depends upon the *presence* of water, but not upon its *movement* (see page 109); you have noted that into living cells and through them these solutes move in accordance with the laws of osmosis and are not carried along in a mass movement of water. But by far the greatest part of the journey from root to leaf is not through living cells. It is through dead cells. Those conductive vessels of the wood through which water chiefly ascends are dead. In this region osmosis ceases to be the controlling force. Here solutes cease to be independent of the solvent as to the rate and direction of their movement. They are carried along in the mass movement of the solvent. Water and the solutes in it move together, and far more quickly than when the movement is by diffusion and osmosis. Now *if there were no transpiration, no loss of water at the top, evidently there could be no such mass movement of water and solutes up through the conductive vessels of the wood.* Without this mass movement the upper parts of the plant would cease to get the supplies they need. Except in the lowliest plants and in those submerged in water, the journey of the solutes from where they enter to where they are used is far too long for the slow movements of osmosis alone to attend to it. Were it not for the mass movement through the wood vessels, plants could never have attained the stature which they now possess; without this mass movement all the living cells in the upper part of the plant would be without the raw materials for doing their work.

Thus we see that transpiration, though attended by obvious dangers, is also attended by obvious benefits. Whether its advantages are as great as its disadvantages is an unsettled question.

B. General Relation of Water to the Plant. — We can hardly consider transpiration without considering the general relation of water to the plant. If you can form a true picture of the way water moves through the plant and passes from it, it will help you greatly in understanding all other features of plant life. Evidently the rôle of water in plant life is very different indeed from its rôle in our own lives. To think of the absorption of water by plants as corresponding in any way to the drinking of animals is to have an idea that is entirely wrong. It is also wrong to think of the absorption of water and of solutes as one and the same thing, even though they take place together. The statement that " water from the soil carries food up into the plant " is wrong in two places: the things referred to on much of the way are not carried and they are not food.

We may think of the uses of water to the plant as at least three: (1) It furnishes the medium by means of which all movement inside of the plant is accomplished; without it, the protoplasm itself ceases its activity. (2) It is itself used in food manufacture. (3) It causes the turgidity of cells without which all the soft parts of a plant collapse and wilt. (See page 75.)

C. Respiration. — This function is not restricted to leaves. It is a function of all living cells whether in plants or in animals. In plants it occurs in stems, in roots, and in seeds, especially in the cambium layers, and in all other living cells as well as in those of the leaves.

Respiration is that process whereby, principally through the oxidation of food, energy is released. Without respiration no work can be done by living things; they quickly perish.

We are considering respiration under the subject of leaves simply because it is principally through the leaves that there occurs that exchange of gases which is the outward evidence of respiration. It is through the leaves that oxygen is principally received into the plant, and oxygen has a great deal to do with respiration, it is through the leaves that carbon dioxide is given off into the air, and carbon dioxide is a waste which results from respiration.

In our own bodies the lungs are the organs whereby oxygen is taken in and carbon dioxide given off. The process whereby these and other gases are inhaled and exhaled is called breathing, but breathing is not respiration any more than eating is assimilation. Breathing simply brings oxygen in contact with cells which can absorb it and carries off the carbon dioxide which they give off as waste. Plants respire, but, strictly speaking, they do not breathe. The leaves are not like our lungs, except that they are the parts in which oxygen is principally received and from which carbon dioxide is principally given off.

Oxygen enters the intercellular spaces of the leaves and there some of it passes through the permeable cell walls. It becomes a solute in the cell sap, for molecules of gases as well as molecules of solids may become solutes. Oxygen is also found as a gas in the plant body. Bubbles of oxygen mixed with other gases are found in the wood cells. Stems of plants which grow under water often have a special form of structure which provides for *aëration, i.e.* oxygen supply; they contain extensive *air passages*. (See *Figures 80* and *81*.) Just as water supply is commonly the hardest problem for leaves, so oxygen supply is often the hardest problem for roots and submerged plants.

Root-hairs absorb oxygen from the air which is present in the soil among the soil grains. The aëration (air supply) of roots is a matter of much importance in agriculture. One of the great advantages of plowing is that it permits more air to enter the soil and thus supply roots with oxygen.

That soil is best for crops which is loose enough to permit air to get into it, and yet compact enough to prevent water from leaving it too rapidly. Sand, for example, is excellent for aëration, but poor for holding water. Lack of drainage, as well as too great compactness of soil, interferes with the proper respiration of roots.

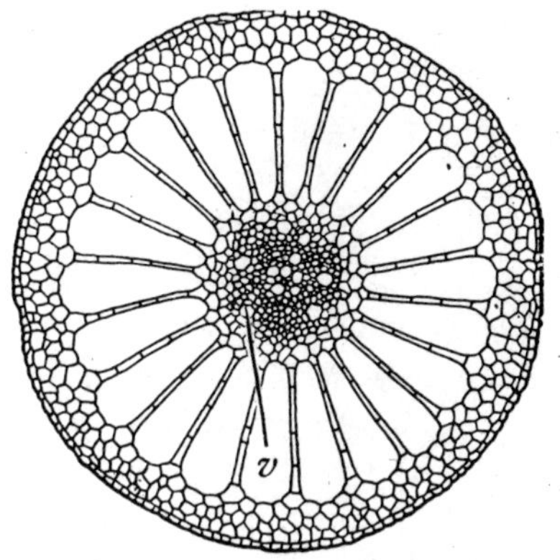

FIG. 80. — Cross section of the stem of water milfoil (*Myriophyllum*), a submerged plant. Note the large and symmetrically placed air passages. Note that the vascular cylinder, indicated by *v*, is in the center, as in roots. Note that its cells are but slightly differentiated. This is common in plants which grow under water; they evidently do not need heavy cell walls in order to maintain an erect position.

Thus, by one means or another, all living cells of the plant receive the oxygen which they must have, and the leaves are the organs through which it chiefly enters. When the plant is not actively at work, not so much oxygen is needed; plants which live through the winter without leaves appear to get along mainly with what oxygen is already in their tissues; so also with the living cells in seeds.

As has been said, we know very little about respiration. We know that it appears to be principally a kind of *oxidation*, which means that it is a chemical change involving the use of oxygen and the giving off of energy. We know, of course, that without it no work can be accomplished by living things. We know that as work increases respiration increases. This you have noticed yourself when exercise gets you out of breath; the body has greater need for oxygen and greater need for the carrying off of wastes. All this we know, but just what happens when molecules of food release their energy and break down into simpler molecules — this still remains a mystery.

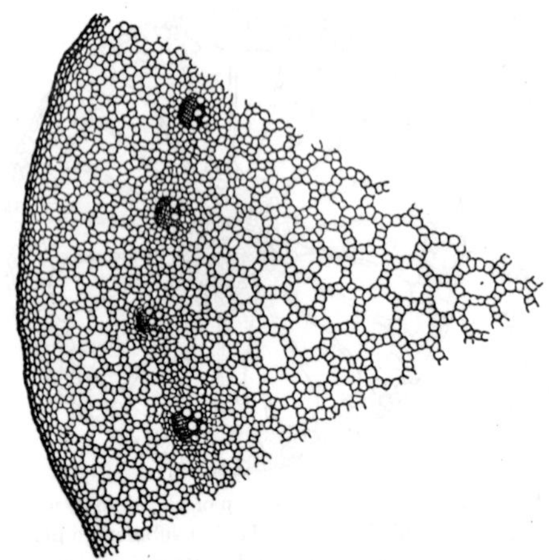

FIG. 81. — Cross section of the stem of hornwort (*Ceratophyllum*), a plant which grows under water. Note the abundant small air passages. Compare with *Figure 80*. In this plant the vascular bundles are small, but they are not in the center.

Respiration is something like what happens when burning takes place, and yet it is not burning. It takes place at low temperatures and water is the medium in which it occurs. This is not true of burning. As in burning, however, oxygen is used in this process (except by some bacteria) and carbon dioxide is given off. This carbon dioxide passes off chiefly through the stomates, but some of it, as you have noted, is given off in solution from the root-hairs. (Carbon dioxide in solution in water becomes carbonic acid, which is formed by the union of a molecule of water with a molecule of carbon dioxide.)

D. Photosynthesis. — The manufacture of food from inorganic materials is the principal function of leaves. There is little question of sharing this function with other parts of the plant, and no question at all as to whether the process is an advantage or a disadvantage. Here we have the thing which appears more than anything else to have made leaves what they are; they nearly all appear to be designed primarily for this work of food making in the light. In photosynthesis, we may say, we have the leaf's principal excuse for existence.

Already you have been told nearly as much about photosynthesis as has a proper place in a book like this. The chemistry and the physics of the process are things to be studied elsewhere than in a first brief course in plants. A few new terms at least are needed now, however, and the principal one is the term *carbohydrate*, the name of that kind of food which photosynthesis produces.

Carbohydrates form one of the three principal classes of foods and photosynthesis is their manufacture. Photosynthesis is sometimes called *carbohydrate synthesis;* it has nothing to do with the manufacture of the other classes of foods. They are made by the protoplasm, and light is not required in the process; they may be made in any living cell. Carbohydrates are, however, the only class of foods which are made entirely out of inorganic materials; in the manufacture of proteins or fats, carbohydrates are necessary as a basis.

The two classes of foods besides carbohydrates are the *proteins* and the *fats;* we shall consider them when we come to the seeds and the foods stored in them (Chapter VIII). In this book the word food is applied only to organic substances, and you can think, of course, of many different kinds of organic substances which we buy in the markets and use for food. Evidently the classification of food into proteins, fats, and carbohydrates is on a different basis from our everyday classification of food into meats, vegetables, etc. It is a chemical classification, which means that it is based on the composition of the molecules. Meat is composed of both fat and protein, vegetables are composed mainly of carbohydrate and protein, and milk is composed of all three. Our bodies appear to be best nourished when our food includes all three of these chemical classes of food, and no one of them in undue proportion.

Sugar and *starch* are the most common examples of the carbohydrate class of foods, and both are abundant in plant tissues. Plants are able to change sugar into starch and starch back again into sugar. This is important. Sugar is soluble and so moves readily as a solute through the plant; starch is insoluble and so does not move through the plant. Since its insolubility makes it independent of the laws of osmosis, starch is an excellent form in which to store food. Storage generally depends upon insolubility, just as movement generally depends upon solubility. The cells of a potato are full of starch, and starch is the principal form of food found in storage organs generally.

Fats and proteins are also quite insoluble, so it is not surprising to find that sugar is the usual transfer form of food in plants. No form of food is more readily soluble or movable than sugar, and out of it any living cell appears to be able to form proteins or fats, just as they are able to transform it into starch.

The molecules of carbohydrates are composed of atoms of carbon, hydrogen, and oxygen. These elements are obtained by the plant from carbon dioxide (CO_2) and from water (H_2O). In the molecules of carbohydrates there are nearly always just twice as many atoms of hydrogen as of oxygen. Since this is so, it is evident that in carbohydrate synthesis there is certain to be an excess of oxygen; that is, if the molecules CO_2 and H_2O are used in equal numbers (as is believed to be the case), there are two atoms of hydrogen for every three of oxygen, and since in the finished product there must be but one atom of oxygen for each two of hydrogen, there is an excess of oxygen under such circumstances.

The oxygen given off by plants as a by-product of active photosynthesis under good illumination is much greater in amount than the oxygen consumed by them in respiration. Since, therefore, the green parts of plants give off more oxygen than they take in, early observers were led to believe that plants breathe in a way which is just the opposite of the breathing of animals. It was believed, since plants absorb carbon dioxide and give off oxygen, and since animals absorb oxygen and give off carbon dioxide, that there is herein an admirable arrangement whereby the supply of both of these gases is maintained in the air. No doubt the oxygen given off by plants does to some extent aid animal life by keeping up the oxygen content of the air, and no doubt animal life aids plant life to some extent by being a source of carbon dioxide, but to say that "plants breathe one way and animals another" is to state the matter very improperly. Plants respire in the way we do just as truly as we do ourselves, but they also give off oxygen in large quantities in connection with a process which does not occur in us.

The real, inner organs of photosynthesis are, as you know, the chloroplasts. They are composed of chlorophyll and protoplasm. (See page 74.) Light is essential to their work. Just what part of this process is done by the chlorophyll and just what part by the protoplasm is not known, but it appears that the chlorophyll is chiefly concerned in the absorption of energy from the sunlight and in the breaking up of the molecules of carbon dioxide, while the protoplasm is more directly concerned in building up molecules of carbohydrates. The protoplasm of the chloroplasts is also continually manufacturing chlorophyll, for the chlorophyll itself is constantly being broken down by the effects of light. It needs constant renewal.

Having found that water and carbon dioxide are the only substances used in carbohydrate manufacture, you may be wondering what becomes of the various substances which come up as solutes from the soil. They are used in the manufacture of proteins, which contain several different elements, and of other substances. Thus, for example, the presence of iron is necessary to the manufacture of chlorophyll. If the roots absorb no molecules of iron, chlorophyll does not appear.

The first product of photosynthesis which can be readily observed is starch. This is principally because starch is insoluble; it collects in the chloroplasts in the form of *starch grains*. The molecules of starch are much larger than the molecules of sugar and other soluble carbohydrates; they contain many more atoms. Thus we note that the solubility of a substance may depend somewhat on the size of its molecules; the larger the molecules, the less likely they are to dissolve. Certain carbohydrate molecules, much smaller than those of starch, are the first products of photosynthesis, and these smaller molecules are gradually built up into larger and more complex ones. Starch molecules may contain hundreds of atoms, but the proportion of two of hydrogen to one of oxygen is never lost; if it were, the resulting substance would not be starch.

Iodine is a substance which turns starch blue. It is common to use a solution of iodine in order to demonstrate the presence of starch in leaves which have been actively at work in sunlight. First, the leaves are soaked in alcohol which dissolves out the chlorophyll. Then they are dipped in iodine. The action of iodine turns such leaves dark, proving the presence of starch, while upon similar leaves which have been shaded for some time iodine produces no such effect. (See *Figure 82.*) The application of iodine to a leaf which has been partly shaded and partly illuminated shows that starch is present in the lighted part and absent in the darkened part, but evidently this experiment does not prove that this result is due to differences in light alone unless care has been taken that the shaded part receives carbon dioxide as well as the lighted part.

FIG. 82. — A geranium leaf tested with iodine for starch. The darker portion had been exposed to light, while the lighter portion had been screened from light. Both parts of the leaf, however, were exposed to air.

A good definition for photosynthesis is the following: "It is the manufacture of carbohydrates by chloroplasts in the presence of light, water and carbon dioxide being used, and oxygen being given off as a waste product." (J. M. Coulter, *Textbook of Botany.*)

E. Summary as to Gaseous Exchanges. — The following sums up what we have been discussing: —

"The chief gas movements in plants are associated with

respiration, carbohydrate synthesis, and transpiration. Respiration, involving the absorption of oxygen and the emission of carbon dioxide, takes place in nearly all plants at all times, though it is slight, or even wanting, in resting organs, such as seeds. Carbohydrate synthesis, involving the absorption of carbon dioxide and the emission of oxygen, is confined essentially to chlorophyll-bearing organs in the presence of sunlight. Transpiration, involving the emission of water vapor, occurs in all aërial organs; principally in the leaves. Transpiration involves much the greatest gas movement, respiration much the least." (Cowles's *Ecology*.)

57. **Structure.** — We are now to consider the inner appearance of leaves. We are to study the construction of that apparatus whereby principally plants are related to air and light, and wherein principally they make their food. We are to note how this apparatus appears to fit these uses.

A. Epidermis.— The leaf of a tulip or hyacinth or some similar fleshy leaf is best for studying the epidermis. From such a leaf the thin, transparent, whitish skin may be easily peeled off.

The cells of the epidermis are usually about the same size. Except for those which border the stomates, they do not contain chlorophyll. Their outer walls, those which form the leaf's surface, are usually a good deal thicker than those walls which are not so exposed. (See *Figure 79*.) It is in connection with this outer wall that there occur those secretions of cutin or wax or those outgrowths of hairs or spines which you have already noted. It is evident that the epidermis is a tissue whose use to the plant appears to be for protection exclusively.

B. Stomates. — By far the most interesting feature of the epidermis is the stomates. These seem to be doing something. In *Figure 83* you see among the cells of the epidermis numerous sausage-shaped cells which occur in pairs. These are the *guard-cells* of the stomates. The stomate itself is a slit-like opening in the surface of the leaf and it lies between the guard-cells. It opens directly into the intercellular spaces of the mesophyll; a rather prominent intercellular space lies just beneath it. (See *Figure 79*, page 220.)

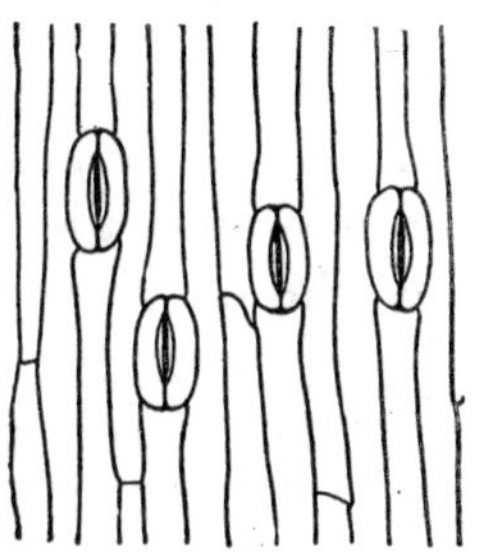

FIG. 83. — Epidermis of a leaf of lily, showing stomates.

Stomates are usually very much more abundant on the under than on the upper side of leaves; often they are not found at all on the upper side. On leaves which stand erect they are about equally distributed on both sides, and on leaves which lie on the surface of water, like those of the water lily, they occur only on the upper side. Usually, on the under surfaces of leaves, stomates are very abundant; about sixty thousand to the square inch of leaf surface is common, and they have been found in such density as to indicate more than four hundred thousand to the square inch. Stomates are not confined to leaves. They are usually found in connection with any green tissues. They occur in the epidermis of young stems and fruits.

Stomates have always been a puzzle to botanists and the puzzle is by no means settled yet; their behavior and the causes which control it are not yet satisfactorily explained. By changes in the condition of the guard-cells, the stomates are opened or closed, but the causes of such changes in the guard-cells are not clear. The chloroplasts in them are thought to have something to do with the matter, but no one has been able to show just what.

The guard-cells are attached to each other at the ends. As a consequence of this, when they are *turgid* (swollen with water), they bow out; they curve away from each other and the stomate thereby is opened. On the other hand, when the guard-cells are *flaccid* (shrunken through loss of water) they straighten elastically and practically close the slit between them. Hence the stomates tend to be open when the water supply is abundant and closed when the water supply is scanty. A natural conclusion from this is that guard-cells regulate transpiration. To some extent they probably do regulate it, but it has also been noted that guard-cells tend to open the stomates in the light and close them in the dark. This evidence cannot be interpreted like the other, for the reason that danger from loss of water is greater when the sun is shining than it is at night; that is, the tendency of light to open the guard-cells would conflict with the tendency of water shortage to close them.

In view of the responses of guard-cells to light, and in view of their possession of chlorophyll, it has been held that their behavior is related to photosynthesis rather than to transpiration. In any case, whatever it is that causes their opening or their closure, it is evident that the principal advantage to the plant in possessing stomates is that thereby those gas exchanges which are necessary for photosynthesis are facilitated. Whether there is much advantage in guard-cells as checkers of transpiration is doubtful; if a plant is in real danger of drying out, it is probable that the straightening of the guard-cells does little to save it. As to their use, then, stomates may be well defined as "organs which facilitate gas exchanges."

An advantage to the plant in having its stomates generally much more abundant on under rather than on upper surfaces of leaves is evident in connection with rain. The work of leaves depends completely upon the free access of air into the mesophyll. Both photosynthesis and respiration would be seriously impeded if the stomates became choked by raindrops, or if the intercellular spaces were filled with water. Even in the absence of stomates, the upper surfaces of leaves are usually so constructed as to cause water to run off rapidly. Depressed veins and ribs often form a sort of drainage system for the upper surface of the leaf, while hairs and waxy coatings prevent water from soaking in as well as interfering with its evaporating out.

Another advantage in having stomates on under rather than on upper surfaces is that the loss of water by transpiration is less under the former condition. Plants which grow in moist places commonly have stomates almost equally abundant on both sides of the leaves, while plants which frequent dry places have their stomates almost exclusively on under surfaces.

C. Mesophyll. — This term refers to all the tissues of the leaf except the veins and the epidermis; the word means *in the midst of the leaf.* Another useful word in connection with this subject is *chlorenchyma.* All tissue which contains chlorophyll, whether it is in the leaf or elsewhere, is called chlorenchyma; the word means *green tissue.* Why is it more accurate to refer to photosynthesis as a function of the chlorenchyma than it is to refer to it as a function of the leaves?

Mesophyll is quite distinctly differentiated into palisade tissue and spongy tissue. (Study *Figure 79*, page 220.) It is believed that the palisade tissue has more to do with photosynthesis, and the spongy tissue more to do with relations to the air. Evidently, on account of their positions, the palisade is better related to light, and the spongy tissue better related to air.

a. Palisade. — In an ordinary leaf the palisade is on the upper and the spongy tissue on the under side. Some vertical leaves have palisade on both sides. The common wild lettuce develops palisade on both sides in intense light, on only one side in diffuse light, and in dense shade it develops none at all. Should we be justified in concluding from this that as a general rule the development of palisade depends upon light? Surely not. In the first place the wild lettuce may be quite an exception among plants in this matter, and in the second place the behavior noted may have been due to other conditions than the differences in light alone, as, for example, differences in the amount of transpiration. Under ordinary conditions of plant growth, transpiration increases as light increases, and decreases also about as light decreases.

It has also been noted that the palisade tissue is much more extensively developed in dry than in wet soil, the exposure to light being equal in both cases. Moisture evidently has an effect upon the development of this tissue as well as light. The greatest development of palisade tissue is in plants which grow in deserts, where intense light goes along with danger from too much evaporation.

So it appears that excess of light and excess of transpiration are both associated with prominent development of palisade. Does it follow from this that palisade tissue protects the plant from too much light or from too much transpiration? Not necessarily. To find the cause which induces a structure is not to prove the advantage of that structure.

As to light, it has been held that the shape of palisade cells is an advantage in that it reduces the number of walls the light has to penetrate in reaching the chloroplasts, and that it permits the chloroplasts to get out of danger in case of too intense light, that is, they can move to the bottoms of these deep cells. Evidence shows that very intense light does injure chlorophyll, but evidence also shows that in some palisade cells the chloroplasts do not move; in others, however, their movements have been observed and they have been seen to assume various positions, apparently under the influence of light.

As to reducing the loss of water by transpiration, evidently the compactness of the palisade tissue is an advantage as compared with the looseness of the spongy tissue. As in the case of the closure of stomates by the guard-cells, however, it is doubtful whether this structure has a protective effect with respect to water loss which actually amounts to much.

b. Spongy Mesophyll. — The air spaces which characterize the spongy mesophyll deserve our attention as much as the cells of that tissue themselves. These air spaces are connected; they form a continuous system in which air can circulate freely throughout the leaf; it is sometimes called the aërating system. In some plants the air spaces are beautifully symmetrical in arrangement. They are particularly well developed in connection with water plants.

The mesophyll of submerged leaves is entirely composed of spongy tissue and air chambers; there is no palisade at all, and the air chambers may be much larger than the cells which surround them. Such plants, of course, do not need protection from transpiration, but do need some arrangement to insure their supply of oxygen; the air chambers constitute such an arrangement.

As you have noted, that place at which the plant has its relations with the soil is the region immediately surrounding the root-hairs. Similarly, that place at which the plant has its principal relations with the air is the region immediately surrounding the cells of the spongy mesophyll, the region of that "internal atmosphere" which the air in the air spaces composes. The root-hairs are exposed directly to the soil; the mesophyll cells do not appear to be directly exposed to the air, but as a matter of fact they are, since the outside air is continuous through the stomates with the air in the aërating system. These two regions — the root-hair region and the air-space region — may be regarded together as the two regions in which the relations of the plant to the outside world are principally established. The exchange of liquids goes on actively between root-hairs and the surrounding soil, and the exchange of gases goes on actively between the mesophyll and the air in the spaces which they border. The root-hairs are the organs of liquid exchanges; the mesophyll cells are the organs of gas exchanges.

D. Veins. — The veins, in a cross section of a leaf, are found to lie in the spongy mesophyll rather than in the palisade; they are often in contact, however, with the inner ends of the palisade cells. (See *Figure 79.*) The structure of a vein is practically identical with the structure of a small vascular bundle, which is natural inasmuch as the veins are the extensions of the vascular bundles into the leaves. The larger veins often exceed in thickness the rest of the leaf; they may fill the cross section from epidermis to epidermis and cause a swelling in it, but the veinlets are embedded in the mesophyll. The larger veins evidently serve as a mechanical framework for the blade as well as serving as paths of movement.

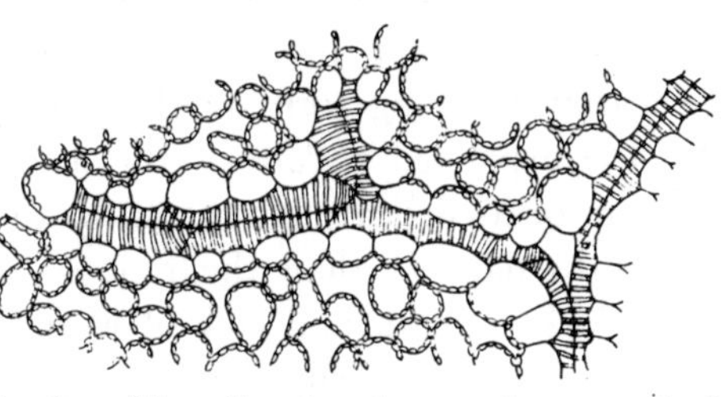

FIG. 84. — The ending of a xylem strand among cells of the mesophyll.

The veins are seen to be so related to the soft tissues of the leaf that their delivery of raw materials and their absorption of manufactured food is not difficult to understand. No transportation system could more thoroughly penetrate the region which it serves than the veins penetrate the leaves. (See *Figure 84.*)

The xylem part of the vein lies above the phloëm part, nearer to the palisade tissue. This is the position we would expect in view of the fact that in the stem the xylem lies within the phloëm.

58. Protection. — Certain features of leaves not yet considered may be grouped under this general title. The *shedding of leaves* and the *movements of leaves* are such features. Protection from excessive transpiration is the principal advantage which the plant derives from both of these phenomena. Similarly, the *edgewise position* of leaves with reference to light is protective with reference to excessive light and with reference to excessive transpiration as well.

Almost every part of a plant has its protective aspect. In leaves we have considered the protective nature of the epidermis and of the palisade tissue, and we may say of the veins that they protect the otherwise soft blades of the leaf from being torn by the winds. We have noted that the expanded nature of the leaf, and the necessary exposure of its delicate inner cells to the air, make it more subject to dangers than the root or the stem. Lack of water, lack of light, and lack of warmth appear to be the principal dangers to which the leaf is exposed, and of these the first named is by far the greatest. Many different means of protection from these dangers are found. Having in mind the kind of protection a leaf needs, as well as the kind of work which it does, we are able usually to perceive the advantages or the disadvantages of the various kinds of structures which we meet.

A. The Shedding of Leaves and their Autumnal Colors. — The regular fall of leaves from the plant is a very important means of protection. It is, of course, not a means of protection to the leaves themselves, but as a means of protection to the plant as a whole this habit is of very great advantage.

The most striking difference between the plants with which we are familiar and the plants of the tropics is that with us plants lose their leaves in the winter, while tropical plants generally bear leaves the year round. Plants which have the habit of losing all their leaves at about the same time are called *deciduous*. Deciduous plants are not confined to temperate regions; wherever there are long dry seasons many plants drop their leaves at the beginning of the dry season, just as with us they drop them at the approach of winter. Evidently the effect of dry weather upon the fall of leaves is about the same as the effect of cold weather.

It should not be understood that evergreen plants keep the same leaves throughout life. They too shed their leaves, but gradually rather than all at about the same time; new leaves appear at about the same rate that the old ones fall. The life of a leaf, save in some annuals, is much shorter than the life of the plant which bears it. The cotyledons and the leaves of some plants which grow under very dry conditions may endure only a few weeks or even days. Evergreens rarely retain any one leaf more than a year or two, though the needle-like leaves of some pines have been observed to endure as long as ten years. The life of most leaves is limited to a few months, varying as the season of active work is long or short.

Why is it that leaves cannot do their work as well in winter as in summer? Evidently it is not on account of lack of light, for the sun often shines as brightly on winter days as on summer days. It is not on account of lack of carbon dioxide, for carbon dioxide is always present in the air. Water too appears to be abundant. On account of the cold, however, the water, especially the water in the soil, is more likely to be solid than to be liquid. Herein appears to be a principal reason why cold weather arrests the work of plants. Water in the form of ice is of no more use to the plant than no water at all. Only in liquid form can water enter the roots or move through the plant body or furnish means for the activity of the protoplasm. So we see why freezing weather and long droughts have practically the same effect on plants; they both affect the water supply. Cold apparently does not affect plants directly in the manner in which it affects us directly. Indirectly, however, through stopping the water supply or by freezing the water already inside of plants, it has an even greater effect upon them than it has upon us.

Under these circumstances we can see why the loss of the leaves is so great a means of protection to the plant as a whole. If the wide, green leaves were present in winter, there would be great loss of water by transpiration, especially on dry days when the temperature is above the freezing point. The roots in the frozen soil would be entirely unable to make good this loss, and the death of all the aërial parts would result.

The following quotation sums up this matter: —

"The shedding of leaves at the inception of a cool or dry period is of inestimable advantage, especially in trees with delicate leaves, because of the enormously reduced transpiration thus resulting. The leafless tree is one of the most perfectly protected of plant structures, since the impervious bud scales and bark cover all exposed portions." (Cowles's *Ecology*.)

Of course when we speak of an advantage to the plant in losing its leaves, we are assuming that it has enough food stored up to nourish it until new leaves become active in the next growing season. The advantage is even more evident for plants which shed their leaves at the approach of a dry season than for those which shed them at the approach of winter.

Probably our evergreens have occurred to you in connection with this subject. You are familiar with trees like the pine, whose needle-like leaves stay on the year round, and you may be wondering how they avoid injury in winter. One glance at a pine needle does much to answer this question. Both by shape and structure it is evidently far better protected than wide, thin leaves are protected. It is in no such danger from loss of water or from injury by the freezing of the water in it. Instead of being a disadvantage, leaves of this kind are a positive advantage to plants in winter. On many days it is warm enough and bright enough for photosynthesis, and on such days the needle-like leaves are at work. (It is to be noted that the roots of the plants which bear such leaves usually penetrate below the frozen zone of the soil.) In summer, however, such leaves are a disadvantage as compared with broad, thin leaves, for the latter have much greater exposure to light and much greater chlorophyll content.

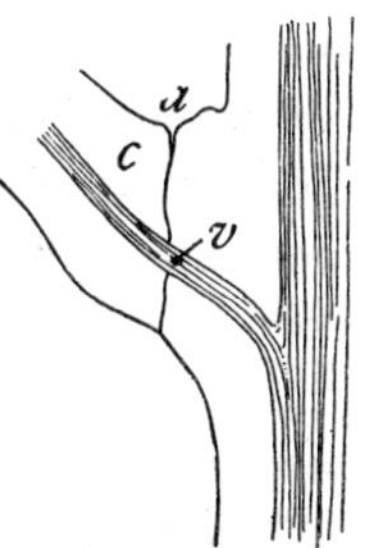

FIG. 85. — Diagram illustrating preparation for leaf fall. A section through the base of the petiole of a mature leaf; *a* indicates the absciss layer. This layer runs through the cortex, indicated by *c*. The leaf is held till the last by its vascular strands, *v*.

Most plants which lose their leaves in the fall, as the time to shed them approaches, form a special layer of cells at the base of the petiole. By means of this layer the leaves are shed by a sort of self-amputation. This special layer is called the *absciss layer.* (The word absciss means *cutting-off.*) It loosens the leaf from the stem, and forms a place at which any gentle breeze or even its own weight is likely to detach the leaf. (See *Figure 85.*) After the leaf falls, the wound which might be expected is often found to be already healed. The absciss layer is responsible for this. It often forms a healing scar even before the leaf falls off. The cells of this layer differ from adjoining cells in that they are more turgid and have thinner walls. Cork sometimes appears in them, and forms the healing scar.

In temperate regions the colors of leaves in autumn form one of the most striking beauties of nature. The reds and golds of the woods in October delight us as much as their soft green of early spring. Maples and oaks, the dogwood and the sumach, give us masses of color which make us exclaim with pleasure. Just what it is that causes this change in color cannot be expressed in a single word. It is common to hear that frost does it, but this is disproved by the fact that the bright colors often come before the first frost. Sometimes a bright-colored branch is found in the woods in midsummer, and young leaves often show a tendency to redness much like that which they show in the fall when they are old. *It appears that the cause of the colors may be anything which reduces the activity of the leaves.* It is not surprising that an early frost has been considered the principal cause. Sometimes it is the principal cause, for it may be the first thing to interfere with the work of the leaves.

Redness in leaves is produced by certain substances called *anthocyans.* The anthocyans are found in solution in the cell sap. They seem to appear when sugar is abundant. Redness has been induced in leaves by growing plants in solutions of sugar. It has been suggested that the absciss layer, which is usually full of starch, checks the migration of sugar out of the leaf and so indirectly causes the shades of red to appear.

The yellowness of leaves is not due to a newly manufactured substance. It is due to a substance called *xanthophyll* which is present in chlorophyll all the time. The yellowness of xanthophyll does not show in active leaves. It is obscured by the green. But as the chloroplast begins to break down in the dying leaf, its greenness disappears, and the color of the xanthophyll begins to show.

B. Movements of Leaves. — The most striking case of the movement of leaves is the case of the sensitive plant. Perhaps you have seen this plant. It is a native of the tropics, but it is common in greenhouses. It is a small creeping plant having somewhat the habit of growth of the strawberry. Its compound leaves are composed of many small opposite leaflets. If the plant is touched or affected by vibration, first these leaflets fold together, and then the whole leaf bends down as on a hinge at the base of the petiole. (See *Figure 86.*) These same movements are produced, but more gradually, by drought.

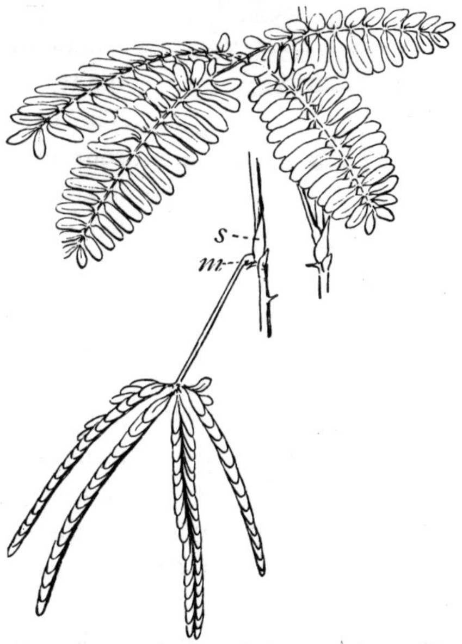

FIG. 86. — Movement of the leaves of the sensitive plant (*Mimosa pudica*); *m*, pulvinus, *s*, stipule. See context.

In its home in the tropics this plant is called the *shame plant.* It seems to shrink in shame. It is a very common weed, and is especially troublesome in the rice. From car windows thousands of these plants may be seen shrinking down in response to the vibration caused by the passing train.

The sensitive plant belongs to that great family (*Leguminosæ*) to which also peas and beans belong. Our own clovers and the locust trees also belong to this family, and they too show leaf movements. Their compound leaves close up at night, especially in cool weather. The little *Oxalis*, called sorrel or sour grass, grows often in lawns; it also has this habit, as you may have noticed at twilight. Its compound leaves resemble those of the white clover. (See *Figure 87.*)

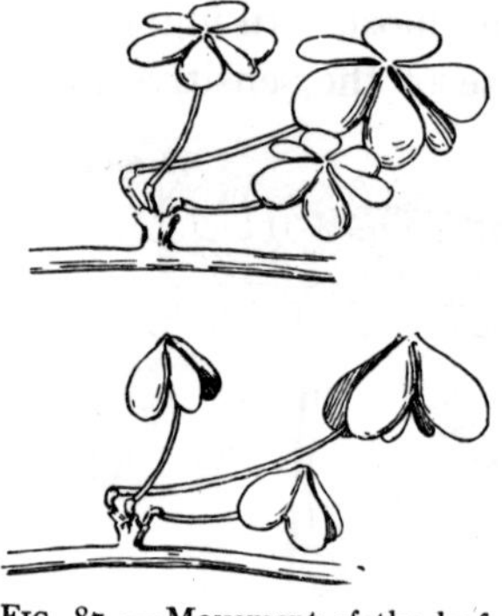

Fig. 87. — Movement of the leaf of wood sorrel (*Oxalis*). The upper picture shows the leaves as they appear in daylight, the lower one shows their position at night.

Evidently the chief advantage of this closure of leaves is that transpiration is checked thereby. But the advantage of closing at night rather than in daytime, or when touched rather than when untouched, is not evident. These may be examples of habits which have no advantages.

The immediate cause of leaf movements is change of turgor in certain cells of the pulvinus. You will recall that the pulvinus is the enlarged part of the petiole at its base. It is evident that if the cells on one side of the pulvinus suddenly lose their water, while the cells on the opposite side remain turgid, the petiole will tend to bend toward the flaccid cells. This is a striking example of the way in which the condition of the protoplast affects turgor. It is the protoplast, of course, which is the sensitive part of the cell and receives the stimulus. The stimulus appears to cause quick alterations in the protoplast, and these in turn cause those physical changes which result in the movement.

C. The Edgewise Position. — Leaves which are crosswise to the light are so much more conspicuous than those which are edgewise that it is natural to think of all leaves as exposing themselves to light as much as possible. One has only to visit a swamp, however, to realize that very many plants have their leaves set edgewise to the sun, edgewise at least when the sun is high in the heavens. The cat-tails and flags which are common in swamps have their sword-like leaves set with the points straight up. Even in a meadow or on lawns you may note that the blades of grass grow so close together that it is usually impossible for them to be crosswise to the light.

It appears that, where there is an abundance of light, leaves tend to avoid the intense light of midday. Intense light often seems to be too much of a good thing. It has been observed that excessive light produces injury to the chlorophyll. A thing which is absolutely essential for photosynthesis may thus sometimes interfere with it. In the open plants get more light than they need, and the supply of carbon dioxide and the temperature are the things which, under this condition, chiefly limit photosynthesis.

The edgewise position results in the loss of less water by

transpiration than the crosswise position. The direct exposure of the leaf blade to the warmest and brightest rays of the sun results in a higher rate of evaporation than if the edge is turned to these rays. In many situations this advantage is probably more important to edgewise leaves than the advantage of protection from excessive light. In some dry regions the leaves of many forest trees and shrubs have the edgewise position. This is particularly true in certain regions of Australia.

Apart from protection from excessive light and excessive transpiration, there is another distinct advantage to be noted in connection with the edgewise position of leaves. This is an advantage, not to one plant, but to a great many plants taken together; it is a collective advantage rather than an individual advantage. Though the edgewise position results in minimum lighting for any one leaf, *it results in maximum lighting for the vegetation as a whole.* The more vertical the leaves, the more numerous may they be in any given space and yet have sufficient light to live. This position, which may seem bad in some respects for one leaf, seems very good for all the leaves; it results in " the greatest food production possible within a given volume of leafage." (See *Figure 88.*)

Certain plants with edgewise leaves are called *compass plants,* for the reason that when they grow in the open their leaves tend to point north and south. It is evident that this position is more favorable for receiving the morning and afternoon rays of the sun than if the leaves pointed east and west. The rosin-weed of the prairies is the best-known compass plant of the United States, and the same tendency in a less degree may be noted in the edgewise leaves of the common wild or prickly lettuce.

59. **Special Forms of Leaves.** — It has already been mentioned that the parts of flowers and, often, thorns and tendrils are like leaves in origin. From this point of view, such organs may be described as "special forms of leaves." It seems much better, however, to discuss the parts of flowers in the chapter devoted to flowers than to discuss them under the present heading.

FIG. 88. — View in a swamp showing the vertical arrangement of leaves, which gives the maximum amount of illumination for the vegetation as a whole.

As to tendrils, it may be noted that the Virginia creeper is a familiar plant whose tendrils arise as leaves arise, while the tendrils of sweet pea are like leaflets in origin. They show the same sensitiveness to contact, and the same general behavior as a consequence of it, that has already been noted in connection with stem tendrils. In the climbing nasturtium the petiole often behaves like a tendril.

As to thorns, the barberry is a commonly cultivated shrub whose thorns are leaves as to origin. The thorns of the locust trees are stipules as to origin.

A. Needle-like Leaves. — Pine trees bear the needle-like leaves which are probably most familiar to you. The common varieties of pine may be distinguished by the character of their needles. The advantages and disadvantages of needle-like leaves as compared with broad leaves have been noted in the preceding section. Since these leaves live and work during the winter, it is not surprising to find in them thick protective layers of tissue surrounding the mesophyll. The stomates are found at the bottom of pit-like depressions. (See *Figure 89.*) These structures, combined with the comparatively small surface exposed and with the small water content, enable these leaves to survive through the severest winters.

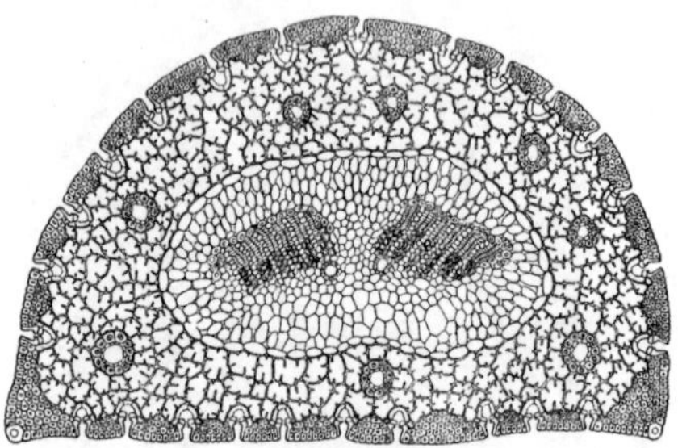

FIG. 89. — Cross section of pine needle. Note the layers of protective cells just under the epidermis. Note that the stomates are at the bottom of little pits. The vascular tissue is in two strands in the center. The rings of cells in the mesophyll indicate cross sections of resin ducts.

B. Leaves as Insect Traps. — You may have heard of carnivorous plants. You may even have seen them in conservatories. They are often kept there as curiosities. Carnivorous means meat-eating. Only a few kinds of plants possess this power of using animals as food. The animals used are insects, and the organs which capture them are leaves. Two principal types of plant insect-traps are found: one is the *pitcher* type, the other the *steel-trap* type.

Pitcher plants are found, usually in swamps, both in the northern and southern United States. (See *Figure 90.*) The southern pitcher plant produces a sweet liquid (*nectar*) at the mouth of the pitcher; it is quite similar to the nectar which attracts insects to flowers. If the insect goes beyond the mouth of the pitcher, it rarely if ever gets out again. Just beyond the rim there is a surface so smooth that even a fly slips on it; it seems to be glazed. If the insect slips on this smooth place, it finds itself in the water which is just beyond. The wetness of its wings prevents it from flying out, and stiff, downward-pointing hairs, as well as the very smooth space, keep it from crawling out. It drowns.

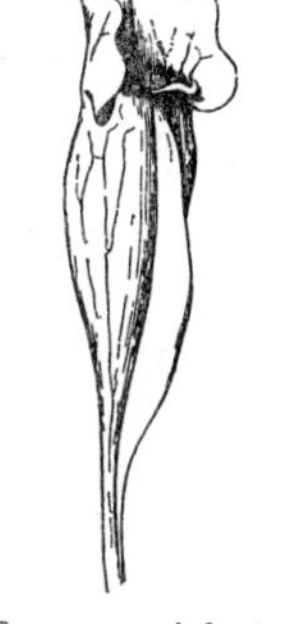

FIG. 90.—A leaf of the pitcher plant (*Sarracenia*). Usually these pitcher leaves are partly filled with water into which insects crawl or fall.

The other type is well illustrated by a famous plant called Venus's-flytrap. This plant has been found only in certain swamps of North Carolina. (See *Figure 91.*) It captures insects by the sudden closure of the trap part of the leaf. This part bears sensitive hairs, and the trap is sprung, like a steel trap, when an insect disturbs these hairs. Another somewhat similar insect-catching arrangement is that of the sun-dew, a swamp plant which is found in various parts of the United States.

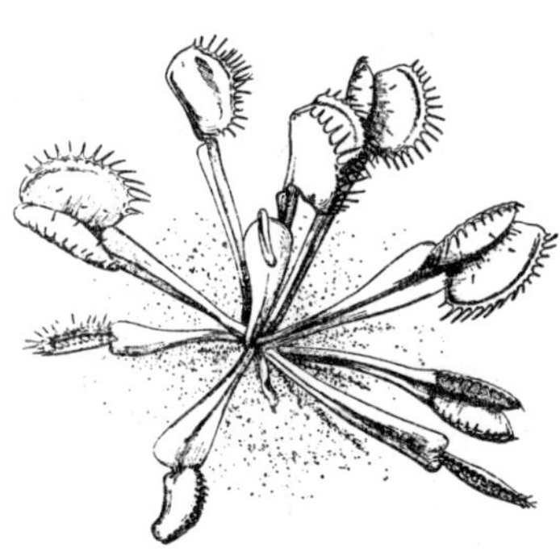

FIG. 91.—A rosette of leaves of the Venus's-flytrap (*Dionea*).

(See *Figures 92* and *93*.) In this plant it is the long outgrowths of the leaf surface and not the leaf itself which close down upon the insect. The insect is first caught, however, by getting its feet entangled in a sticky substance which is produced at the swollen ends of the outgrowths. The leaf acts somewhat like a bit of sticky fly paper. This plant gets the name *sun-dew* from the dew-like appearance of the sticky drops; they do not evaporate in the sun like ordinary dew.

FIG. 92. — Plant of sun-dew (*Drosera*).

How does the food derived from the captured insects get into the body of the plant? As you already know, it can enter only in liquid form. The change of food from a solid into a liquid condition is accomplished by the process called digestion. The captured insects are digested where they are caught. Certain fluids which the plant manufactures act upon them much as the digestive fluids act upon the food which you eat. After this process, the molecules of food enter the plant as solutes.

These insect-catching plants all possess chlorophyll. They appear to be as capable of photosynthesis as many of their neighbors which do not catch insects. It is estimated that the food which they derive from insects is small in amount as compared with the food which they obtain by the more usual methods. Insect catching appears to supplement their diet; it does not appear to be their main source of food. The conditions which have caused this remarkable habit and the advantage which it may be to the plant have not yet been satisfactorily explained.

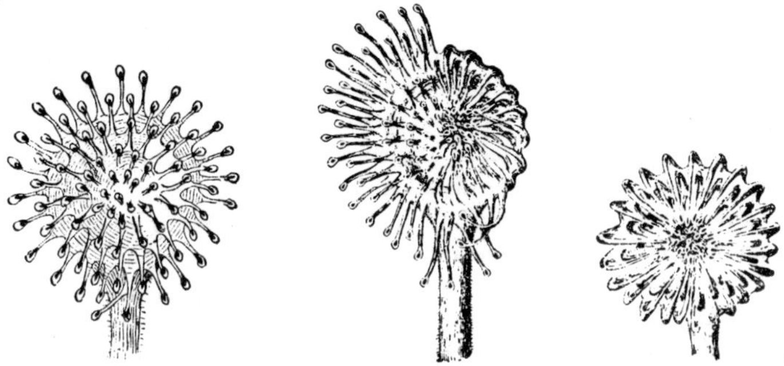

FIG. 93.—Leaves of sun-dew magnified, showing the method of capturing an insect.

QUESTIONS AND SUGGESTIONS

SECTION 54. 1. Define leaves as to structure, function, and origin.

A. 2. Discuss the value of leaves to man.

B. 3. Describe a leaf externally and internally, naming its principal organs and tissues and indicating their functions.

SECTION 55. *A.* 1. Describe the venation of leaves, comparing monocotyledons with dicotyledons. 2. Define pinnate and palmate leaves, giving examples.

B. 3. Illustrate by examples of leaves the principle that organs vary in their power to vary. 4. Define and give examples of linear and ovate leaves. 5. Define and give examples of entire, toothed, and lobed leaves. 6. Distinguish between pinnately and palmately compound leaves, giving examples.

C. 7. Define pulvinus, stipule, sessile, clasping, and sheathing. 8. Describe the opposite, alternate, and whorled arrangements of leaves and give examples. 9. Describe gradation in the length of petioles, explaining the advantage thereof, and defining rosette.

D. 10. Describe glaucous and pubescent leaves, giving examples. 11. Discuss the advantage of hairiness to leaves. 12. What are three characteristics of leaves of the heath family, and in what regions are members of this family most abundant?

E. 13. Contrast large leaves with small leaves as to advantages and disadvantages. 14. State the advantage and disadvantage of large and small foliage area, indicating what limits the amount thereof.

Suggested Study. — Select some common weed which is familiar to you and write a description of what seem to you the particularly advantageous features of its roots, stems, and leaves.

Section 56. *A.* 1. Define transpiration and explain why it is not a function. 2. State the disadvantages of transpiration. 3. State the advantages of transpiration. 4. Explain how transpiration may be an advantage at one time and a disadvantage at another.

B. 5. Describe the general relation of water to the plant, making a comparison with animals, and stating its threefold use to plants.

C. 6. Define respiration. 7. To what extent are leaves like lungs? 8. Explain the difference between respiration and breathing. 9. Explain how air reaches parts of plants that are under water. 10. State similarities and differences between respiration and burning.

D. 11. Name the three principal classes of food, and distinguish between them as to how and where they are made. 12. Name the materials out of which carbohydrates are made, and state the source of each. Are these materials organic or inorganic? 13. What is the basis of the classification of food into the three classes indicated? 14. Discuss the physical properties of sugar and starch as related to movement through the plant body. 15. Discuss the composition of the molecules of carbohydrates. 16. Explain why the belief arose that the breathing of plants is the reverse of that in animals. 17. Distinguish between the functions of chlorophyll and protoplasm in photosynthesis. 18. State the use of the mineral solutes which enter the plant, and give an example of a mineral and its use. 19. Describe the first observable product of photosynthesis. 20. Describe an experiment to show that an illuminated part of a leaf will produce starch while a darkened part will not. 21. Define photosynthesis.

E. 22. Give a summary as to the gaseous exchanges of the plant.

Section 57. *A.* 1. Describe the structure and function of the epidermis of a leaf.

B. 2. Describe stomates and guard-cells. 3. Describe the action of guard-cells in opening and closing the stomates, defining turgid and flaccid. 4. Define stomates as to function. 5. Are the guard-cells of advantage to the plant in regulating transpiration? Explain your answer. 6. Explain the advantage in having most of the stomates on the under surface of the leaf.

C. 7. Define mesophyll and chlorenchyma. 8. Describe the structure of mesophyll, indicating the advantages of the palisade structure. 9. Indicate what goes on in the spongy mesophyll, contrasting the work of this region with that of the root-hair region.

D. 10. Describe the structure and position of the veins.

SECTION 58. 1. What are the principal dangers to which the leaf is exposed? 2. Name the subtopics of this section and indicate how each has to do with protection.

A. 3. Define deciduous plants. 4. Discuss the length of life of leaves. 5. Explain why cold and drought have similar effects upon plants. 6. Explain the advantage in the loss of leaves before a cold or a dry season. 7. Compare a pine tree with a maple as to the advantage or disadvantage of their foliage habits. 8. Describe the shedding of leaves, defining absciss layer. 9. Discuss the cause of autumnal colors of leaves, defining anthocyans and xanthophyll.

B. 10. Describe the movements of the leaves of the sensitive plant. 11. Give other examples of plants with motile leaves. 12. Describe the process in the pulvinus which causes leaf movement.

C. 13. Discuss the advantages of the edgewise position of leaves, indicating in what kind of localities it is most advantageous. 14. What are compass plants, and what advantage is there in the compass habit?

SECTION 59. 1. Give examples of plants whose tendrils are leaves in origin, and of those whose thorns are leaves in origin.

A. 2. Describe the internal structure of needle-like leaves.

B. 3. Describe the pitcher plant. 4. Describe the sun-dew. 5. Explain how insects are made available for use as plant food. 6. Discuss the advantage of the insect-catching habit.

CHAPTER VII

FLOWERS

60. **Introductory.** — You think of flowers as the bright-colored parts of plants, and you have learned that their work is the production of seed. This idea as to flowers is only partly true. It is true that flowers lead to the production of seed, but it is also true that many kinds do not have bright colors. To define flowers as to function is easy. We can call them *organs of seed production* and be done with it. But to define them as to their structure is difficult. In structure they are exceedingly various.

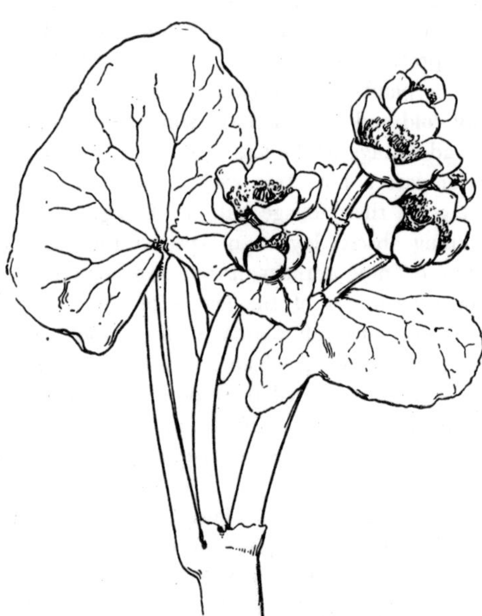

FIG. 93 *A*. — Flowers and leaves of marsh marigold (*Caltha palustris*). Common in swamps and wet meadows in early spring.

As to function, flowers are much more alike than roots,

stems, and leaves are alike, but as to structure they are much less alike than are these other organs. Because of this variety in structure it is the flowers which are principally used in telling the kinds of plants apart. Often it is hard to identify a plant by means of its stems and leaves alone, but it is comparatively easy to do so when you have also the flowers. Thus there is a special need to understand flowers — the need to understand them in order to identify plants.

FIG. 93 *B.* — Bluebells (*Mertensia virginica*). Common in spring, especially along the banks of streams.

Many plants which produce seeds do not produce what are ordinarily considered flowers. Grass and most trees do not produce what are ordinarily considered flowers. Yet grass and all trees produce seed. Evidently, if all seeds are produced by flowers, we need a broader definition for the word than simply " the bright-colored part of a plant." We need it also because there are some bright-colored parts which are not flowers at all. So, for our purpose, we may define flowers as *those organs or groups of organs, having to do with seed production, which appear before the seeds and from a part of which the seeds develop.* According to this definition, all plants which produce seeds also produce flowers, and you may take it as a rule that *all fruits and seeds are preceded by flowers.*

Fig. 93 C.—*A*. Flowers and compound leaf of Dutchman's breeches (*Dicentra cucullaria*). Common in woods in early spring. *B*. The cluster of grain-like bodies which form the peculiar bulb of this plant. *C*. Flower of squirrel corn (*Dicentra canadensis*). In foliage and flowering habit this plant closely resembles *D. cucullaria*. *D*. The underground part of *D. canadensis* from which it gets its common name. The scattered little tubers resemble grains of corn.

According to the definition just given, it is evident that gymnosperms (page 183) have flowers as well as angiosperms. That is, a pine tree has flowers as well as the monocotyledons or dicotyledons. A young pine cone is composed of many small flowers placed closely together.

As to its structure, we may say of a flower that it is *composed of the end portion of a stem and of outgrowths from that stem.* These outgrowths arise from the stem just as leaves arise from stems; in fact, as to origin, they *are* leaves. It is common to speak of these outgrowths as *specially modified leaves.* The most conspicuous colored parts of flowers are usually the parts called petals, and petals are commonly much like leaves in shape as well as in the way in which they arise from the stem.

FIG. 93 *D.* — Dogtooth violet, or yellow adder's-tongue (***Erythronium Americanum***). One of the most graceful of spring flowers. The bulb is found from four to seven inches below the surface of the soil.

A. Pollination. — You have learned as to leaves that their character is chiefly determined by photosynthesis. This is the principal function of leaves and appears to have the most to do with determining their forms and positions. As to flowers there is also something which appears to have the principal influence in determining their forms and their positions. There is a certain thing which must be accomplished before good seeds can be produced, and it is chiefly for the accomplishment of this certain thing that the visible parts of flowers appear to be designed. That certain thing which must be accomplished is *pollination.*

You have already heard of pollen (see page 60). It is that dust-like substance, usually yellow, which comes from the inner part of the flower. You know that before seed can be produced, this pollen must be transferred; it must go from the anthers in which it is produced to the stigmas. Once upon the stigma, the pollen grain produces a structure which is very important in seed making, as you shall see. *It is this transfer of pollen from anthers to stigmas which is called pollination.* To secure this transfer appears to be the principal aim of the flower. Other processes are equally important in seed production, and with them the flower is also concerned, but these other processes are hidden. Pollination is very plain. It is the evident, external process to which flowers appear to be devoted. It is the principal key to an understanding of the forms of flowers and of the positions which they assume.

FIG. 93 *E.*—Chokeberry (*Prunus arbutifolia*). One of the common spring-flowering shrubs of the rose family.

Pollination is of two kinds. Pollen may be transferred

from the anther to the stigma of the *same* flower; this is called *self-pollination* or *close-pollination*. It may also be transferred to the stigmas of *other* flowers; this is *cross-pollination*.

This cross-pollination is one of the most wonderful things about plants. The thing about it which makes us wonder is the way in which it is accomplished. It is a thing which the plant cannot accomplish for itself. It is a thing which is accomplished entirely outside of the plant. It is a process which is essential to the perpetuation of most seed plants, and yet it is a process which the plant alone is powerless to accomplish. It is a process over which the plant itself or the pollen itself has no direct control. And yet this cross-pollination,—this transfer of pollen from one plant to another, this long movement to a small and distant goal by bodies which themselves are entirely without power to move,—occurs each season as a regular feature of the plant's life, as regular as the respiration or the photosynthesis which goes on inside of it.

There are two agencies of the outside world by which principally pollen is carried from flower to flower. *One of these agencies is the wind; the other is insects.*

It is the fragrance and the bright colors of flowers which make them conspicuous. These are the features which make them attractive to us, and it is believed that these features also make them attractive to insects. Many flowers contain *nectar*, a sweet liquid which is a favorite food of insects. Bees make honey of it. Some insects eat pollen, and in eating it they are sure to carry some of it about from flower to flower. It is believed that the odors and bright colors of flowers act as signals to insects that nectar or pollen may be found, and are thereby of advantage to the plant; they attract pollinators. In our own climate the humming birds, and in the tropics many other small birds, are also attracted to flowers to get nectar, and they aid in carrying pollen about. Among insects the bees are by far the most important pollinators. Butterflies and moths are also of service to plants in this way.

FIG. 94.—Flowers and leaf of a common willow. The two middle pictures show the clusters of flowers which give rise to the name pussy willow. Clusters like these are called catkins. The upper catkin is composed of flowers which produce pollen but do not receive it (staminate flowers). The flowers of the lower catkin receive pollen but do not produce it (pistillate flowers). Of the two small pictures at the left, the upper one shows a single staminate flower, the lower a single pistillate one.

As to the wind, it is evident that no odors or bright colors are needed to attract it. You would not expect wind-pollinated flowers to be fragrant or brilliant, and they are not. They are usually inconspicuous, and, if they are noticed, they may not be considered flowers at all. (See *Figures 100 B* and *106*.) Now you can see why it is common to think that grass and trees, except fruit trees and a few others, do not have flowers. The flowers of grass and of most trees are pollinated by wind, and are inconspicuous.

Pollination is a subject which needs a section to itself. It is discussed in more detail in *Section 63*. It is mentioned now simply because it *explains* flowers more completely than any other single thing explains them.

B. Protection. — There are certain parts of the flower, parts that are usually green, which are concerned with protection more than they are with pollination. They are the outermost parts of the flower and they protect the inner parts. These outermost parts are the sepals. They form the covering of the flower buds, and, in flowers which are open, they may be found beneath the petals. Sepals usually look like little leaves. You have already noted the injury which may come to the tender inner parts of fruit blossoms when a frost comes after the sepals have opened (*Section 51*).

C. Evolution of Flowers. — You cannot understand much about flowers without understanding at least a little about evolution. Pollination helps us to understand the advantages of the many different kinds of flowers and evolution helps us to understand why there are so many different kinds.

Evolution is the name of an idea which is very big, but not very complex. It is an idea which is neither too big nor too complex for you to grasp, and it will help you understand living things as no other idea can. It helped the scientists of the past as no other idea ever did, and the scientists of the present accept it as a great fundamental truth. *All things keep changing more or less. The sum of such changes makes up what we call evolution.*

We may say of flowers that they have been evolved in connection with reproduction. By this we mean that all plants have come from simple forms without flowers which existed remote ages ago; we mean that these ancestral forms had the habit, as plants have the habit now, of changing somewhat as time goes on and as conditions alter; we mean that some of these ancestral plants gradually became more complex in structure, and that those forms whose new structures were of advantage were more likely to live than those forms whose new structures were not of advantage; we mean that among the new structures which were gradually evolved the flower was one. All this and a great deal more is involved in the idea of the evolution of flowers.

FIG. 94 *A*.—Lady's slipper, or moccasin flower. One of the orchid family (*Cypripedium parviflorum*). One of the most peculiar of spring flowers, found in boggy places.

It is in connection with this process of evolution that the great variety of flowers has appeared. The nutritive organs which you have been studying are also the result of long evolution; in fact, everything which exists to-day has evolved from what existed in the past. It may occur to you to wonder, since flowers and the nutritive organs evolved together, why there is such general similarity among leaves, for example, and such general dissimilarity among flowers. The answer is that, to do their work successfully, leaves need to have one general kind of structure, while, as to flowers, many different kinds of structure have proved to be successful.

That method of reproduction with which flowers are concerned is, as you know, the sex method. Almost all of the lower plants (ferns, mosses, mushrooms, etc.), though they do not produce seeds, do possess this sex method of reproduction. They possess structures whose appearance and behavior is like the appearance and behavior of certain parts of the flowers. To study them gives a clear idea of the way in which the flower probably has evolved, for some of these lower plants which exist to-day are much like the remote ancestors of the flowering plants. To study them throws much light upon the evolution of the *seed habit*; for the seed habit has evolved from the simpler reproductive habits of plants which did not produce seeds.

The making of seeds is an elaborate process. In this chapter we can tell you how the various parts of the flower are concerned in that process, but the real meaning of the process you will understand only after you have read the chapters about plants which do not produce seeds.

D. Flowers as Sex Organs. — You have been told that the flower is not a sex organ (page 58). But you have also been told that it is concerned with sex reproduction;

you have been told that the embryo in the seed is the direct result of a sex process. (See page 58). Why, then, is a flower not a sex organ?

This matter would not be worth discussing if it were merely a question of how we are to use the expression *sex organ*, but it is a question of much more than that. It is a question of your understanding or misunderstanding a very fundamental thing in plant life.

The *sex process* is the union of two cells into one which develops into a new individual, and a *sex organ* is an organ which produces one or the other of these two kinds of sex cells whose union constitutes the sex process. Now those cells from whose union the embryo of seed plants arises are not cells of the flower. If they were, the flower would be a sex organ. Since they are not, the flower is not a sex organ.

Any single cell of a plant which gives rise to a new individual is called a *spore*. All plants produce spores. The young grains of pollen are spores.

An individual which arises from a single cell of its parent is a new *generation;* it is not, strictly speaking, a part of its parent, even though it remains in the body of its parent. Spores give rise to new generations.

The new generation to which the pollen grain gives rise is contained within the grain; it consists of but a few cells, but it is none the less a new generation. The pollen grain after pollination gives rise to a structure called the *pollen tube.* This pollen tube grows down into the very center of the flower, and the contents of the pollen grain move down through it. (See *Figure 95.*) In the center of the flower the pollen tube reaches another individual, another hidden generation, which has also resulted from the germination of a spore. This second kind of spore never escaped from its parent, but it was a spore none the less. It remained within at the very point where it was produced, instead of being shed as pollen is shed. The new individual or generation into which this second kind of spore grew also remains within, never escaping from the tissues of its parent. It derives its nourishment from the tissues of its parent, somewhat as embryos derive their nourishment from the tissues which surround them in the seed. It is, none the less, an individual distinct from its parent except for its position; it is a different generation from its parent.

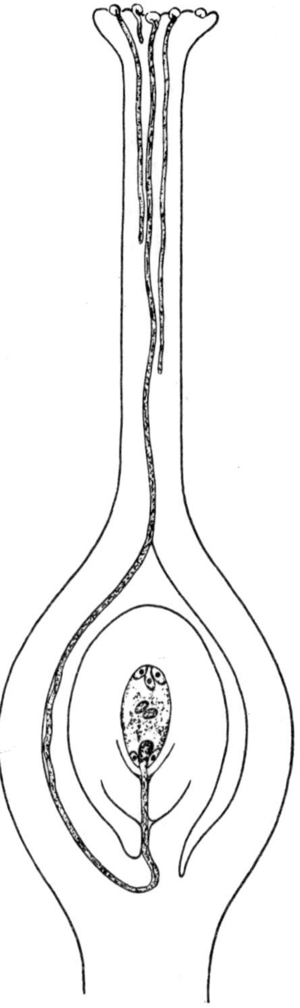

FIG. 95. — Diagram illustrating the way in which pollen tubes grow down into the center of the flower and reach the "hidden generation" which grows there. See context, and page 276.

It is these two individuals which have just been described which produce those cells whose union is the sex process. Neither of them may be properly called parts of the flower, though both develop within the tissues of the flower.

Thus we see why the flower should not be called a sex organ. Though very much concerned with sex reproduction, it is not, strictly speaking, a sex organ itself. This matter is more fully explained in Chapter X.

61. **General Structure.** — You have already learned (page 59) the names of what may be called the "principal parts" of flowers. You have learned that some of these parts are directly concerned with seed making and that others are not. Stamens and pistils are directly concerned with seed making, while sepals and petals are more directly concerned with protection or with the attraction of insects.

A. Essential and Accessory Parts. — Upon the basis just indicated, all flower parts are classified either as *essential* (to seed making) or as *accessory* (to seed making). Some flowers are composed of essential parts only.

The familiar wild flowers are composed of both essential and accessory parts. The outer, conspicuous parts are the accessory ones; the inner, inconspicuous parts are the essential ones. Many cultivated flowers, such as roses, carnations, and chrysanthemums, do not have well-developed essential parts; the outer parts have been cultivated at the expense of the inner ones, and these showy flowers have lost the power of producing seed.

FIG. 95 *A*. — Wild geranium (*Geranium maculatum*). The open flower shows plainly both essential and accessory parts. This is one of the common and handsome spring flowers.

Most of the common wild flowers have four kinds of parts; two kinds are outer and accessory, the other two are inner and essential. All these parts are usually arranged in *whorls*. (A whorl is a group of leaves or leaf-like structures which arise at the same node.) In a flower the innermost whorl usually forms a single structure.

B. The Outer Parts. — Generally there are one or two whorls of those bright-colored leaf-shaped parts which have been already called petals. They form the corolla. Usually the corolla is the most conspicuous part of the flower. Often, instead of being composed of entirely separate petals, it is a tube or funnel-shaped structure which appears to be composed of united petals, separate only at the top. (See *Figure 107*, page 286.)

Outside and under the petals it is usual to find a whorl of those protective green parts which we have already called sepals. They form the calyx. The calyx often appears to be composed of more or less united sepals just as the corolla appears to be composed of more or less united petals. (See *Figure 112*, page 290.)

All the outer parts taken together form the *perianth*. (The word means *around the flower*, which expresses the fact that the perianth is not an essential part of the flower.) Sometimes the perianth is composed of only one whorl. In such cases its parts are regarded as sepals whatever their color may be, and petals are said to be lacking. Sometimes there is no perianth at all. In such cases the flowers are said to be *naked*.

C. The Inner Parts. — These are more complex than the outer parts. Some flowers possess only one kind of inner part. In these inner parts originate those hidden generations which produce the sex organs.

a. The Stamens. — As has been noted, there are two kinds of inner parts. There is usually a central structure called the pistil, and it is usually surrounded by other structures called the stamens. The stamens are slender organs at whose upper ends are borne little sacs which contain pollen. (See *Figure 96.*) Usually there are one or two whorls of stamens. Taken all together, the stamens form the *andrœcium.* The pollen-containing part at the tip of the stamen is called the *anther;* it is usually composed of two distinct lobes or *pollen sacs.* (See *Figure 96.*) That part of the stamen which supports the anther is called the *filament.* Usually the filament is quite slender.

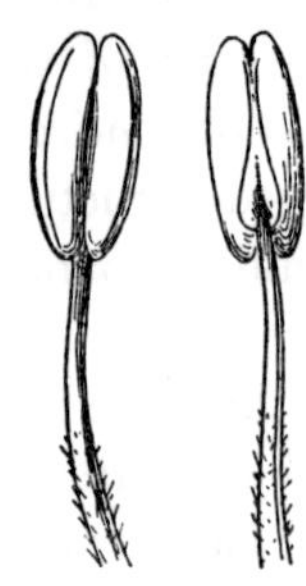

FIG. 96. — Stamens, showing *filament*, *anther*, and the two *pollen sacs* of which the anther is composed.

b. The Pistil. — The pistil is the most complicated of all the parts of the flower. It is the part within which the seed is produced. It is the part which develops into the fruit. It is the part into whose tissues the pollen tube penetrates, carrying with it one of the hidden generations. It is the part within whose tissues develops that other hidden generation which is also concerned with the sex process. It is the part within which one of the cells of the pollen tube unites with one of the cells of the other hidden generation.

From top to bottom pistils are usually divided into three distinct regions. (See *Figure 97.*) At the top there is an expanded sticky part which catches and holds the pollen brought by insects or otherwise. This part is called the *stigma*. Beneath the stigma there is an elongated and usually slender part called the style; the function of the style appears to be to get the stigma in a position which is favorable for receiving pollen. Some pistils have no style, as in the water lily. (See *Figure 108*, page 287.)

The base of the pistil is an enlarged part called the ovary or *seed vessel*. (Seed vessel is a preferable name for this structure for the reason that the ovary of animals is an entirely different structure from the ovary of plants. But this structure was called an ovary before its real nature was understood and that name is still generally given to it.) The ovary is that part of the pistil which develops into the fruit and contains the seeds.

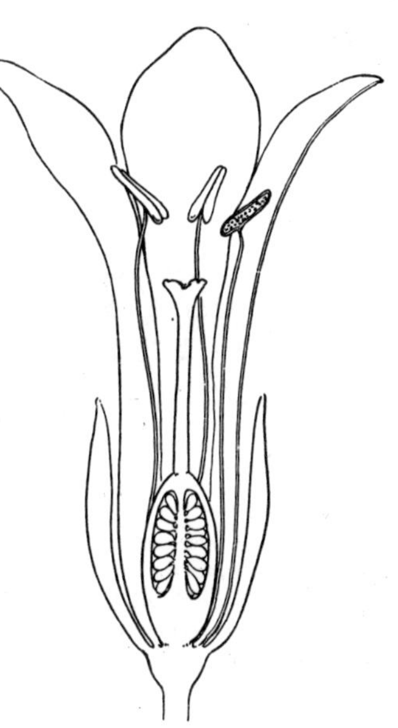

Fig. 97. — Lengthwise section through a flower, showing the relation of parts. See context.

Sepals, petals, and stamens are all, as has been noted, leaves as to origin; they arise as leaves arise; when they are very young they look exactly like very young leaves; they are sometimes referred to as "modified leaves." Thus the calyx, the corolla, and the andrœcium may all be regarded as sets or whorls of modified leaves. Similarly the pistil may be regarded as a modified set of leaves, but since it is usually a single structure, the parts which compose it are not so well known or so evident as the parts of the calyx, corolla, and androecium. These parts of the pistil which usually do not appear as parts are called *carpels.* In some flowers the carpels are separate, and in nearly all flowers the number of carpels is indicated by the number of divisions or lobes of the stigma. (See *Figures 114* and *116*.)

Buttercup and hepatica are flowers which have separate carpels; they are sometimes described as having many pistils. Thus it is evident that the word pistil sometimes refers to one carpel, sometimes to many. Like the word ovary, it is a name which was used before the real nature of flowers was understood. The carpels taken together are called the *gynœcium.* Having these two words, carpel and gynœcium, the word pistil is not really needed at all. It is a very commonly used word, however, and it would be inconvenient to try to get along without it.

FIG. 98. — Lengthwise section through a flower of peony; *K* indicates the calyx of separate sepals; *C* indicates the corolla of separate petals; *A* indicates the androecium of separate stamens; *G* indicates the gynœcium of separate carpels (or simple pistils).

Within the ovary the undeveloped seeds are found. They commonly resemble small pearls. They are called *ovules.* The ovules are attached to the sides of the ovary or to walls which sometimes divide the ovary into separate compartments. The number of such compartments indicates the number of carpels of which the pistil is composed. (See *Figure 97*.)

c. Fertilization. — It is within the ovule that the sex process occurs. It is here that there develops that hidden generation which the pollen tube seems to seek and finds. That hidden generation produces a cell called the *egg*. The pollen tube contains two cells called *sperms*. (See *Figure 99*.) Sperms are male cells. The egg is a female cell. Both are sex cells, or *gametes*. The egg and one of the sperms unite. This union of egg and sperm is called *fertilization*. Fertilization occurs within the ovule. (See *Figure 95*.) It is the sex process. It has already been described as the union of "portions from the bodies of separate parents." (See page 58.) As you now see, the portions referred to are the sperm and the egg. The cell which results from the fusion of egg and sperm is called the *fertilized egg*. The fertilized egg develops into the embryo. (See *Figure 99 A*.)

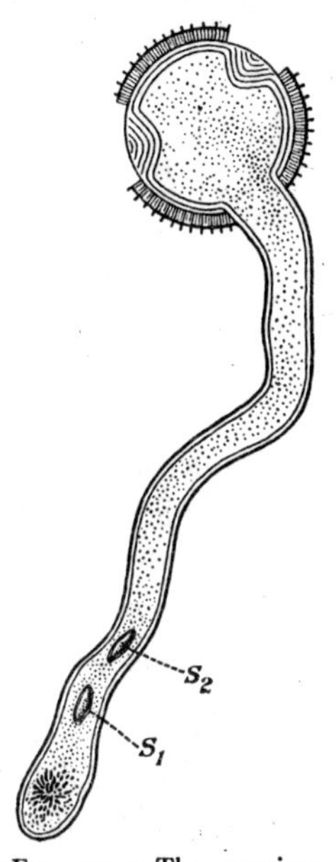

FIG. 99. — The germination of a pollen grain, the pollen tube being well developed. Much magnified. S_1 and S_2 are sperms.

Fertilization is, you see, a very different thing from pollination. Pollination takes place on the stigma. Fertilization takes place within the ovule. The growth of the pollen tube separates them. The pollen tube must grow down the style and reach an ovule before fertilization occurs. (The difference between pollination and fertilization is thus emphasized chiefly for the reason that pollination is often wrongly called fertilization.)

One pollen tube can accomplish the fertilization of only one ovule. Since ovaries often contain many ovules, it is evident that many pollen grains are often needed on the same stigma.

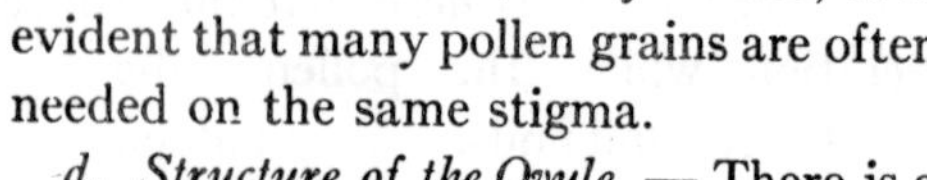

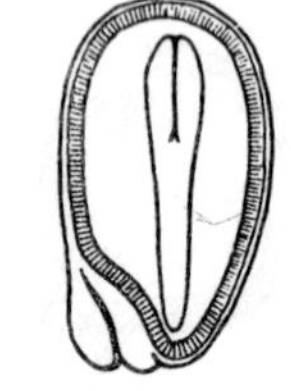

FIG. 99 *A*. — Diagram of the seed of violet. The embryo lies in the center.

d. Structure of the Ovule. — There is a little opening into the ovule called the *micropyle.* (The word means *little gate.*) The pollen tube usually enters through the micropyle. This little opening may sometimes be seen in mature seeds. In a bean it has the appearance of a hole made with the point of a pin.

The micropyle leads to the central part of the ovule, which is called the *nucellus*. It is within the nucellus that the egg-producing generation lies embedded. (See *Figure 95*.) The tissue around the nucellus forms a sort of coat called the *integument;* sometimes there is one integument, sometimes two.

D. The Receptacle. — That part of the stem from which the parts of the flower arise is called the *receptacle.* The receptacle is really a part of the flower. In some flowers the ovary, and even parts of the calyx and corolla, appear to be formed by the receptacle. It is often hard to tell where the receptacle ends and the other parts begin.

If we regard sepals, petals, stamens, and carpels as modified leaves, evidently we may regard the whole flower as a modified shoot, the receptacle being the stem part.

62. Variations in Structure. — You yourself have noticed that there are many kinds of flowers. There are thousands of perfectly distinct kinds, all alike as to function, but all different as to structure. The kinds of flowers are not only more easily distinguished from one another than the kinds of leaves and other organs, but flowers of the same kind of plant do not vary as the leaves and other organs of the same kind of plant vary. (See *Figure 100.*) On different kinds of plants, flowers vary much more than leaves, but on the same kind they vary much less. On dry, sandy soil a wild rose has small leaves, short stems, and many thorns; on rich soil it has larger leaves, longer stems, and fewer thorns. But wherever it grows the flower is practically the same. It is a general principle that surrounding conditions affect the structure of nutritive organs more than they affect the structure of reproductive organs. Evidently, then, reproductive organs, being less variable, are more useful in classification

FIG. 100. — Variation in the dandelion. Both plants are mature. The smaller one was grown from a portion of the larger one, but under less favorable conditions. The stems and leaves are very different in the two specimens, but the flowers are about the same.

than nutritive organs are. This is true for plants without flowers as well as for plants with them.

The differences between kinds of flowers are differences between their sepals, petals, stamens, and carpels. These organs vary in number, in shape, in color, in size, in relative position, and in various other ways.

A. The Evolution of Flower Forms. — We are far from understanding all the laws of evolution, but we do understand a good deal about the evolution of flowers. We can see clearly that certain kinds have evolved from certain other kinds, that the more complex forms have all come from simpler forms, and that all forms which are successful appear to have certain peculiar advantages of their own.

It is no simple matter to determine the true relationships of plants even by means of their flowers. To tell which plants are related to which is a great puzzle upon which botanists are still at work. They are trying to find out the true genealogy or "family tree" of plants. In this work the flowers have been of great aid, and the real kinships of flowering plants are better understood than the kinships of groups which do not bear flowers. Yet it has taken many years of work by many men to determine even the general relationships of flowering plants to each other.

Linnæus, a Swedish botanist, was one of the first great classifiers of flowering plants. He recognized the flowers as the best basis for classification, but he had no means of knowing which flowers were derived from which. So he divided them into two great groups on the basis of the number of stamens. He put in one great group all those plants whose flowers have ten stamens or more; he put in the other all with fewer than ten stamens. This, as you can see, was a convenient classification, but not a natural one. It was an artificial classification; it was like classifying horses by the white spots on them; it did not indicate true kinships. True kinships are based on descent from the same ancestors.

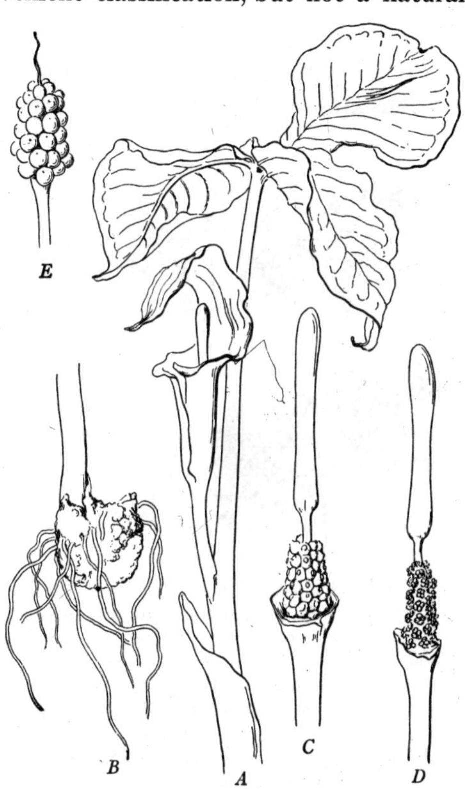

FIG. 100 *A*.— Jack-in-the-pulpit (*Arisæma triphyllum*). *A*, top of the plant showing *spathe* with *spadix* protruding from it. *B*. The corm, which is called Indian turnip. (See page 160.)

This plant is usually diœcious (see page 284) and the flowers are very simple. *C* shows a spadix bearing pistillate flowers, the spathe having been removed. *D*, a spadix bearing staminate flowers. *E*, the fruit, which turns scarlet; note the withered remnant of the club-shaped part of the spadix.

Gradually botanists have discovered the principal relationships of flowering plants. We know that similar flowers indicate similar ancestors, but we no longer classify simply on the basis of the number of stamens. The characteristics of the flowers and of other organs as well are considered together in determining the relationships of a plant. The flowers, however, are the most important and convenient means of identification.

B. The Simplest Flowers. — We can understand the various kinds of flowers best by considering first the simplest kinds. The more complex kinds have evolved from simple kinds, from ancestral kinds which were much like the simple kinds which exist to-day.

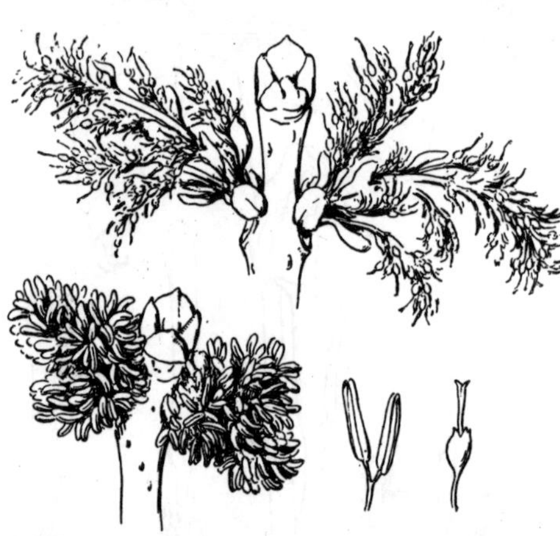

FIG. 100 *B.* — Flowers of the white ash (*Fraxinus Americana*). A common and valuable tree whose flowers appear in early spring. They are very simple and are of two kinds, nearly always borne on different plants, *i.e.* diclinous and diœcious. The picture shows the end of a pistillate twig above and the end of a staminate one below. Of the two smaller pictures, the one at the left is a staminate flower, the one at the right a pistillate one.

The simplest flowers have no perianth, and they are composed of essential parts only. The two kinds of essential parts are, as you know, stamens and carpels. The simplest flowers of all are those which have only one kind of essential part. In other words, the simplest flowers of all are borne by plants which have two kinds of flowers; one kind is composed only of stamens, the other kind only of carpels. In some cases, as in the duckweeds, a single stamen or a single carpel is all that there is to the flower.

Flowers which have only stamens or only pistils are called *diclinous;* those which have both stamens and pistils are called *monoclinous.*[1] The flowers with which you are most familiar are monoclinous. Diclinous flowers with stamens are called *staminate;* those with pistils are called *pistillate.* Not all diclinous flowers lack a perianth; there are many which have both sepals and petals.

FIG. 101. — A group of cat-tails, showing the dense spikes of very simple flowers. The upper, woolly part of the spike is composed of pollen-producing (staminate) flowers which die when the cat-tail is ripe. The lower, dark part is composed of seed-producing (pistillate) flowers. These flowers are wind-pollinated and the pistillate ones ripen before the staminate ones of the same plant.

Diclinous flowers are more primitive than monoclinous ones; that is, flowers which possessed only stamens or only pistils are believed to have preceded those possessing both stamens and pistils. The flowers of nearly all gymnosperms are dicli-

[1] Diclinous flowers are also called *unisexual* or *imperfect*, and monoclinous ones *bisexual* or *perfect*. These terms are objectionable. Flowers are not, strictly speaking, sexual; and diclinous flowers, while more simple, are not less perfect than are monoclinous ones.

nous. They, and diclinous flowers generally, are pollinated by wind.

Note that if diclinous flowers are more ancient than monoclinous ones, then cross-pollination is more ancient than close-pollination, for the latter evidently could not occur when all the flowers were either staminate or pistillate.

The cat-tail is a common plant in marshes. (See *Figure 101.*) Its flowers are about as simple as any. They have no perianth and they are of the two kinds, staminate and pistillate. That part of the plant which is supposed to look like a cat's tail is composed of separate carpels and stamens which grow very close together. The stamens are on the upper part of the tail. They die after they have shed their pollen. You may have noticed this dead part sticking out of the top of a mature cat-tail.

FIG. 102. — The tassel of corn. Note the hanging anthers which are full of pollen. The pollen is scattered by the wind. — *After Hunter.*

Corn is another common plant which has diclinous flowers. The tassel of corn, which appears at the top of the plant, is composed of staminate flowers (see *Figure 102*), while the undeveloped ear is composed of pistillate flowers. The threads of the silk, which hang out from the young ear, are nothing but unusually long styles and stigmas. (See *Figure 103.*) Each style leads to an ovary. These ovaries are the undeveloped grains of corn. Pollen is produced in great abundance in the tassel. Some of it is quite sure to blow or fall upon the silk. A pollen tube must grow down each thread of the silk if all the ovaries are to become well developed grains of corn. If the days when the pollen is ripe are rainy, the corn crop is sure to be injured thereby. Dry days are needed when the pollen is flying. Rain makes wind-pollination very uncertain. The ovary of corn becomes completely filled by the one seed which develops within it. Thus, strictly speaking, a grain of corn is not a seed; it is rather a fruit completely filled by the one seed which it contains.

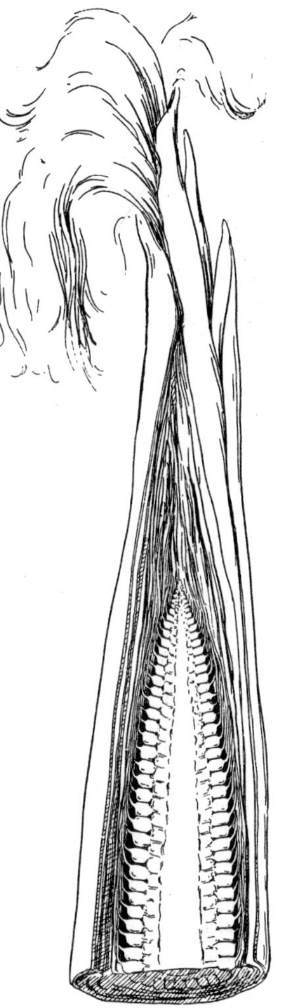

FIG. 103. — The end of a young ear of corn, the silk being exposed at the end. Note that a thread of the silk runs to each of the young grains.

Plants with diclinous flowers bear these flowers in two ways. One way is to bear both staminate and pistillate flowers upon the same individual; the other is to bear the staminate flowers upon some individuals and the pistillate flowers upon others. Plants bearing both kinds of flowers upon the same individual are called *monœcious*. (The word means *one household*.) Cat-tail and corn are both monœcious. (Study *Figure 104.*) Plants which bear staminate and pistillate flowers on different individuals are called *diœcious*. (The word means *two households*.) Many common trees are diœcious. The box elder, the poplar, and the willow are examples. (See *Figures 105* and *106*.) It is evident that only the pistillate box elders or poplars can produce fruit, and this fruit will produce good seed only if staminate trees are in the same vicinity with pistillate ones.

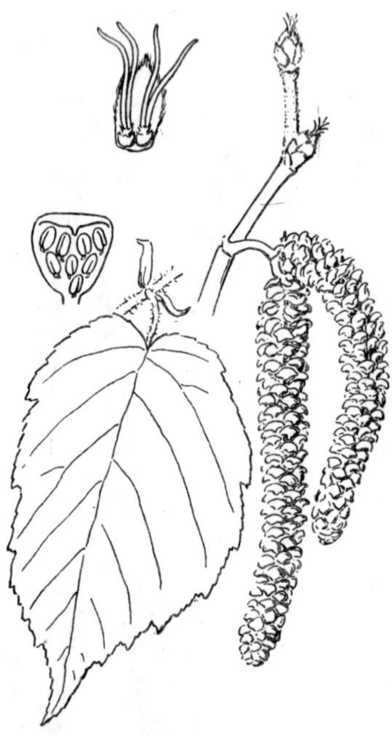

FIG. 104. — A flowering twig of hazel, a shrub which has monœcious, wind-pollinated flowers. Note that the staminate flowers are lower than the pistillate ones; this tends to prevent pollen from falling on stigmas of the same plant. The catkins sway in the breeze, the pollen grains often being blown from them in little clouds. The uppermost small picture shows a single pistillate flower; the one below it, a single staminate flower.

There are some plants, some kinds of maple trees, for example, which produce both monoclinous and diclinous flowers. Such plants are called *polygamous*.

C. Kinds of Perianths. — It is believed that plants with perianths came from plants without them; that is, naked flowers are more primitive than flowers with perianths. The more primitive perianths appear to be those whose parts are spirally arranged on the stem; they are not arranged in definite rings or whorls. The white water lily illustrates this point very well. In it we find the many petals spirally arranged. We also find parts which are partly petal and partly stamen. (See *Figure 108.*) The stamens and carpels as well as the petals are, as to number, numerous and indefinite. This latter characteristic is considered more primitive than the low and regular number of parts which characterize most of our familiar flowers. In all these respects the water lily is regarded a rather primitive type of flower.

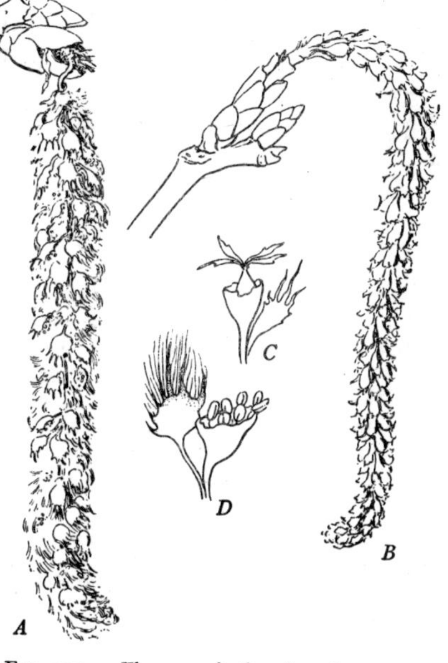

FIG. 105.—Flowers of the American aspen (*Populus tremuloides*); sometimes called the trembling aspen. It is diœcious. *A* shows a catkin composed of staminate flowers. *B*, pistillate catkin. *C*, a single pistillate flower with the characteristic scale which accompanies it. *D*, the staminate flower and scale.

Petals and sepals are either separate or more or less united. Flowers with these parts separate are considered more primitive than those in which they are united. This applies especially to the petals. Some flowers have united petals, but separate sepals. The bluebell, the phlox, the morning-glory, and the petunia are common plants whose petals are united. (See *Figures 93B, 107,* and *111.*) Flowers with petals more or less united are said to be *sympetalous.* (The word signifies *petals together.*) Flowers with separate petals are described as *polypetalous.* (This word signifies *many petals.*)

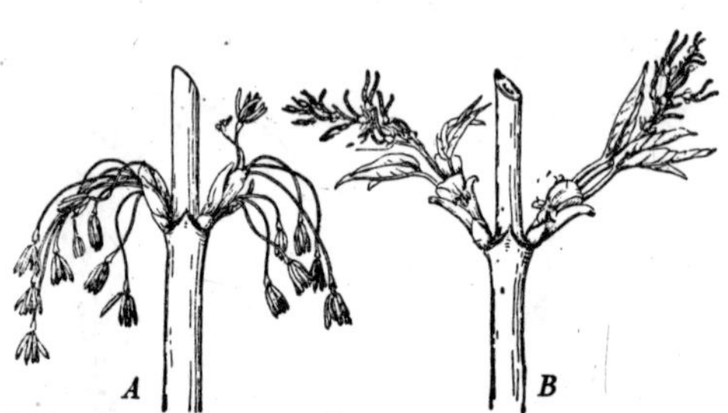

FIG. 106. — Diœcious, wind-pollinated flowers of the box elder. *A*, staminate flowers; each one consists of a group of stamens borne at the end of long, drooping flower stems (pedicels); each stamen is composed of a long anther and a very short filament, too short to be seen in the picture. *B*, pistillate flowers; note the prominent stigmas extended as two branches from each flower; these flowers have a small perianth.

As to their general form, corollas and calices (plural of calyx) are either *regular* or *irregular.* Regularity of form is thought to be a more primitive characteristic than irregularity. By a *regular* corolla is meant one that is *radially symmetrical.* (See *Figure 108.*) A wheel is radially symmetrical. You can divide it into halves along the line of any radius and the halves will be alike. The same thing is true of regular corollas. *Irregular* corollas are *bilaterally symmetrical*, which means that there is one way and only one way in which they can be divided into halves which are alike. (See *Figure 109.*) Our own bodies, so far as we can tell from their outward appearance, are examples of

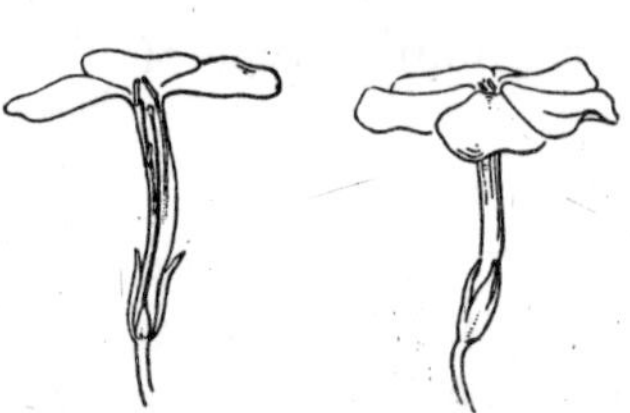

FIG. 107. — Flowers of *Phlox*, showing petals united and, by their union, forming a tube. Note that the stamens are attached to the corolla tube.

bilateral symmetry. This kind of symmetry of body is very general among animals. A sweet pea is a familiar example of an "irregular" flower; a vertical plane passing through the center of it divides it into halves which are alike.

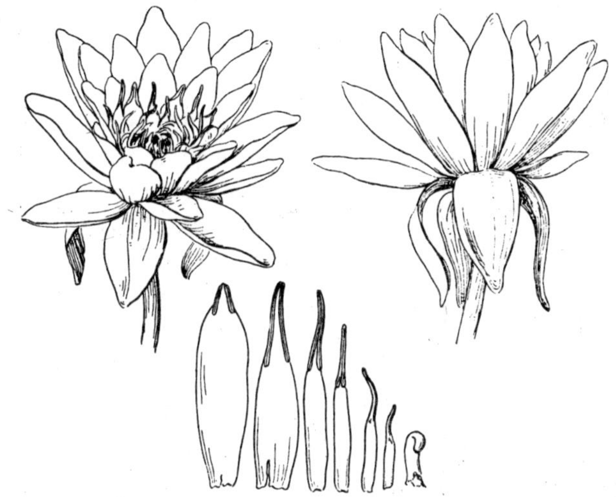

FIG. 108. — The water lily (*Castalia odorata*). Note the indefinite numbers of the spirally-arranged petals and stamens. Note also the gradations of petals into stamens. Note that the pistil has no style.

The so-called irregular corollas, since they are really not irregular at all, are better called *zygomorphic* (*yoke-formed*). Similarly, regular corollas are called *actinomorphic* (*ray-formed*).

D. Variations of the Andrœcium. — Usually the stamens are all alike and separate. It is common to find two whorls or sets of them. The anther is usually attached to the filament at one end and stands in line with it. Sometimes, however, the anther is attached near its middle and swings free at both ends.

There is a good deal of variation in the way in which the pollen sacs open. The most common way is by a lengthwise slit. Sometimes, however, the slit is only at the top of the anther, and sometimes the pollen is discharged through a round pore. (See *Figure 110*.) There are still other variations in this feature of the andrœcium.

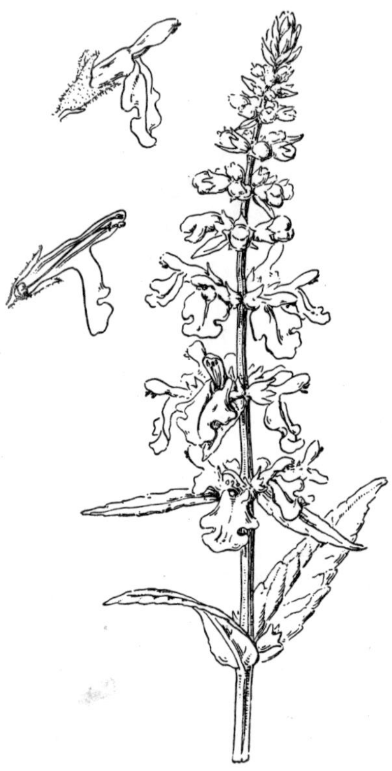

FIG. 109. — Flowers of a mint (*Stachys*). These flowers are irregular, that is, zygomorphic. See context.

Stamens vary as to where they are attached to the rest of the flower. In flowers with separate petals they are usually attached to the receptacle like the other parts, but in flowers with united petals it is common to find them attached to the corolla. (See *Figures 107* and *111*.)

Sometimes the stamens are not all alike. A striking case of this is found in the sweet pea. In its flower you find nine stamens united and one which is free. A similar arrangement is found in the flower of clover.

Sweet peas and clover belong to the same great family (*Leguminosæ*), and this arrangement of stamens occurs in the flowers of very many members of that family. (See *Figure 112.*) Another great group of plants is characterized by the fact that the stamens form a tube which closely surrounds the pistil. From this tube the stamens branch out irregularly and in indefinite numbers. The common mallow, the hollyhock, the hibiscus, and the cotton plant are members of this group. (See *Figure 113.*)

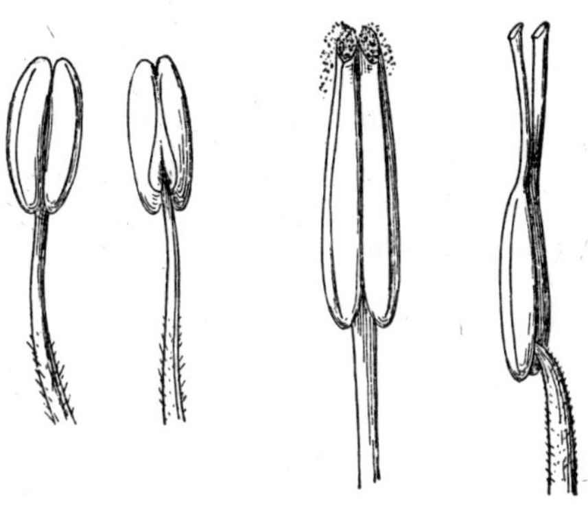

FIG. 110. — Types of anthers, showing various ways in which pollen is discharged.

FIG. 111. — Flower of morning-glory (*Ipomea purpurea*), showing a sympetalous corolla and the attachment of the stamens to it.

E. Variations of the Gynœcium. — The carpels are usually united, while some flowers, as these of the pea family, have only one. Many flowers, however, have separate carpels, and this condition is known to be more primitive than the united condition. Flowers with

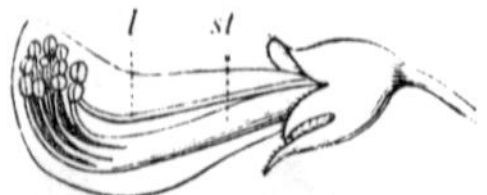

FIG. 112. — Diagram of a lengthwise section through the flower of sweet pea; *l*, the one free stamen; *st*, the nine stamens whose filaments are united.

united carpels evolved from flowers with separate carpels. (See *Figure 98*, page 274.)

A flower with separate carpels is said to be *apocarpous*, while one with united carpels is said to be *syncarpous*. The flowers of the buttercup family (*Ranunculaceæ*) are mostly apocarpous; besides the buttercup, the anemone and the hepatica are examples of this. (See *Figure 113 A*.)

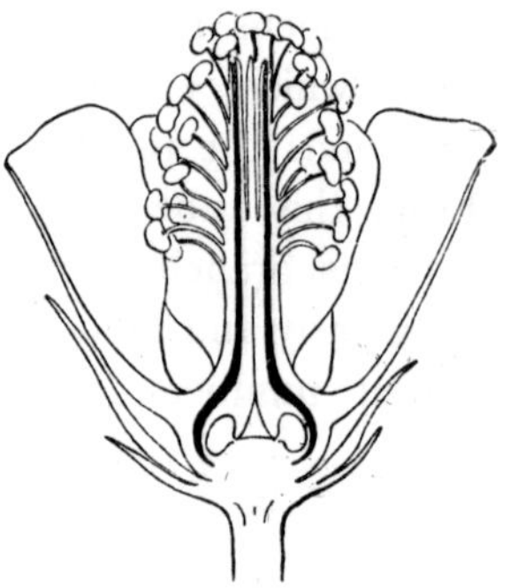

FIG. 113. — Section of a flower of one of the mallow family, showing the way in which the stamens form a tube which incloses the style.

As you have noted, the word pistil is sometimes synonymous with gynœcium and sometimes synonymous with carpel. (See page 274.) When the carpels are united into a single structure, pistil is synonymous with gynœcium. When the carpels are separate, each pistil is but a single carpel. Thus we may define a pistil as *any structure composed of one or more carpels which appears to be a single organ.* A pistil composed of one carpel is called a *simple pistil* (see *Figure 98*); one composed of more than one carpel is called a *compound pistil*. In a compound pistil the divisions or lobes of the stigma, or the number of compartments of the

FIG. 113 *A*. — Mature fruits which arise from a single flower of the anemone, a flower with separate pistils. The fruits of hepatica are similarly arranged.

ovary, usually indicates the number of carpels which compose it. (See *Figure 114.*) Some ovaries of compound pistils have but one seed compartment. In such cases the number of carpels is usually indicated by the number of rows of ovules along the walls of the ovary. The walls of the ovary are usually raised in little ridges to which the ovules are attached. These ridges are called *placentæ* (singular, *placenta*).

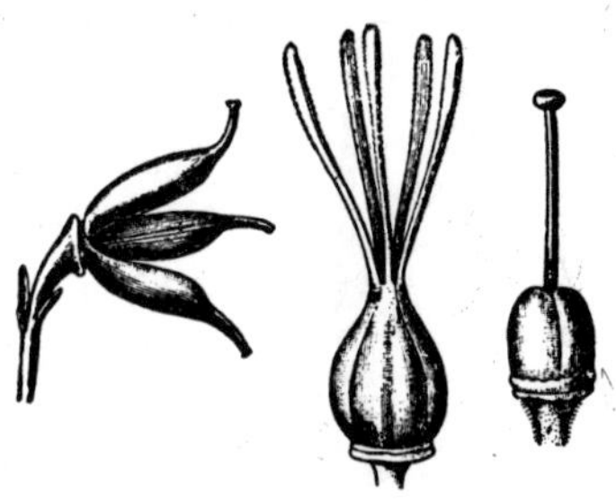

FIG. 114. — Three types of pistils. The figure farthest at the left shows three simple pistils composed of separate carpels. The other figures show compound pistils composed of united carpels.

The most striking variation of the gynœcium is the variation in the position of the ovary with reference to the rest of the flower. (Study *Figure 115.*) You observe that ovaries have three distinct positions with reference to the rest of the flower.

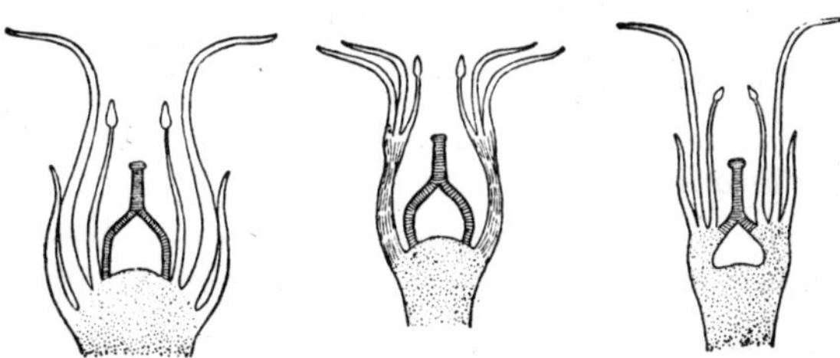

FIG. 115. — Diagrams illustrating *hypogyny*, *perigyny*, and *epigyny*. See context.

In the first of these the other parts of the flower are all attached to the receptacle beneath the ovary. This arrangement is called *hypogyny*. (The word signifies *under the gynœcium.*) In the second arrangement the other parts arise from the top of a ring of tissue which surrounds the ovary. This arrangement is called *perigyny.* (The word signifies *around the gynœcium.*) In the third arrangement the tissue from which the other parts arise grows right around the ovary so that it forms the ovary wall; the other parts appear to arise from the top of the ovary itself. This arrangement is called *epigyny.* (The word signifies *upon the gynœcium.*) Of these three arrangements, hypogyny is considered the most primitive, and epigyny the most advanced. Perigyny is somewhat intermediate; it is commonly found in flowers of the great rose family to which peach, pear, apple, cherry, and plum belong. The great majority of flowers, however, are either distinctly hypogynous or distinctly epigynous. An hypogynous flower is sometimes described as having a *superior ovary,* while the expression *inferior ovary* refers to epigyny. (See *Figure 116.*)

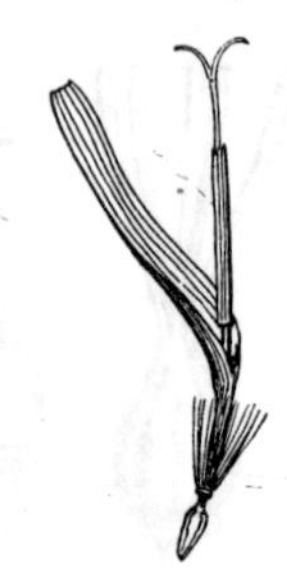

FIG. 116. — A single flower (magnified) from the flowering head of dandelion. Note that the ovary is *inferior* (*epigyny*), all the other parts of the flower arising from above it.

F. Variations in the Number of Parts. — A lily has three sepals, three petals, six stamens, and three carpels. An apple blossom has five sepals, five petals, an indefinite number of stamens, and not more than five carpels. A buttercup has five sepals, five petals, and an indefinite number of both stamens and carpels. The flower of a dandelion has a five-notched corolla, five stamens, and two carpels. (What seems to be the flower of dandelion is really a dense cluster of many small flowers.)

As these examples show, the parts of flowers occur in

both definite and indefinite numbers. Sometimes the numbers are both definite and indefinite in the same flower. Thus a flower may have a definite number of petals,

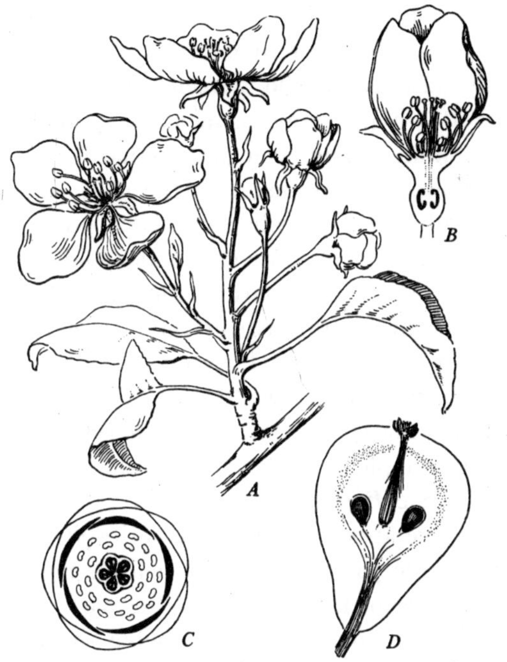

FIG. 117. — *A*, a flower-bearing twig of pear. *B*, lengthwise section through the flower. *C*, diagram, showing the number and relation of the flower parts as revealed by a cross section. *D*, lengthwise section through the fruit.

for example, along with an indefinite number of stamens, as is the case with the apple blossom and other flowers of the rose family. (See *Figure 117.*) Generally, definite numbers appear in some of the sets, as in the buttercup, or in all of them, as in the lily and the dandelion. Indefiniteness as to the number of parts is considered a more primitive character than definiteness.

It is evident that a flower may be primitive as to some whorls and advanced as to others. In other characteristics, as well as in this one of the numbers of the parts, flowers often show a mixture of primitive and advanced features. Thus, actinomorphic flowers are sometimes epigynous, as in wild parsnip, while zygomorphic flowers are often hypogynous, as in snapdragon or members of the mint family. Thus you can see that it is impossible to arrange flowers in a series from the most primitive up to the most advanced. Evolution has evidently proceeded by many branching roads rather than along one main highway. Also it is not necessarily true that the simplest flowers are the most primitive or that the most complex flowers are the most advanced, for evolution may result in an increase in simplicity as well as in an increase in complexity. Thus flowers of the dandelion type, though apparently quite simple, are believed to be the most advanced type of all. (See *Figure 116*.)

Three and five appear to be the favorite *floral numbers*. That is, when the parts *do* occur in definite numbers it is usual to find three or five of each kind of part, except that stamens are quite commonly just double the number of the other parts. You have noted this in the case of the lily.

G. Monocotyledons and Dicotyledons.—Three is the floral number which is characteristic of the monocotyledons, and five is the floral number which is characteristic of dicotyledons. (See page 183.) This may be noted in the examples given above among which the lily is the only monocotyledon; the others are all dicotyledons. This rule as to floral numbers is by no means invariable. In some families of dicotyledons the parts of the flowers occur in threes and fours.

This fundamental difference between the flowers of monocotyledons and of dicotyledons gives us additional evidence that these great groups are natural groups. The greater the number of fundamental differences that we find between them, the more remote may we regard their common ancestry to have been. You have now noted of monocotyledons in general that they have one cotyledon, scattered vascular bundles, leaves with parallel veins, and the parts of the flowers in threes or multiples of three. As to dicotyledons in general, you have noted that they have two cotyledons, vascular bundles cylindrically arranged, net-veined leaves, and the parts of the flowers usually in fives or multiples of five.

63. **Inflorescences.** — Plants differ greatly in the manner in which they bear their flowers. Often you can tell a plant by the manner in which its flowers are arranged. In some you find the flowers on little stems which arise from the axils of ordinary leaves, no particular part of the shoot being devoted to flower bearing. In others, as in some of the lilies, you find single flowers borne on the ends of rather long stems. In most plants, however, you find the flowers in clusters. These clusters are called *inflorescences.* (The word signifies *manner of flowering.*)

In gathering flowers it is often an inflorescence which you break off rather than a single flower. It is so with clover and goldenrod, with butter-and-eggs and queen's lace (wild carrot). The beauty of lily-of-the-valley is in its graceful, drooping inflorescence quite as much as it is in the individual flowers. The brightness of geraniums is due to the massing of their small flowers into good-sized inflorescences.

A. Advantages. — In connection with pollination, inflorescences are an advantage; that is, it is of advantage to have the flowers in clusters rather than widely separated. Whether pollination be accomplished by wind or by insects, the same thing is true. As for wind-pollinated flowers, evidently they need to be exposed in such manner that the pollen is likely to be blown from them and upon them. Evidently, also, it is good economy for the plant to expose flowers in clusters rather than singly. It requires more of stem growth to expose flowers borne singly than to expose an equal number borne in clusters. A further advantage is evident when the stigmas of a number of clustered flowers form a pollen-catching arrangement which would not be possible if the flowers were not close together. This is illustrated by corn. (See *Figure 103.*) Evidently the many threads of the silk catch pollen better when they are together than they possibly could if they were separate.

FIG. 117 *A.* — Inflorescence and leaf of the waterleaf (*Hydrophyllum canadense*); one of the spring flowering plants often very abundant in woods.

As for insect-pollinated plants, the nearer the flowers are together, the more visits can be made by an insect in the same period of time, and thus the more pollination will be accomplished. Also a cluster of flowers, being more conspicuous than single ones, may serve better to attract the insects. The most successful of all the families of seed plants is that family whose flowers are most closely clustered together. They are pollinated by insects. This family is named *Compositæ* from the fact that what appears to be the flower is really a composite of many small flowers clustered very closely together. Daisies and dandelions are examples of these "flowers" of *Compositæ* which are really inflorescences. They are composed of many small flowers arranged in that compact type of inflorescence which is called a *head*. (Other things besides the close clustering of their flowers have undoubtedly had to do with the great success of the *Compositæ*.)

The advantage in connection with pollination is not the only advantage of inflorescences. Other advantages appear in connection with the scattering of seeds.

B. Kinds. — To classify all the kinds of inflorescences is not simple and it is not important. But the names given to the principal kinds are names you should understand. They are frequently used in descriptions.

All inflorescences belong to one or the other of two general types, the *determinate* type and the *indeterminate* type. The flowers of an inflorescence do not all open at the same time; usually buds may be found along with the open flowers. If the youngest flowers or buds are in the center or at the tip of the inflorescence, it is of the indeterminate type. Most inflorescences are of this type. They can go on blooming somewhat indefinitely or indeterminately; shepherd's-purse and the common peppergrass are examples; their inflorescences go on flowering throughout the season.

If the youngest flowers are at the sides or at the bottom of the inflorescence, the older ones being at the center or top, it is of the determinate type. The extent of its flower production appears to be determined beforehand. The growing tip ends with the oldest flowers. The inflorescences of geranium are determinate.

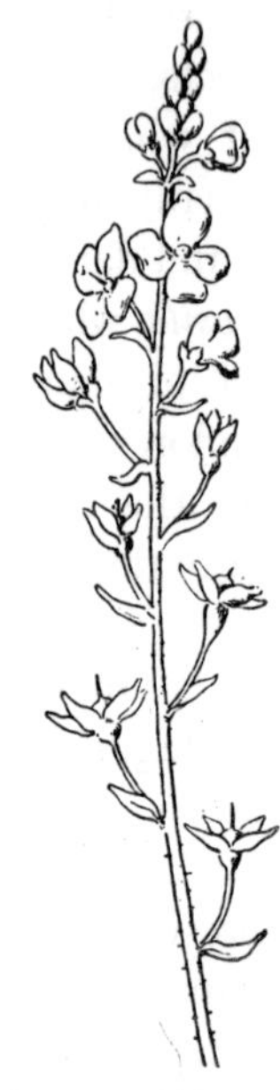

FIG. 118. — A raceme of *Veronica*. Note the *pedicels*, each one arising from the axil of a *bract*.

When it comes to describing inflorescences, certain terms not used before are needed. You have noted that a flower is composed of the end of a stem and of certain modified leaves which arise from it. It arises usually, as any other branch arises, from the axil of a leaf. In inflorescences those leaves from whose axils the flower branches arise are usually very much smaller than ordinary leaves. They are called *bracts*. (See *Figure 118*.) Any small, leaf-like organ found in an inflorescence, but not a part of a flower, is properly called a bract. You have noted those green structures which cover dandelion heads when they are closed. These structures are bracts. They form what is called the *involucre*. (See *Figure 124*.) The heads of all *Compositæ* are surrounded by involucres which, in function, correspond to the calices of other kinds of flowers. The stem which bears a single flower is called the *pedicel;* its outer end is the receptacle. Evidently the flowers of *Compositæ* do not have pedicels, since many of them arise from the top of a single stem. A flower stem of that kind, or one which subdivides into a number of pedicels, is called a *peduncle.*

a. Indeterminate Inflorescences. — Of these the following are the most common: —

FIG. 119. — The upper part of a spike of plantain. Note that the stigmas, *g*, ripen before the anthers, *a*. This plant is wind-pollinated, and, like other wind-pollinated plants, its stigmas are prominent. They have the appearance of small white plumes. The fact that they ripen before the anthers of the same flower prevents close-pollination; the pollen which is received must come from a different flower. The ripening of stigmas and anthers at different times is called *dichogamy*.

The *spike.* (See *Figures 101* and *119.*) This kind of inflorescence is illustrated by the common plantain, at least when the flowering stems are not kept cut off by a lawn mower. A much more dense kind of spike is that of the cat-tail. In spikes the flowers are *sessile*, that is, they have no pedicels. They arise directly from the main flowering stem and they usually grow close together.

The *catkin.* (See *Figures 94, 104,* and *105.*) Catkins are spikes whose flowers are borne in the axils of structures called *scales.* Scales are like leaves in origin. Catkins are commonly flexible and swing freely, like tassels, in the wind. As is almost equally true of spikes, the plants which bear them are wind-pollinated.

The *raceme.* (See *Figures 109* and *118.*) This type is well illustrated by the graceful, drooping inflorescence of lily-of-the-valley. It may be described as a spike in which the flowers have pedicels. The inflorescences of peppergrass and shepherd's-purse, already referred to, are also racemes. It is a common type.

The *panicle*. (See *Figure 120*.) This kind of inflorescence is found especially among various kinds of grasses, all of which are wind-pollinated. It is very graceful. It may be described as a raceme whose flower-bearing branches have subdivided; that is, it is a sort of compound raceme.

FIG. 120. — A panicle of a common kind of grass. Note that it might be described as a compound raceme.

The *corymb*. (See *Figures 117, 121*.) This kind of inflorescence is of frequent occurrence. It may be described as a shortened raceme in which the pedicels are of unequal length, the lowermost being the longest, with the result that the flowers (or heads) are borne at about the same level.

The *umbel*. (See *Figure 122*.) This is an inflorescence whose pedicels arise from the same place, and are usually of about equal length. It is a kind which is very easy to recognize. Milkweed affords a familiar example of it. Cherry blossoms are also borne in umbels. The inflorescences of the wild carrot (queen's lace) and wild parsnip are compound umbels; the large family to which these plants belong is called *Umbelliferæ;* its flowers are always in umbels.

FIG. 121. — A corymb of the yarrow, a common weed, and one of the *Compositæ*. Note that in this case, as is common among *Compositæ*, we have an inflorescence made up of inflorescences.

The *head*. (See *Figures 123* and *124*.) This kind of inflorescence has been mentioned as characteristic of the *Compositæ*. It might be described as an umbel whose flowers are sessile. In addition to dandelion and daisy, the asters, the goldenrods, sunflowers, wild lettuce, thistle, burdock, and many other common weeds belong to the *Compositæ*. Though all *Compositæ* have their flowers in heads, not all plants with heads are *Compositæ*. White and red clover have their flowers in heads, and they belong to the pea family (*Leguminosæ*). The buttonbush and the sycamore also bear their flowers in heads.

FIG. 122. — An umbel of milkweed. Note that the pedicels all start from the same place. Note the whorl of bracts at the top of the peduncle.

b. Determinate Inflorescences. — Of this type the *cyme* is the only one of importance. (See *Figure 125*.) It is illustrated by the geranium and by the syringa. In a cyme the terminal or central flower blossoms first, and after that only buds already formed open out into flowers.

64. More about Pollination. — Pollen escapes from the anthers which produce it. After its escape any one of a number of things may happen to it. It may be blown by the wind, it may be carried away by an insect, it may fall to the ground and perish. It may reach a stigma or it may not. Most of it does not.

A. The Waste of Pollen. — For nearly all of the pollen which is produced there is no such thing as pollination; not one grain out of a hundred reaches a stigma. This "waste" of pollen is much greater in wind-pollinated plants than it is in insect-pollinated ones. The chance that a wind-blown grain of pollen has to reach its destination is even less than that of one carried on the body of an insect. Of wind-blown pollen probably not one grain in many thousands ever reaches a stigma, yet the quantity produced is so enormous that few of the ovules fail to become seeds through lack of pollination.

FIG. 123. — Flowers and inflorescences of *Rudbeckia*, the coneflower or brown-eyed Susan.

B. Kinds of Pollination. — As to that pollen which does reach stigmas, there is not at all the same story in all cases. Different kinds of stigmas are reached, and different kinds of means are used in reaching them. Thus pollination may be classified as to the means by which the pollen is transferred, or as to the stigmas which are pollinated.

a. As to the Pollen Carrier. — Of the means by which pollen is transferred you have already noted wind, insects, and birds. Of these the first two are by far the most important; they are the means by which more than ninety per cent of all cross-pollination is accomplished. Some of the minor agents are worthy of mention. In some close-pollinated plants the pollen simply falls on the stigma; gravity is the agent of transfer. In others, as in the fireweed (*Epilobium*), the style turns as it grows so that it rubs against the anthers. There are some plants whose cross-pollination is accomplished by means of water. (Study *Figure 126.*) Snails, where they are very abundant, have been observed to act as pollinators.

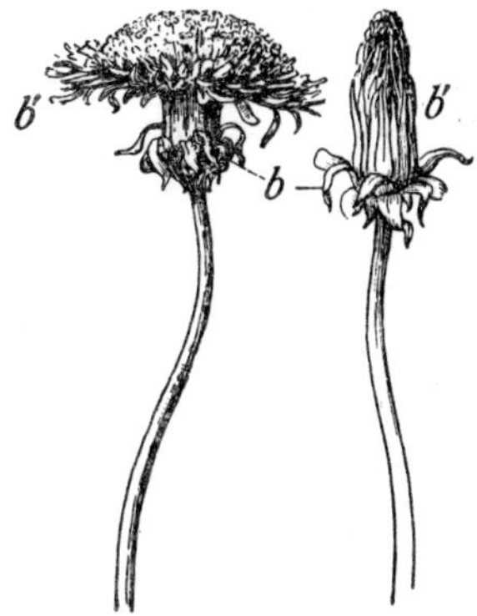

FIG. 124. — Heads of dandelion. At the left is an open head as seen in the sunlight. At the right is the same head as it is at night. The bracts of the involucre are of two kinds, short outer ones (b) and long inner ones (b'). The opening and closing is due chiefly to the movements of the inner bracts.

b. As to the Stigma Pollinated. — For some reasons it is more important to note *what stigmas are pollinated* than it is to note what it may be that carries the pollen. If it is the stigma of another flower that the pollen reaches, that is cross-pollination; if it is the stigma of its own flower, that is close-pollination. Evidently these are two quite distinct things. Close-pollination is also called *autogamy*. Of cross-pollination there are two kinds; three if the pollination of one kind of plant by the pollen of another kind be included.

FIG. 125. — Cyme of syringa, or mock orange. The terminal flower blossoms first. Note that the calyx is composed of sepals which, for about half their length, are united.

(1) *The pollen may cross to the stigmas of flowers of the same*

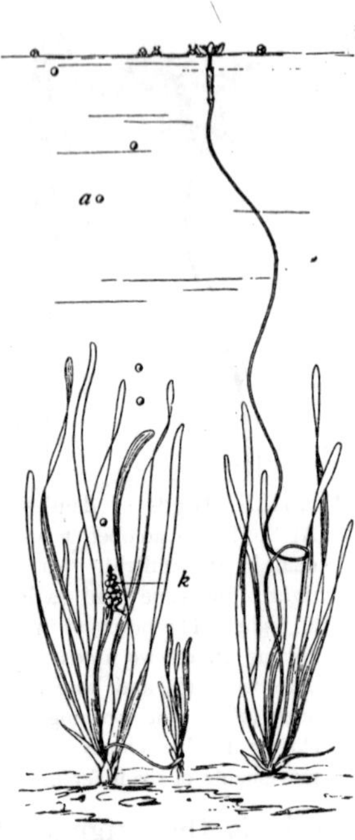

FIG. 126. — Pollination by means of water as illustrated by the tape grass (*Vallisneria*). This plant is diœcious. The individual at the left bears a spike of staminate buds (*k*). These spherical buds (*a*) become detached and rise to the surface, where, as they float, they open and expose the stamens. The pistillate flowers are borne on long stems which come just to the surface of the water. Pollination is accomplished by those staminate flowers which float against the pistillate ones. After pollination the stem of the pistillate flowers coils into a spiral, withdrawing the ovary below the surface. The fruit develops under the water.

plant. This is called *geitonogamy*. It is the commonest kind of pollination accomplished by insects. The principal pollinating insects usually explore thoroughly all the flowers of one plant before going to another.

(2) *The pollen may cross to the stigmas of flowers of other plants of the same kind.* This is called *xenogamy.* Evidently this is a more complete kind of crossing than geitonogamy is; in fact, the latter is more closely allied to close-pollination than it is to xenogamy. *Close-pollination and geitonogamy both mean that the pollen parent and the ovule parent of the seed are the same, while in xenogamy the pollen and the ovules have different parents.* It is evident that all diclinous flowers are cross-pollinated, also that all diclinous flowers borne by diœcious plants (see page 284) *must* be *xenogamic.* That is, since the staminate and pistillate flowers of diclinous plants are borne on different individuals,

there evidently must be a crossing from one plant to another if pollination is accomplished at all.

(3) The pollen may cross to the stigmas of flowers of a different *kind* of plant. Usually in such a case a pollen tube is not produced, but sometimes it is. Plants of just the same kind form what are called *species*. (The singular and plural of this word have the same form.) When the pollen of one species germinates upon the stigma of another, it is more than ordinary cross-pollination; it is the *crossing* of one species with another. Only nearly related species cross in this manner; crossing occurs frequently, for example, among the different species of oak. Plants produced from seeds which have resulted from the crossing of species are called *hybrids*. Hybrids are sometimes unable to produce fertile seeds; crossing appears to weaken their reproductive power. The production of hybrids has been of much importance in the cultivation of plants, especially in floriculture.

C. The Stigma. — Stigmas are not definite organs so much as they are "regions for the reception of pollen." They show much variety in form. Think of the small knob-like stigmas of many common flowers as compared with the silk of corn or with the feathery stigmas of some grasses. (See *Figures 127* and *128*.) The stigmas of wind-pollinated flowers are as a rule larger and more prominently exposed than those of insect-pollinated ones. Stickiness, on the other hand, is more characteristic of stigmas which are pollinated by insects than it is of those which are pollinated by the wind.

When flowers are abundant there is much pollen in the air. It is the flying pollen of some kinds of plants which is believed to be one of the causes of rose cold and hay fever. With the wind blowing it and many kinds of insects carrying it, pollen may frequently be caught by other stigmas than those of its own kind. Since in such cases the pollen usually does not germinate, it is believed that stigmas generally stimulate their own kind of pollen to germinate and do not so stimulate the pollen of other

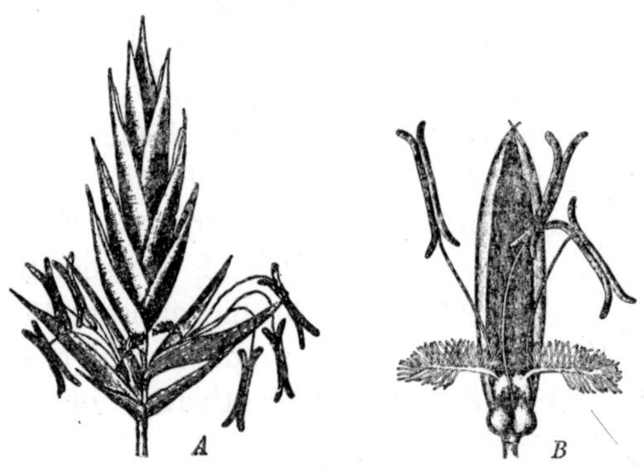

Fig. 127. — A common meadow grass (*Festuca*). *A*, the end of a spike showing the bracts from whose axils the flowers arise. *B*, a single flower and bract much magnified; note the prominent swaying stamens and the feathery expanded stigmas. — *After Strasburger*.

species. This power of the stigma is probably due to varying properties of the secretions which are found upon its surface. It has been found that on some stigmas pollen from the same flower will not germinate, while pollen from other flowers of the same kind will germinate. This is true of buckwheat and of the day-lily. This fact evidently prevents close-pollination in such plants. Much more common, however, are cases in which pollen does germinate upon its own stigma, but not so well as it does on other stigmas. Thus if pollen from the same flower and pollen from other flowers are both upon the same stigma, that from other flowers will produce pollen tubes of more vigorous growth than those produced by the pollen from the same flower. The more vigorous pollen tubes will be the ones to reach the ovules. Thus cross-pollination is favored even though close-pollination is not absolutely prevented.

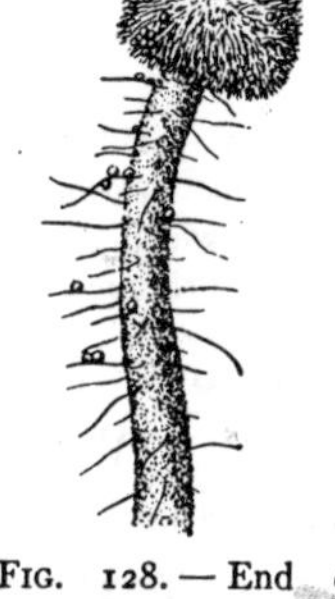

FIG. 128. — End of the pistil of *Hibiscus*, much enlarged. Note the hairy surface of the stigma to which pollen grains are adhering.

D. Wind-pollination. — This is the simplest form of pollination. It is also the most ancient. Insect-pollinated plants came from ancestors that were wind-pollinated. Those structures in lower plants which correspond to the pollen of seed plants are scattered by the wind.

The pollen of wind-pollinated plants differs from that of insect-pollinated plants in two general features. It is more buoyant and it is more abundant. It is generally light and smooth and dry, whereas insect-carried pollen is generally heavy and rough and moist. In some cases the buoyancy is increased by wing-like outgrowths. (See *Figure 129*.)

The flowers of wind-pollinated plants have other features which are distinctly favorable to this process. It is common to find both stigmas and anthers prominently exposed. (See *Figures 106, 119, 127,* and *130*.) It is also common, among those which are diclinous, to find the staminate flowers arranged in slender, pendent catkins from which, when ripe, even the gentlest breeze may start a cloud of pollen. (See *Figures 104* and *105*.) Such catkins are characteristic of many trees; poplars, oaks, birches, and pines are examples. In most wind-pollinated trees the flowers appear before the leaves, another feature which is evidently favorable to the process.

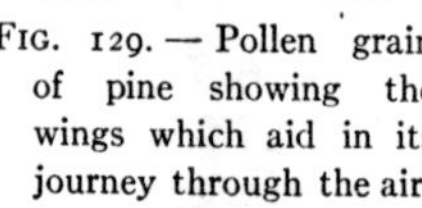

Fig. 129. — Pollen grain of pine showing the wings which aid in its journey through the air.

Many monoclinous flowers and nearly all diclinous ones are wind-pollinated; insect-pollination, with a few exceptions, is confined to monoclinous flowers. (Why is monocliny more favorable to insect-pollination than dicliny is?) As you have noted, close-pollination is a physical impossibility for diclinous flowers; for them only geitonogamy and xenogamy are possible. Thus it is interesting to note that many wind-pollinated plants have structures and habits which favor xenogamy and make geitonogamy difficult or impossible. Poplar, ash, box elder, juniper, and meadow rue insure xenogamy by being diœcious, while in the hazel (see *Figure 104*) and in the pine xenogamy is favored by the fact that the pistillate flowers are borne higher up than the staminate ones. Also in monœcious forms, the pistillate flowers usually ripen before the staminate ones of the same individual. In alders and in cat-tails this difference in the time of flower ripening may amount to several days.

Fig. 130. — Part of a panicle of a meadow grass. Two of the lower flowers have opened. Note that each of these has two plumose stigmas and three stamens whose long and slender filaments expose the anthers to the wind.

Wind-pollinated flowers, as you have noted, are usually much less conspicuous than are insect-pollinated ones. This has the natural consequence that the latter are much more familiar and include practically all the flowers that are cultivated for their beauty. Wind-pollinated flowers are not generally recognized as flowers at all. They lack bright colors, odor, and nectar.[1] The perianth is usually inconspicuous or absent.

So far as may be judged by the number of plants which use this method, pollination by wind appears to be just as successful as pollination by insects. Of course wind-pollination involves an enormous waste of pollen. On the other hand, the chance of a favorable wind may be greater than the chance of transfer by insects, especially if only a particular kind of insect is able to make this transfer. Oaks, pines, and grasses are very successful plants which are wind-pollinated. The abundance of these plants may, however, be due to other features than their pollination. Pollination is only one of many factors which determine the success or non-success of plants.

E. Insect-pollination. — The relations between flowers and insects is one of the most wonderful things in nature. Some insects are absolutely dependent upon certain flowers for food and for the rearing of their young, and some flowers are absolutely dependent upon certain insects for pollination and the production of seed. The structures of some flowers and of some insects so correspond to each other that only these insects can get the pollen from these flowers. Insects are the chief agents of pollination for

[1] Cottonwood and field sorrel are exceptions to this in that, though wind-pollinated, their flowers are bright colored.

nearly all conspicuous or odorous flowers, though "corresponding structures" between flowers and insects are the exception rather than the rule. Most insect-pollinated flowers may be pollinated by any one of a considerable number of kinds of insects. In the case of dandelion, for example, over one hundred different kinds of insects have been observed visiting its flowers. Quite commonly one kind of insect appears to be the favorite visitor of one kind of flower, even though pollination by other insects is possible. This is the case with red clover, whose favorite visitor is the bumblebee. One condition necessary for a good crop of clover seed is an abundance of bumblebees in the clover field. Evidently bumblebees are of commercial value to the farmer who grows clover for seed.

As you have noted, nearly all insect-pollinated flowers are monoclinous. Willows (see page 264) and some of the *Compositæ* are exceptions to this. Evidently monocliny is an advantage in insect-pollination; for one reason, to effect an equal amount of pollination, only half as many visits are necessary to monoclinous flowers as to diclinous flowers.

a. Why do Insects visit Flowers? — Pollen and nectar are the two things for which insects visit flowers. Both are used for food, though many insects use one and not the other. Nectar is a sweet liquid which is secreted by a part of the flower called the *nectary*. It is usually so placed that in order to reach it the insect must brush the stamens or stigma or both. In some cases the nectar drips from the nectary and collects in a protected extension of the corolla called the *spur*. (See *Figure 131*.) Insects also visit flowers for shelter, and, in some cases, to lay their eggs.

Bees take nectar and make honey from it. Honey is nectar which has been swallowed by bees and has been partly digested. They also take pollen which is largely used as food for the young bees. A certain part of the hind legs of bees is called the *pollen basket;* it is coated with stiff hairs and becomes covered with a mass of pollen; it may be easily observed upon bees which are at work on flowers.

FIG. 131. — Lengthwise section through a flower of nasturtium. The nectar (*n*) is at the base of the spur (*s*), which is a prolongation of the zygomorphic corolla. Note that to reach it the insect must come in contact with the stamens and pistil.

Moths and butterflies take only nectar. They have a long *proboscis* by means of which they reach into corolla tubes or spurs and suck the nectar. They are not so generally useful in pollination as are the bees; the bees work at the flowers in a much more businesslike and thorough way. The night-flying hawk moths (see *Figure 132*) are an exception to this rule. "They visit flowers rapidly and with the precision of bees, thus contrasting with the more languid and haphazard movements of the butterflies." (Cowles's *Ecology.*)

b. Color and Odor. — It is believed that both the colors and odors of flowers serve as means of attracting insects. Of the two, odor is doubtless much more important in this connection. Insects are short-sighted and are thought to be usually color-blind; the honeybee is the only insect which has been positively proved to have a sense of color.

As to the sense of smell, however, insects are very keen. Fragrant flowers which are inconspicuous are visited much more than are showy ones which have no odor. Night-flying moths locate flowers readily by their fragrance, and there are some flowers which open and give out their fragrance only at night; the moonflower and night-blooming cereus are famous for this. There is reason to believe that many insects detect odors which we are quite unable to perceive. The skunk cabbage and the carrion-flower have decidedly disagreeable odors. These odors, like those of decaying flesh, are attractive to flies.

FIG. 132. — A flower of *Petunia* visited by a hawk moth. Note the long corolla tube of the flower, and the long mouth part of the insect.

c. Examples. — In their relation to insects there is a general difference between flowers with actinomorphic (regular) corollas and those with zygomorphic (irregular) ones (see page 287). The former generally have a considerable variety of insect visitors, while the visitors of the latter are more limited as to kinds. It is among zygomorphic flowers that we find those highly specialized forms which make pollination possible by certain insects only.

Flowers with regular corollas and spreading perianths, like buttercups or wild roses or apple blossoms, are open to all comers. Their pollen and their nectar may be obtained by any visiting insect. Regular flowers whose corollas form a tube are somewhat more select as to their visitors; only insects with elongated mouth parts are able to reach the nectar without breaking the wall of the tube. (See *Figure 132*.) Sometimes bees, wasps, or ants reach

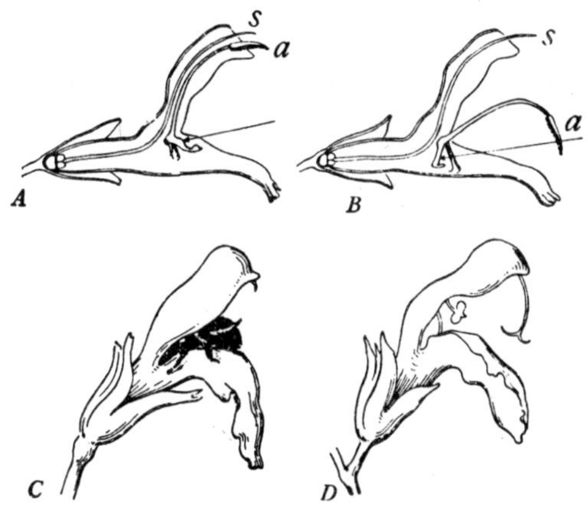

FIG. 133. — Pollination in *Salvia*. *A*, lengthwise section of the flower; *s*, the unripe stigma; *a*, the ripe anther; the arrow points to a lower arm of the stamen; by pushing back this arm the visiting bee forces the anther down upon its own back. *B* shows the position of the anther when the lower arm of the stamen is pushed back. *C* shows a bee at work. *D* is an older flower in which the stigma, now ripe, is ready in its turn to be brushed by the back of the bee. — *After Kerner and Avebury.*

the nectar of such flowers by biting holes through the corolla tubes near the bottom. Such holes are often seen in the corolla tubes of the trumpet creeper, a plant whose flowers are frequently visited by humming birds.

Flowers with zygomorphic corollas are, as a rule, dependent for pollination upon certain kinds of insects, and it is the bees upon which such dependence is principally placed. Zygomorphic flowers are characteristic of the legumes (pea family), the violets, and the mints, and these flowers generally do not set seed unless pollinated by bees. The natural distribution of clover, for example, is confined to those parts of the world which are inhabited by bees.

FIG. 134. — An inflorescence of one of the species of yucca. Note that the flowers hang downward, thereby protecting their inner parts from rain.

The flower of *Salvia* illustrates a structure favorable to pollination by bees. This structure, or one much like it, is quite common among zygomorphic flowers. (Study *Figure 133*.) *Salvia* is a member of the mint family. It is grown commonly in beds near houses and walls, and is familiar on account of its bright red flowers which bloom in late August and September.

The snapdragon is another cultivated flower with a zygomorphic corolla. It is usually tightly closed and small insects cannot enter it. But when a bee alights upon the lower lip of the corolla, its weight depresses the lip and opens the flower.

The violet has a zygomorphic corolla and a spur. The

pollen is concealed in the throat of the corolla and the nectar is in the bottom of the spur, but both can be easily reached by the tongue of bees.

Yucca and Pronuba.—This is one of the most famous cases of mutual dependence between a plant and its pollinating insect; there is a sort of partnership in their reproductive process, each being necessary to the other.

The yuccas grow chiefly in deserts, especially in southwestern United States and Mexico; the Spanish bayonet is a commonly cultivated species. (See *Figure 134.*) The pronuba is a small moth most of whose life is spent upon yucca plants.

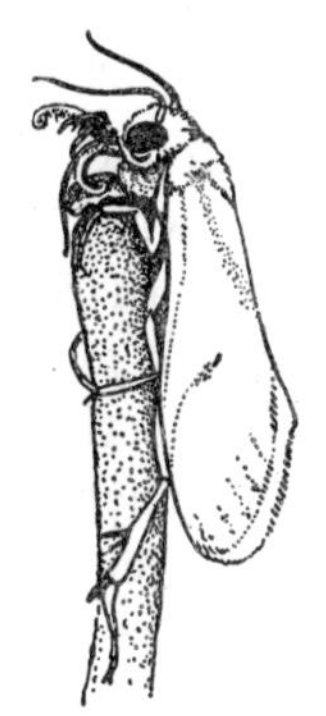

FIG. 135. — Pronuba collecting pollen from a stamen of yucca. The stamens open at the top.

Pollination is performed by the female moth. During the day she remains hidden within the flower. At twilight she begins her work. First she visits the stamens and collects a mass of pollen. (See *Figure 135.*) Still holding the pollen, she moves about the flower and finally comes to rest astride of one of the stamens and with her head toward the stigma. (See *Figure 136.*) While in this position she thrusts her sharp-pointed egg-depositing organ (the ovipositor) down between the stamens, through the wall of the ovary, and in among the ovules. There she deposits an egg. As soon as this is done, she rushes to the top of the hollow pistil and vigorously thrusts pollen down into it. She then repeats the performance, going to the top of the pistil and thrusting pollen into it after each egg is deposited. In this way the same pistil is usually pollinated several times. It is as though the moth were following an instinct which impels it in this way to provide food for its young.

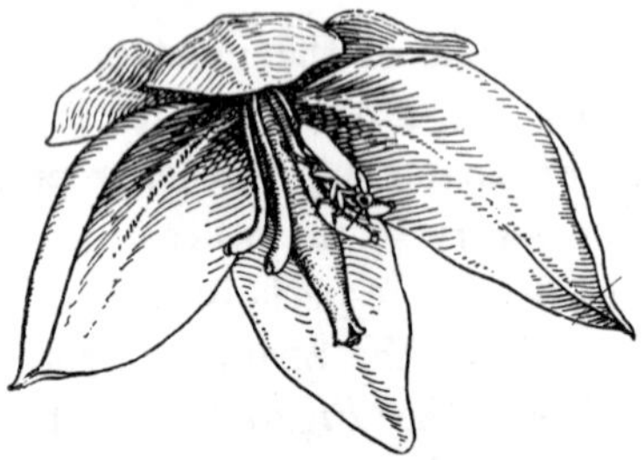

Fig. 136. — Pronuba depositing eggs in the ovary of yucca.

The ovary of yucca contains many ovules. While they are developing into seeds, the egg of the moth is developing into a grub, the larva. The grub eats a number of the young seeds, and then bores its way out of the ovary and drops to the ground. Ripe ovaries of yucca show a number of holes from which pronuba grubs have escaped. (See *Figure 137*.) In this process a good many seeds are sacrificed, but still more come to maturity, which, without the aid of pronuba, they are quite unable to attain. On both sides there seems to be fair payment for the service rendered.

Fig. 137. — Fruit of yucca showing holes made by escaping grubs of pronuba.

The pronuba usually thrusts the pollen into the pistil of the same flower in which she collects it. Note that this is therefore a case of insect-pollination which is not cross-pollination; also that in this case close-pollination is not synonymous with self-pollination. Sometimes, however, after collecting pollen, the moth flies to another flower before beginning to lay eggs and to pollinate the pistils. In this way the pronuba may effect geitonogamy or even xenogamy.

Fig and Wasp. — Here is an even more complicated

case. Figs are fruits which do not result from single flowers. They result from the growth of whole inflorescences and of the tissue which surrounds the inflorescences. Note the lengthwise section of a green fig as shown in *Figure 138*. The inside of the fig is lined by a large number of small flowers. Entrance to these flowers is possible only by a small opening at the top. All kinds of figs are diclinous and the fig of commerce is practically diœcious, the pistillate flowers being sterile on some trees, while the staminate flowers are sterile on others. This kind of fig is pollinated by a small wasp whose life history is most extraordinary.

The female wasps force their way into the inflorescences. Usually in doing so they scrape off their wings on the scales which guard the small opening. (See *Figure 138, B.*) After laying eggs, they die without escaping. Those which enter pistillate inflorescences (which are the figs we eat) have no progeny. The styles of the fertile pistillate flowers are so long that the wasp cannot deposit its eggs in a favorable place. (See *Figure 138, C.*) But in the staminate figs (called *caprifigs*) there are numerous sterile pistillate flowers with short styles. (See *Figure 138, A.*) In these are deposited eggs which later hatch into wasps. The male wasps die within the caprifigs in which they are born, but the females escape, and carry with them the pollen with which they became dusted as they crawled about within the floral cavity. Of the escaping females, those that later go into pistillate inflorescences are of service in pollination, but produce no progeny; those which enter staminate inflorescences produce progeny, but are of no service in pollination.

Centuries before this process was understood, the fig growers of the old Mediterranean countries used to cut branches of caprifigs from the staminate trees and place them among the branches of the pistillate figs. They knew this process improved the quality of the fruit, though they did not know why. It was because it made more certain the entrance of the weak-flying, pollen-bearing female wasps into the pistillate inflorescences.

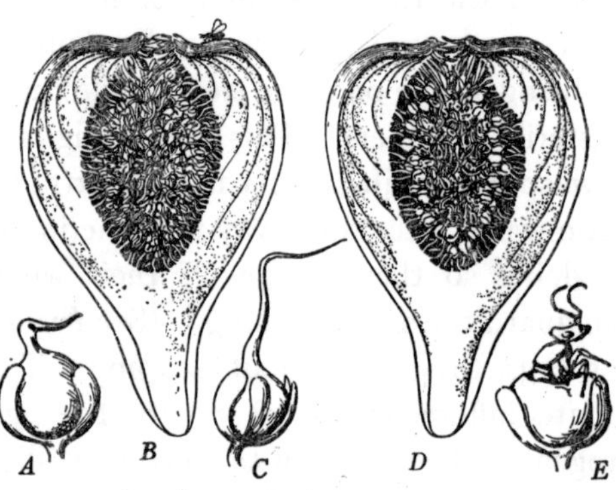

FIG. 138. — Pollination of the fig. *B*, lengthwise section through a seed-producing (pistillate) inflorescence; note a wasp on top and another which has just crept inside. *D*, a staminate inflorescence with numerous sterile pistillate flowers in which young wasps have hatched; note one near the opening crawling out. *A*, one of the short-styled sterile pistillate flowers. *C*, a long-styled fertile pistillate flower. *E*, young wasp just emerging from one of the sterile pistillate flowers within which it was hatched.

Another remarkable thing about the fig is that it will mature its fruit (though not its seeds) even if pollination does not occur. Pollination, however, especially in the Smyrna fig, improves the fruit in plumpness, juiciness, and flavor. These figs are now grown in California, where their culture has been greatly improved by the introduction of the proper pollinating wasp. Commercial figs are propagated by cuttings, not by seeds, so pollination is not necessary in their culture.

F. Prevention of Close-pollination. — Some plants with monoclinous flowers possess devices which prevent close-pollination; for diclinous flowers, of course, close-pollination is impossible anyhow. You have already noted (see page 306) that in some monoclinous flowers the pollen will not germinate upon the stigma of the same flower; this is one such device.

Another thing which prevents close-pollination is what is called *dichogamy*. This is the ripening of the stamens and the pistils of the same flower at different times. Many flowers have this habit. Among wind-pollinated plants, the common plantain furnishes an example of it. (See *Figure 119*.) Among those which are insect-pollinated, the figwort (*Scrophularia*) is one of a number of common dichogamous forms. (See *Figure 139*.) Sometimes the stigmas ripen before the anthers, and sometimes the reverse is true.

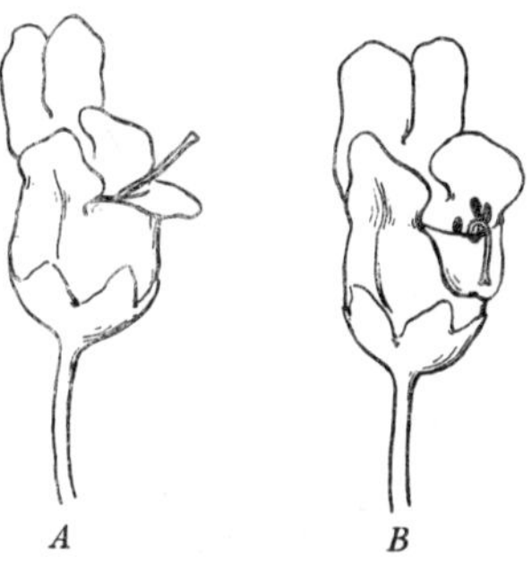

FIG. 139. — Flowers of the figwort (*Scrophularia*), illustrating dichogamy. In *A* the unripe anthers are not yet visible, while the ripe stigma is prominently exposed. In *B*, which is older, the anthers can be seen, and the top of the old style has drooped down, out of the way of visiting insects.

Some plants have two kinds of monoclinous flowers; in one the anthers are above the stigma, in the other the stigma is above the anthers. This is called *dimorphism*. (See *Figure 140*.) The primrose, the forget-me-not, and the bluets have such flowers. That part of the insect which brushes the higher anthers will later come in contact with a high stigma in another flower, while that part which is dusted by low anthers will probably pollinate low stigmas. Usually, in such plants, pollen from low anthers will germinate only on low stigmas, and that from high anthers only on high stigmas. And the two kinds of flowers are usually borne on different individuals. Thus we have three devices of the same kind of plant, all tending to prevent close-pollination and to produce xenogamy.

Fig. 140. — Flowers of the Chinese primrose, illustrating dimorphism. Note that that part of the insect which strikes the anthers of one flower will be likely to strike the stigma of the other.

G. Cleistogamy. — Some plants produce flowers which never open and are necessarily self-pollinated. Such flowers are called *cleistogamous*. The violet is the most familiar plant which regularly produces cleistogamous flowers. They are produced some weeks or even months after the open, colored ones. They are colorless and are borne on short stalks, very near the ground. Often they are actually subterranean, the seeds which they produce thus being in a position which is very favorable for their germination. One of the milkworts (*Polygala polygama*) also regularly produces cleistogamous flowers and these are distinctly subterranean. (See *Figure 141*.)

Sorrel (*Oxalis*) and touch-me-not (*Impatiens*) are examples of plants which *sometimes* produce cleistogamous flowers. Low temperature is believed to be the principal cause which induces their production. *Lamium*, one of the mint family, produces cleistogamous flowers in spring and autumn, and open flowers in the summer.

FIG. 141. — One of the milkworts (*Polygala polygama*), bearing both aërial open flowers and subterranean cleistogamous flowers. The subterranean flowers are self-pollinated and produce pods which contain many seeds.

H. Which Kind of Pollination is Better? — The test of pollination is the quality of the progeny resulting from the seeds produced thereby. It is commonly believed that the progeny resulting from cross-pollination is in general better than that resulting from close-pollination. Numerous experiments, conducted through many years, tend to show this to be true. But there is something to be said in favor of the other method. In some plants, close-pollination has been shown to result in progeny of at least equal vigor with that resulting from cross-pollination. Tobacco, petunia, and California poppy are examples of this.

It is evident that close-pollination is much simpler than cross-pollination. It does not require such elaborate structures. It does not involve any waste of pollen. (Some cleistogamous flowers have only one anther and only a few grains of pollen in that.) It is certain, not requiring the plant to depend upon more or less uncertain outside agencies for its performance. In some cases (some cleistogamous flowers) it allows the placing of the seeds in a position favorable to germination. Evidently, if equally good seeds *can* be produced by close-pollination, then this method has distinct advantages over the other.

But cross-pollination is the more common method. Diclinous plants necessarily use it, and the great majority of monoclinous ones not only appear to take pains to secure it, but also appear to take pains to avoid the other kind of pollination. With corn it has been proved that even geitonogamy results in diminished vigor; corn requires xenogamy. Wheat, on the other hand, appears to do very well with close-pollination, and some of our common weeds make use of this method constantly and successfully. Pigweed, knotgrass, chickweed, and mallow are examples. The flowers of trillium and geranium are usually self-pollinated. The evening primrose, though often cross-pollinated, may be self-pollinated even before the flowers open; in its yellow corollas the stamens and the pistil may be seen growing together in such manner that the pollen is fairly rubbed off on the stigma.

It is evident, as stated in Cowles's *Ecology*, that "the benefits of cross-pollination and the disadvantages of close-pollination have been too much emphasized. Close-pollination and its essential equivalent, geitonogamy, are extremely common in nature, nor must it be forgotten, also, that many of the important plant and animal races utilized by man have reached their present state of commercial perfection by the most careful inbreeding," that is, by careful prevention of cross-pollination.

QUESTIONS AND SUGGESTIONS

SECTION 60. 1. Define flowers as to structure and function. 2. Why are flowers used as the principal means for the identification of plants?

A. 3. Define pollination and indicate the apparent extent of its influence upon flower forms. 4. Discuss the two kinds of pollination. 5. What are the agencies by means of which cross-pollination is chiefly accomplished? 6. Define nectar and indicate its use to the plant. 7. Describe wind-pollinated flowers and give examples.

B. 8. Describe the protective organs of the flower.

C. 9. Explain what is meant by evolution, using flowers as an illustration. 10. Why are leaves more alike than flowers are? 11. Give reasons for the study of non-seed plants.

D. 12. Define spore, generation, and pollen tube. 13. Explain why it is inaccurate to call a flower a sex organ.

SECTION 61. *A.* 1. Explain what are the essential and what the accessory parts of flowers, and why they are such. 2. Give examples of flowers whose essential parts are undeveloped. Explain this lack of development.

B. 3. Describe the perianth, and define naked flower.

C. 4. Describe stamens, defining andrœcium, anther, filament, and pollen sacs. 5. Describe the pistil, defining gynœcium, ovary, style, stigma, and ovule. 6. Describe fertilization, defining gamete, sperm, and egg. 7. Contrast fertilization with pollination. 8. Describe the ovule, defining micropyle, nucellus, and integument.

D. 9. Describe the receptacle.

SECTION 62. 1. Why are nutritive organs less useful in classification than reproductive organs are?

A. 2. Discuss the work of Linnæus. 3. What is meant by the "real kinships" of plants?

B. 4. Define diclinous and monoclinous, staminate and pistillate. 5. Give examples of diclinous flowers other than those given in the book. 6. Describe the flowers and the pollination of corn. 7. Define monœcious, diœcious, and polygamous, giving examples.

C. 8. Explain in what respects the water lily is thought to be a primitive type of flower. 9. Define sympetalous and polypetalous, giving examples. 10. Define zygomorphic and actinomorphic.

D. 11. Give examples of variations of the androecium.

E. 12. Define pistil. 13. Distinguish between simple and compound pistils, and define placenta. 14. Define hypogyny, perigyny, and epigyny, giving examples.

F. 15. Are the most complex flowers the most advanced in evolution? Explain your answer. 16. Discuss floral numbers.

G. 17. State four fundamental differences between monocotyledons and dicotyledons.

SECTION 63. *A.* 1. Define inflorescences and discuss their advantages. 2. Give examples of *Compositæ* other than those in the book, and explain why the family is so called.

B. 3. Distinguish between determinate and indeterminate inflorescences. 4. Define bract, involucre, pedicel, and peduncle. 5. Name seven kinds of indeterminate inflorescences and give examples of each. 6. Explain and illustrate what is meant by a cyme.

SECTION 64. *A.* 1. Discuss "waste" of pollen.

B. 2. Mention seven different means by which pollination is accomplished, indicating which of these are more common. 3. Define autogamy, geitonogamy, and xenogamy, indicating which two of these are most nearly alike. 4. Explain the crossing of species, defining species and hybrids.

C. 5. Define and describe stigmas. 6. Discuss the response of pollen tubes to stigmas.

D. 7. Discuss the structures that appear to be related to wind-pollination. 8. Why is close-pollination impossible for diclinous flowers? 9. Describe ways in which wind-pollination is compelled to be xenogamic. 10. Describe the flowers of wind-pollinated plants and illustrate by examples.

E. 11. Discuss the causes of the visits of insects to flowers. 12. What kinds of insect visitors are of chief importance to flowers? 13. Define nectary, spur, and pollen basket. 14. Give examples of apparent correlation between insect forms and flower forms. 15. Describe the relationship between yucca and pronuba. 16. Discuss the case of the fig and wasp.

F. 17. Discuss the prevention of close-pollination, defining dichogamy and dimorphism. *G.* 18. Explain what cleistogamy is, giving examples. *H.* 19. Compare close- and cross-pollination as to advantages and disadvantages.

CHAPTER VIII

FRUITS AND SEEDS

65. Functions of Fruits and Seeds. — Protection, dispersal, and germination — these are the three great functions with which fruits and seeds are concerned. Fruits are usually concerned only with the first two of these functions; seeds are concerned with all three. The thing which is to be protected and nourished, the thing which needs to be dispersed, to be carried away from its parent to a place favorable for its future life, the thing whose successful germination is so important — that thing is, of course, the young plant which lies dormant within the tissues of the seed, the young plant whose existence it has been the whole work of the flower to accomplish. That dormant young plant you know as the embryo, and the renewal of its growth is what is commonly called the *germination of the seed.*

Strictly speaking, however, seed germination is not germination at all. If it were, then seed plants would have two germinations. Germination, strictly speaking, is the beginning of the growth of a single cell into a many-celled individual. It occurs when the fertilized egg divides and redivides to form the embryo; it does not occur when that embryo simply starts to grow again. The embryo is simply an arrested stage in the development of the young plant. To speak of the germination of seeds is not objectionable, however, if it is realized that this is a process entirely different from the germination of one-celled reproductive bodies, as, for example, the germination of the pollen grain upon the stigma, or the germination of the spores of lower plants, as the ferns. It is more accurate to speak of seeds as sprouting than to speak of them as germinating. The "germination" is really the sprouting of the embryo; to sprout is to give off branches.

Since the seed, strictly speaking, does not germinate, then it is evident that, strictly speaking, the seed is not a reproductive body. It does not reproduce anything. It simply protects and nourishes for a time a resting stage in the growth of the young plant. It may be regarded as an organ of protection, nutrition, and dispersal rather than as an organ of reproduction. However, it is usual to regard it as the latter, and here again there need be no objection if there is no misunderstanding.

When the young plant renews its growth, it draws upon the food which is stored in the seed. As you have noted (see page 66), this food may be stored in a tissue of the seed called endosperm, or it may be stored in the cotyledons which are a part of the embryo itself. It may be large or small in amount, the amount of food determining the size of the seed. Seeds vary in size from those which are almost microscopic up to the coconut, which is the largest of all seeds. The milk of the coconut as well as the meat is rich in food. Evidently smallness of seeds tends to be an advantage in dissemination, at least by wind, while largeness tends to be of advantage in germination. (See page 86.)

You have already considered the advantage it is to the young plants to be scattered abroad. You have also considered some of the means by which this scattering abroad (dispersal) is accomplished. (See page 63.) In very many cases it is the seeds alone which are dispersed. Just as frequently, it is the fruits containing the seeds which are dispersed, while in some cases, as in tumbleweeds, whole shoots of the parent plant are scattered about. (See *Figure 142.*)

FIG. 142. — Mature plants of the winged pigweed, which is one of the tumbleweeds. Tumbleweeds are plants which, upon maturity, break from the soil and "tumble" along before the wind. Wherever they lodge, they drop seeds. Plants of this kind are usually frequent in treeless regions. The Russian thistle and the amaranth are other well-known examples of tumbleweeds.

In some plants it is not simple to tell whether the part scattered is seed or fruit. Thus in the dandelion or in the sticktight (*Bidens*) or in others of the *Compositæ* it is common to think of the scattered parts as being seeds, whereas they are more than that. They are fruits, one-seeded fruits, in which the fruit wall (*pericarp*) grows closely around the seed wall (*testa*) so that they are practically one structure. One-seeded fruits of this kind are called, in the dicotyledons, *akenes*, and they are particularly characteristic of the *Compositæ*. Among monocotyledons they are called ***grains***, and they are particularly characteristic of the grass family (*Gramineæ*), to which wheat, corn, and the other cereals belong.

The principal agencies by which dissemination or seed dispersal is accomplished are wind, water, and animals.

These are discussed in the section devoted to this topic (see page 353). It must not be understood, however, that dissemination, like pollination, is a process for which all flowering plants must make definite provision. A vast number of fruits and seeds have no regular means of dissemination. They simply drop to the ground beneath the plant which bore them. Nuts, acorns, and many other heavy fruits or seeds are examples of this.

66. The Nature of Fruits. — Fruits and seeds begin where flowers leave off. They fulfill a work of reproduction to whose first stages the flowers are devoted. Fertilization is followed by changes in the organs of the flower. The most conspicuous of these changes is the growth of the ovary. Its wall becomes the pericarp of the fruit. The ovules develop into seeds. The corolla and the stamens, their work having been accomplished, soon disappear, but the calyx and the receptacle often enlarge with the ovary. *The structure which results from this growth that follows fertilization is called the fruit.*

Young fruits are usually as green as the leaves. They are able to manufacture much of the food used in their own growth. In some trees, as in the elm, the green fruits are prominent before the leaves appear, and in such cases the work of photosynthesis which they do seems especially important.

A. Aid in Dispersal. — Fruits aid in the dispersal of seeds in three principal ways: —

a. By their Edibility. — Many kinds of fruits are eaten by animals through whose alimentary tracts the hard-coated seeds pass without injury. This is not true, however, of all fruits which are eaten; the seeds of some are digested as well as the tissue which surrounds them. Birds and grazing animals are especially helpful to plants in connection with seed dispersal. They are agents by which seeds are scattered widely and are deposited in conditions which are favorable for their germination. Edibility of the seeds themselves is, however, a disadvantage, as in the case of nuts eaten by squirrels.

b. By their Buoyancy. — Buoyancy both in the air and on the water is important. The fruits of many *Compositæ*, buoyant in the air, and often aided in this buoyancy by parachute-like structures, furnish the most familiar and abundant examples of dissemination by wind. (See *Figure 143.*) All seeds and fruits, even the lightest, are heavier than air, so their movement through it is limited and dependent chiefly upon wind. But very many seeds and fruits are lighter than water, and so may be carried by it for great distances. The principal drawback to this means of dispersal is that water-soaked seeds after a time lose their vitality, their power to germinate. For shore-growing plants, however, water is of very great importance in dispersal.

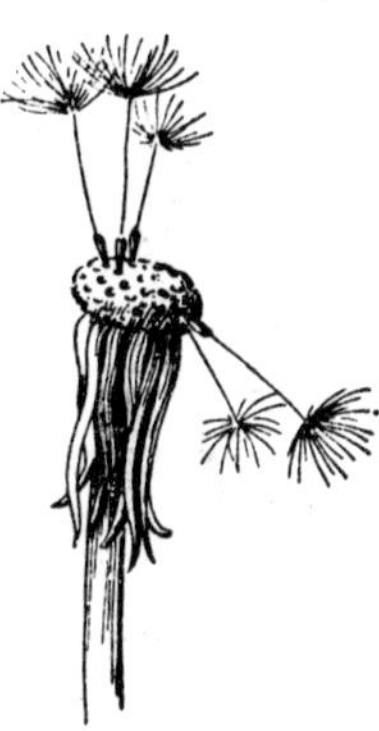

Fig. 143. — Fruits (akenes) of dandelion.

c. By their Clinging Power. — Any one who has walked through woods or weed patches in late September or early October can testify abundantly to the clinging power of several kinds of fruits. (See *Figure 144.*) This, and the two other properties of fruits which make for dissemination, are discussed further in the section on agencies of seed dispersal. (Page 353.)

B. Classification of Fruits. — There is no such classification of fruits as there is of flowers. They are not so useful as flowers are in identifying plants. This is not because they are more variable than flowers; in fact, the contrary is true. It is chiefly because they are much more simple than flowers, and so do not afford in equal degree the means for writing identifying descriptions. They lack those numerous separate organs which flowers possess and upon whose variations the classification of plants is principally based. Fruits are, however, second only to flowers as a means of identification. Often, in case of doubt, it is the character of the fruit which determines the identity of the plant. In collecting specimens, it is always important to obtain the fruit if possible.

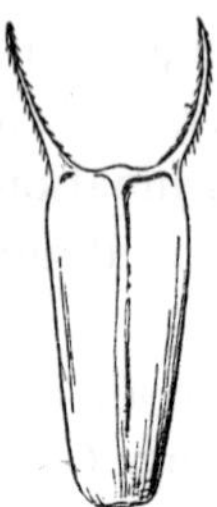

FIG. 144. — Fruit (akene) of stick-tight (*Bidens*).

C. Dehiscent and Indehiscent Fruits. — Fruits, considered apart from the plants which bear them, may be grouped in various ways. One way is to group them into those which are *dehiscent* and those which are *indehiscent.* Fruits that open on maturity, permitting the ready escape of the seeds, are called dehiscent; those which do not open are called indehiscent. In the former it is generally the seed that is scattered, while in the latter it is generally the fruit as a whole that is scattered.

Pods, such as those of beans and peas, are examples of dehiscent fruits. (See *Figure 145, A.*) *Capsules* are like pods in general appearance, and are commonly called pods, but the technical difference is that pods come from simple pistils, while capsules come from compound pistils and thus may have more than one seed chamber. (See *Figure 145, B.*) The parts into which the walls of pods and capsules separate are called *valves;* pods have two valves, while capsules have as many valves as there are carpels. In some cases the valves separate suddenly, twisting as they separate, and so throw the seeds out with some violence. (See *Figure 145, A.*) Such fruits are said to be explosive. The fruit of touch-me-not is a familiar example. The capsules of evening primrose, morning-glory, and larkspur are examples of those which remain on the plant for a long time after they dehisce; they allow the seeds to fall out gradually or to be scattered by the wind. (See *Figure 145, B.*) Pods are especially characteristic of the pea family (*Leguminosæ*) and of the mustard family (*Cruciferæ.*) Sometimes dehiscent fruits open by pores, as in the poppy.

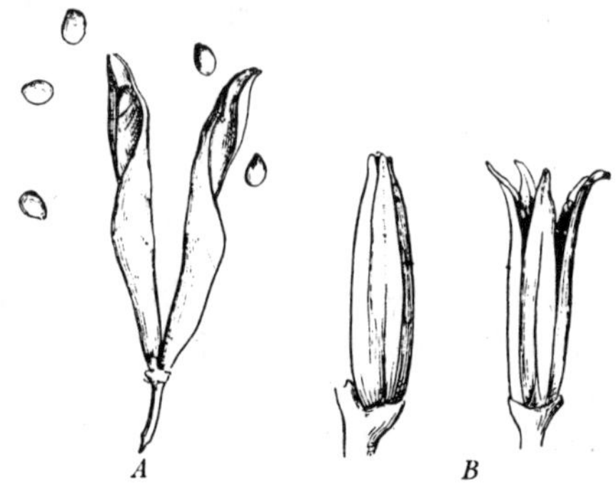

Fig. 145. — *A*, pod of the lupine, one of the pea family; the valves have twisted as they separated, thus scattering the seeds. *B*, capsule of the evening primrose; the four valves begin to split at the apex.

Indehiscent fruits are more common than the dehiscent ones. Akenes and grains are examples. (See page 327.) *Berries, stone fruits* (*drupes*), *acorns, nuts, pomes,* and *samaras* are other examples of indehiscent fruits. In such fruits the seed escapes by the gradual decay of the tissue around it, or the sprouting embryo grows out through the wall of the old fruit. Samaras are the familiar winged fruits of many kinds of trees. (See *Figure 146.*) Pomes are produced by many members of the rose family (*Rosaceæ*). Apples and pears are examples of pomes, while peaches, cherries, and plums, which also belong to the rose family, are examples of stone fruits (drupes). In stone fruits the ovary wall ripens into two layers, the outer layer forming the pulp and the inner one the hard coat which surrounds the seed. In pomes, as in the apple, it is the cup-like receptacle which surrounds the ovary that develops into the edible part of the fruit; the ovary itself develops into the core.

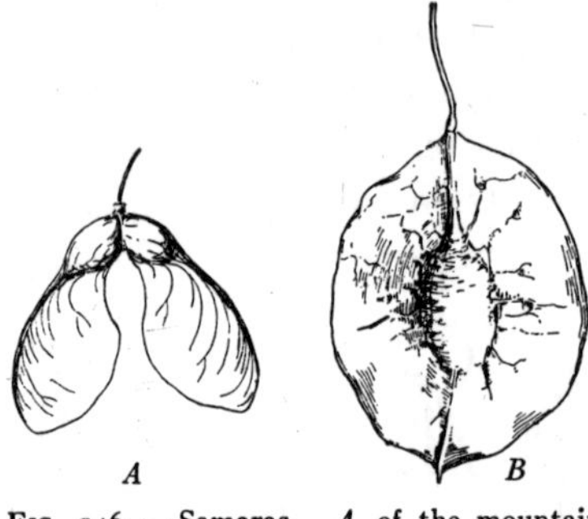

FIG. 146. — Samaras. *A*, of the mountain maple. *B*, of the hop tree.

A berry is a many-seeded fruit in which the whole ovary has become a pulpy mass in which the seeds are embedded. In this scientific sense of the word, some fruits called berries are not berries, while some fruits are berries which are not so called. Thus, in the botanical sense, grapes, currants, gooseberries, and tomatoes, and even oranges and lemons, melons and pumpkins, are to be considered forms of berries. Strawberries and blackberries, on the other hand, are not true berries. The small brown " seeds " on the surface of a strawberry are akenes, while the pulpy mass which bears them is the enlarged receptacle of the flower. Each lobe of a blackberry or raspberry is really a stone fruit, or drupe; it is a mass of pulp surrounding a single seed. In the raspberry the part which we eat slips off of the enlarged receptacle, while in the blackberry we eat receptacle and all.

It is evident that flowers with more than one pistil (that is, with separate carpels) will produce fruit of different character from that produced by flowers with a single pistil. Each carpel or simple pistil may be said to produce a fruit of its own. Blackberries and raspberries are examples of this. Buttercups and hepatica also show several fruits resulting from a single flower. (See *Figure 113 A*, page 290.) The fruit of the tulip tree is a cone-shaped structure composed of many separate carpels which are closely adherent to each other.

D. Dry and Fleshy Fruits. — Fruits may be grouped also into those which are *dry* and those which are *fleshy*. Dry fruits may be either dehiscent or indehiscent, but fleshy fruits are generally indehiscent. The banana is an exception to this. It is a fleshy fruit which dehisces; the wild banana peels itself when it grows old. Those plant products called fruits in the market belong to the fleshy fruits. Apples, pears, grapes, melons, oranges, bananas, and grapefruit are examples. Dry fruits, though not so called, are even more important commercially than the fleshy fruits, for they include wheat, rice, and corn.

E. Inflorescence Fruits. — You have noted that one flower, if its carpels are separate, may produce a number of fruits, as in the case of hepatica. (See *Figure 113 A.*) It is also to be noted that one fruit may be produced by a number of flowers, as in the case of the fig. (See page 318.) The pineapple is another example of a fruit which results from the enlargement of a whole flower cluster, including in this case the stalk and the bracts as well. A pine cone is also a fruit which results from many flowers, bracts, and the stalk which bears them.

F. Self-planting Fruits. — Certain fruits have the interesting habit of planting themselves. The fruit of the peanut, which is a pod, is an example of this. The flowers of the peanut plant are borne under the leaves, near the ground. After fertilization, the flowerstalks begin to grow again, and downwards. They thrust the young pods beneath the soil, and there the seeds ripen and germinate. (See *Figure 147*.)

FIG. 147. — Peanut plant showing the downward growing stems which thrust the pods when young beneath the soil. The roots bear numerous tubercles. See page 377.

The fruit of the porcupine grass (*Stipa spartea*) plants itself after it is separated from its parent. It bears a long, spirally-twisted appendage, the awn. (See *Figure 148*.) This awn absorbs moisture at different rates in its different parts. This causes it to twist and untwist as atmospheric conditions change. The twisting process tends to drive the heavier, sharp-pointed part of the fruit down into the soil, and the upward pointing bristles prevent it from being withdrawn when untwisting occurs. These fruits bore their way through paper or cloth. They will make their own way out of a pocket or envelope in which they may be placed, and they have even been found deep in the flesh of grazing animals. The fruit of the stork's-bill (*Erodium*) has similar powers.

G. Seedless Fruits. — In some plants fertilization is not necessary for the production of fruit. You have noted this in the case of fig, in which not even pollination is necessary. (See page 318.) Usually fruits so produced are said to be *seedless*, though the rudiments of seeds are commonly found in them. Seedless oranges are familiar. The cultivated banana never produces seeds. Certain varieties of grapes, apples, pears, and gooseberries are also seedless. The seedless condition of the fruit may be only an occasional condition, as in the seedless persimmons found growing wild. The seedless orange and other varieties of cultivated seedless plants can be propagated, of course, only by means of cuttings from the parent plant. Seedless varieties have, in some cases, been produced by selecting forms with few and small seeds, and using in propagation only those forms showing the greatest tendency toward seedlessness.

There is a parasitic flowering plant (*Balanophora*) which, without either pollination or fertilization, produces fruits containing seeds which germinate.

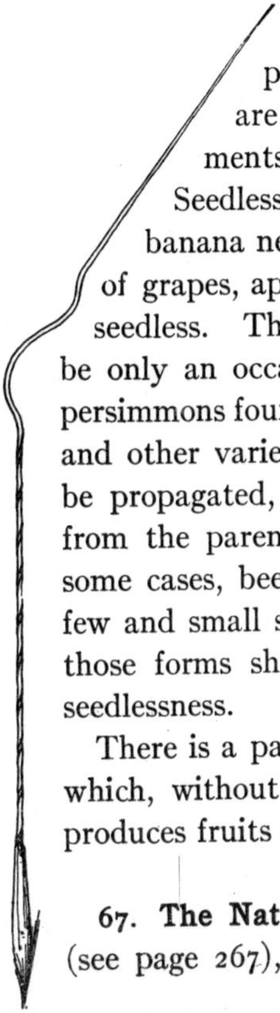

Fig. 148. — Mature fruit of the porcupine grass (*Stipa spartea*). Note the spirally twisted awn and the upward pointing bristles behind the lower apex of the fruit. See context.

67. The Nature of Seeds. — As previously stated (see page 267), the seed is one of the most complex of plant organs, and its nature cannot be fully comprehended until those structures of lower plants are studied which indicate its evolution. It is an enlarged and matured ovule. After fertilization, the ovule, containing the developing embryo, grows to a certain size. Upon attaining this size, the outer part of the ovule (integument) develops into the seed coat or testa, and the growth of the young plant within is checked. It is the appearance of the testa which marks the transformation of the ovule into the seed.

The size of seeds varies exceedingly among different kinds of plants, but it is strikingly uniform in any single species. In fact, in plants of the same kind, no other organs exceed fruits and seeds in uniformity, both as to size and as to structure.

The size of seeds is significant in connection with germination. If the seed is small and poor in food, the young root must quickly find soil water and the shoot must quickly reach the sunlight. The nursing tissue alone cannot sustain them long. If the seed is large and rich in food, the little plant has a better chance. Yet most plants come from small seeds. Small seeds are much cheaper for the parent to produce and are produced in far greater numbers than large seeds. They are much more successful in dissemination than large seeds are. Their disadvantage as to food supply appears to be more than offset by their advantages in respect to numbers and distribution.

A. The Importance of Seeds to Mankind. — Think of what seeds mean to men. Think how different human history would be if it had not been for the evolution of the seed habit. It is the nourishment stored in seeds which is our main source of food. The seeds of rice and corn and wheat are the " staff of life " for human beings. And if it were not for seeds, how would the millions of acres of the farms be brought to harvest? Not only are seeds of great advantage to plants in nature as a means of reproduction; they are also a great advantage to man as a means of agriculture. They are absolutely necessary to it. Without seeds, the cultivation of plants as it is to-day would come to a full stop. Without seeds the life of all animals would be imperiled, and most of the plants which we see to-day would disappear. Seeds thus form a necessary link in the chain of life. Food, whose manufacture began in the leaves, comes to be stored in them. To us they are more than the harvest of this season. They are the assurance of harvest in the seasons which are yet to come.

B. The Importance of Seeds to Plants. — Many flowering plants reproduce their kind abundantly by means of structures other than seeds. You have noted that horizontal stems are especially efficient in this direction. (See page 157.) Reproduction of this kind is especially characteristic of the grass family. Bulbs are especially characteristic of the lily family. Duckweed and bamboo are successful plants which may go for many years without producing flowers, while many cultivated plants are propagated successfully without the aid of seeds.

Most flowering plants, however, depend mainly or exclusively upon seeds for their perpetuation. Consider that extremely large class of plants which are annual, those which die down at the approach of winter. Throughout the world, wherever winter comes, these plants cease to exist for several months except as seeds. Yet, when spring comes again, they grow up by the million, convincing evidence of the ability of their seeds to protect the tender embryos within them. Many of our most common weeds are plants of this character.

Most trees spread only through the agency of seeds. Pines, for example, have no other means of propagation. Yet the survival of trees does not depend upon seeds to the degree that the survival of annuals depends upon them. In order that annuals may survive, *every* crop of seeds must be successful, while trees, if a seed crop fails, have other chances of success in later seasons.

C. Protection of and by Seeds. — It is evident that embryos have much need of protection. They themselves are delicate structures with slight capacity to endure unfavorable conditions. The seeds which contain them are, however, as a class, the most resistant of all plant structures. They resist drought, cold, and heat as no other plant parts do. Moisture is the thing which is most likely to rob them of their vitality, for moisture starts sprouting, and then, unless the young plant is surrounded by conditions favorable to its continued growth, its life is soon ended. Farmers long ago learned to take particular pains to keep dry those seeds which they saved for planting.

Young seeds are protected from excessive transpiration and other dangers by the pericarp, which is the hardened ovary wall. Edible fruits, when young, are usually sour, bitter, or hard, and this is a means of protection for the immature seeds within them. The pericarp, in some cases, produces spines, such as are seen on the gooseberry, the chestnut, or the prickly pear. Protection by the pericarp is particularly noticeable in nuts. The hulls of walnuts and hickory nuts are the enlarged and fleshy calyx of the pistillate flowers, but the bony, hard part is the pericarp. The kernel itself is the seed.

Protection of the embryo by the seed is principally

afforded by the testa, the compactness of the tissues within the testa being an additional protection of minor importance. If there are two integuments, it is the outer one which forms the testa. (See page 276.) At maturity the testa is usually hard and bony, being composed of layers of cells with greatly thickened walls. In most one-seeded fruits, as in grains and akenes, the pericarp closely surrounds the seed and, in such cases, it, rather than the testa, affords the chief protection.

D. Vitality of Seeds.— Some seeds are able to germinate and produce good plants many years after their formation, and after long exposure to conditions which ordinarily put a stop to life very quickly. Nothing shows the capacity of seeds to protect the embryos within them so well as their length of life and their power to endure conditions unfavorable to life.

Some seeds, those of willow for example, die unless they germinate almost immediately, but short-lived seeds are the exception rather than the rule.

Certain popular stories about the length of life of seeds are without foundation. Such, for example, is the story of mummy wheat. It has been widely told and believed that wheat found in the wrappings of ancient Egyptian mummies was successfully sprouted. There is no doubt that the mummies were many hundreds of years old, but there is also no doubt that the wheat was much younger than the mummies. It was put there to fool people. The facts, however, are wonderful enough without exaggerating them. Probably the longest-lived seeds are among those produced by the pea family. There is good evidence that some such seeds may retain their vitality as long as two hundred years. The seeds of water lilies and of some of the mint family have been known to exceed a century in their life.

Equally astonishing with the longevity of seeds is their ability to withstand extremes of temperature. A striking experiment which shows this was the following. Seeds of alfalfa, mustard, and wheat were used. Their seed coats were perforated before the experiment began. They were then thoroughly dried for six months, placed in a vacuum for a year, subjected for three weeks to a temperature of $-190°$ C., and for three days to $-250°$ C. After that, they were successfully germinated.

One almost wonders whether seeds which can endure such very unfavorable conditions might not, under favorable conditions, live forever. There is no satisfactory explanation of how life is maintained under such conditions. To say that the embryo is in a state of suspended animation simply expresses our ignorance of the subject, yet that is about all we can say.

The vitality of seeds is a very important matter to the farmer. He needs to know that the seed he sows is sure to sprout. Careful farmers test their seed by actually sprouting a number of samples before the sowing is done. More than ninety per cent of the samples should show good sprouting power if the seed is to be considered good.

E. Structure of Seeds. — Necessarily, to understand the structure of seeds, the structure of ovules must be understood. In the chapter on the flower you learned of the nucellus, the integument, and the micropyle. (See page 276.) The relation between the integument and the testa you have just noted. The micropyle may usually be observed in mature seeds; usually, as in the bean, it looks like a hole made with the point of a needle. The nucellus disappears. Its place is taken by the tissue called endosperm, or, in the absence of endosperm, by the enlarged cotyledons. (See page 66.) Cotyledons are sometimes called seed leaves. At first they do not have the appearance of

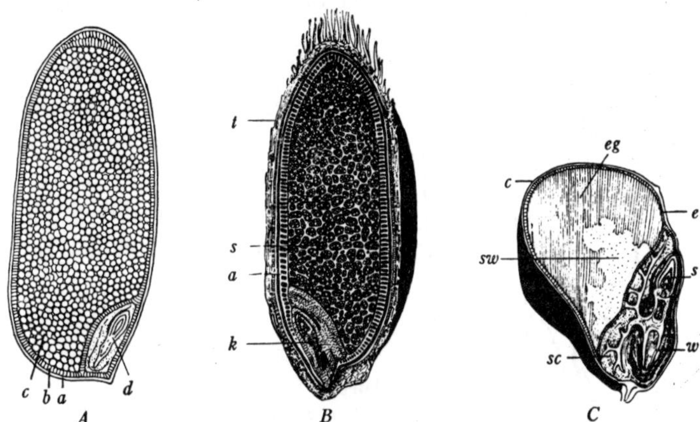

FIG. 149. — Diagrams of lengthwise sections of the seeds of rice, wheat, and corn, the three most important sources of the food of man. *A*, rice; *a*, the seed coat; *b*, the aleurone (protein) layer of the endosperm; *c*, the starchy part of the endosperm; *d*, the embryo. *B*, wheat; *k*, the embryo; *a*, the gluten (protein) layer; *t*, layers of the husk which is partly pericarp and partly testa; *s*, the starchy part of the endosperm. *C*, corn; *c*, the seed coat; *e*, the aleurone (protein) part of the endosperm; *eg*, the yellowish, hard part of the endosperm which contains both protein and starch; *sw*, the whitish, starchy part of the endosperm; the region which contains the embryo is rich in both oil and proteins; the cotyledon is extended into a branched, absorbing organ, *sc*; *s*, the plumule; *w*, the radicle.

true leaves, but, as to origin, they are leaves, the first leaves of the embryo. In some cases the cotyledons, after germination, turn green and manufacture food for the seedling, as well as contributing that which was manufactured in the parent plant and stored in them. (A seedling is a young plant which is still drawing nourishment from the

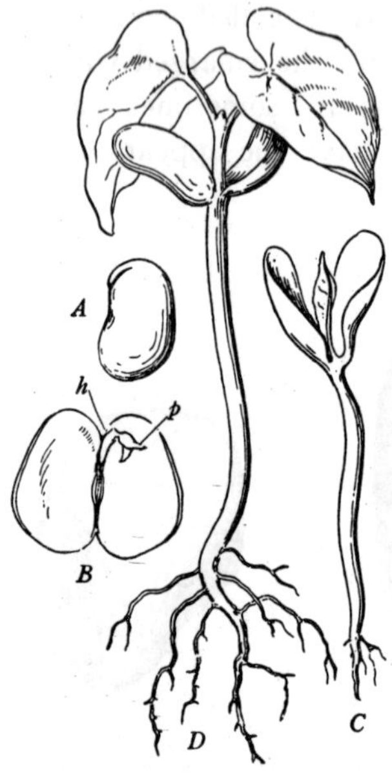

FIG. 150. — Embryo and seedlings of bean. The two large cotyledons entirely fill the seed; there is no endosperm. *A*, embryo removed from the testa; the colorless testa slips off easily after beans have been soaked in water. *B*, the embryo split open; it separates readily into the two cotyledons, revealing the rest of the young plant lying between them; *h*, hypocotyl; *p*, plumule. *C*, young seedling, showing cotyledons, hypocotyl, plumule, and radicle; the hypocotyl is the part of the embryo below the attachment of the cotyledons; the plumule is the bud between the cotyledons; it is sometimes called the epicotyl; its outer part unfolds and forms the first true leaves; the root part of the embryo, indicated by the little rootlets arising from it, is called the radicle; it is a part of the hypocotyl. *D*, an older seedling showing the first internode (above the cotyledons), and the first true leaves; nourishment is still being drawn from the cotyledons.

seed.) In some cases, as in corn, the cotyledon may develop into an absorbing organ, drawing nourishment from the rest of the seed. (See *Figure 149*.)

Usually a scar may be found on the surface of a seed indicating the point at which it was attached to the ovary. This scar may be readily found in the bean. It is called the *hilum*.

When the seed is mature, the embryo is usually differentiated into the following parts: *cotyledons*, *hypocotyl* (the word means *below the cotyledons*), and *plumule* (or *epicotyl*, that word meaning *above the cotyledons*). (Study *Figure 150* and its explanation.)

F. Cotyledons and the Divisions of Seed Plants. You have noted that one of the principal characters which determines the division of angiosperms into two great groups is the character of their embryos as to the number of cotyledons. This is the character which gives name to these great groups. One is called the monocotyledons, the other the dicotyledons. (See page 183.) The two cotyledons of dicotyledonous plants are lateral organs, that is, they arise from the sides of the embryo, below its apex, one on each side. This becomes evident when the embryo enlarges. (See *Figure 150.*) The one cotyledon of monocotyledonous plants is, on the other hand, a terminal organ. It arises at the apex of the embryo; in fact, it *is* the apex of the embryo, and, naturally enough, is but one in number.

The embryos of monocotyledons are, as a rule, much smaller than those of dicotyledons. Their cotyledons are never enlarged and rich in food like those of a bean. The embryo usually lies at one end of the seed, while the much larger endosperm fills the rest of it. (Study *Figure 149.*)

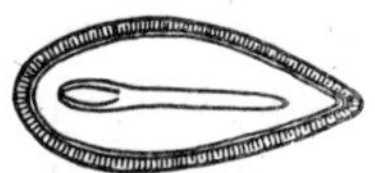

Fig. 151. — Diagram of lengthwise section of seed of pine. Note that the embryo has several cotyledons. It lies embedded in the endosperm. Note the thick testa.

The embryos of all but one of the families of gymnosperms (see page 183) have two cotyledons. Most of the members of that family (*Pinaceæ*) have embryos with several cotyledons, and it happens that our commonest gymnosperms belong to that family. (See *Figure 151.*)

G. Seeds and Foods. — In the chapter on leaves you learned that foods, chemically considered, are divided into three great classes, the carbohydrates, the proteins, and the fats. You learned that the products of photosynthesis are of the carbohydrate class, and that, by use of these products, and the addition of other substances, all other foods are built up. You learned something of the nature of carbohydrates, starch and sugar being described as examples of this class of substances. (See page 230.)

Further consideration of the classes of foods was postponed till now, for the reason that it is in seeds that the most conspicuous accumulation of foods occurs. Foods are present in all living parts of plants, and often they are conspicuously stored in stems and roots, but it is the seeds that are more generally filled with food than any other plant organ.

Life is best nourished by the use of all three kinds of foods. This you have probably already learned in your study of the human body. Both carbohydrates and proteins are essential to life, and, under certain conditions, fats are also necessary. Without proteins, protoplasm, the living stuff, cannot be manufactured.

Proteins, unlike the fats and carbohydrates, contain nitrogen. Nitrogen is absolutely essential to their formation. Evidently, then, nitrogen is an element of great importance both to plants and animals; their life is impossible without it. Nitrogen is abundant. Four fifths of the air is composed of it. But very few forms of life, only some of the smallest and simplest plants, are able to make use of this atmospheric nitrogen. Green plants obtain their nitrogen, not from the air, but from the soil. They get from the soil certain soluble compounds called *nitrates*. Molecules of nitrates are absorbed by root-hairs along with the water which they take in. Atoms of nitrogen are present in the molecules of nitrates. It is largely in order to maintain the supply of nitrates in the soil that farmers use fertilizers. Manure, for example, is rich in nitrates and ammonia, both of which contain nitrogen.

In that part of the book which deals with the lower forms (Chapter X) you will learn of certain microscopic plants that live in the soil and are very important in keeping up the supply of nitrates. This is because they are able to use the nitrogen of the air which penetrates the surface of the soil. From it and other common substances they manufacture nitrates which are used by the higher plants.

Proteins, then, are *nitrogenous* foods, while fats and carbohydrates are *non-nitrogenous* foods. Fats differ from carbohydrates in that they contain a much lower percentage of oxygen than the latter do, and a much higher percentage of carbon. They are like carbohydrates, however, in that carbon, hydrogen, and oxygen are the three elements of which they are composed. Just as carbohydrates usually exist in plants in the form of sugar and starches, so fats usually exist in plants in the form of oils. Very small drops of oil are frequently found in plant cells, and in certain tissues they are of regular occurrence. Proteins exist in various forms. Like starch, they are insoluble, a property which is evidently desirable for a storage form of food.

The *gluten* of wheat and the *aleurone grains* of corn and rice are almost pure protein. It is common for the outermost layer of the endosperm of grains to be particularly rich in proteins. (See *Figures 149* and *154*.) It is the gluten in them which makes the grains of wheat form a gummy mass when chewed. It is the gluten in wheat flour which prevents the gas which arises from the yeast from escaping, and so gives lightness to wheat bread. Since corn does not contain gluten, light bread cannot be made from corn flour.

Of all the kinds of foods, it is starch which is accumulated in the greatest abundance, both in seeds and in other plant organs. Potatoes are filled with starch, and the bulk of the seeds of grain is made up of it. (See *Figure 149.*) Starch occurs in the form of grains. These are microscopic structures marked by concentric rings which are rings of growth, in this respect being somewhat like the rings in wood. (See *Figures 152* and *153*.) Small starch grains are sometimes found in green leaves. Their presence indicates that the starch has been manufactured more rapidly than it has been transformed into sugar. Usually it is transformed into sugar almost as fast as it is manufactured, and in that soluble form the carbohydrate food moves through the plant body. The starch grains begin to form within the chloroplasts.

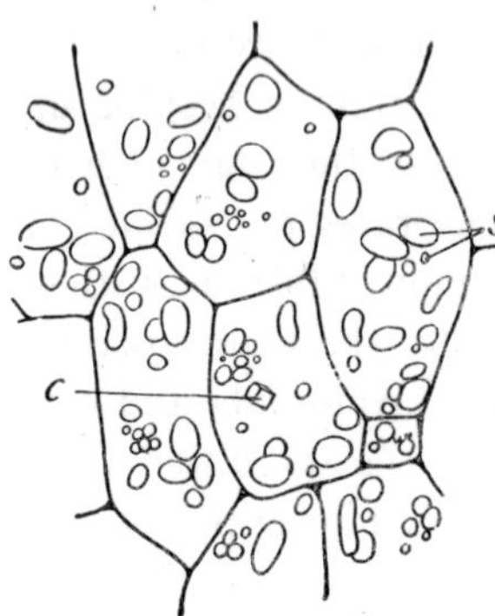

FIG. 152. — Cells of a potato showing starch grains, *s*; *c* is a protein crystal.

Many seeds are rich in fat in the form of oil drops. Such are the seeds of the castor bean, of cotton, and of the sunflower. From all of these oil is extracted for commercial purposes.

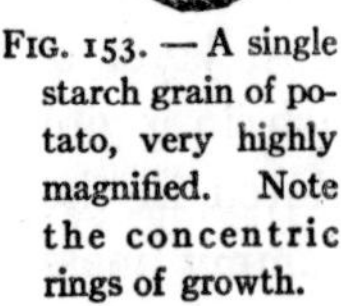
FIG. 153. — A single starch grain of potato, very highly magnified. Note the concentric rings of growth.

A third form of non-nitrogenous food is the substance of which cell walls is composed. That substance is cellulose. Like starch and sugar, it is one of the carbohydrates. Wood, chemically speaking, is cellulose. It is composed chiefly of the walls of dead cells. The cellulose of most plant tissues is not ordinarily used as food, but in some seeds the cell walls of the endosperm as well as of the testa become very thick. This thickening is composed of what is called *reserve cellulose*. (See *Figure 155*.) Upon germination this reserve cellulose is used as food for the seedling. The seeds of persimmons and dates have much food stored in this way. Reserve cellulose gives exceptional hardness to the endosperm; seeds which have much of it are said to be *horny*.

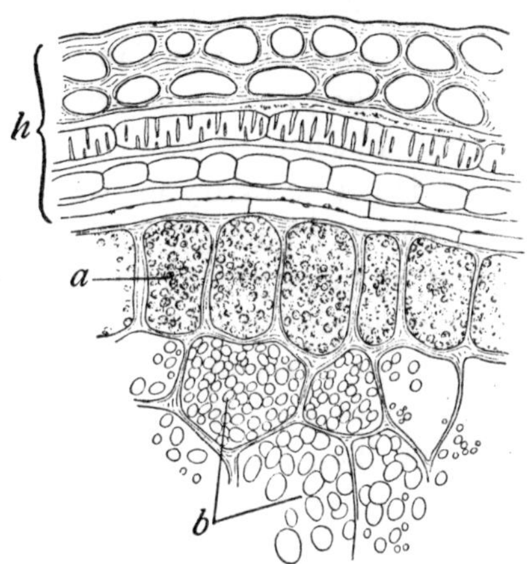

FIG. 154. — Cross section through the outer part of a grain of wheat, much magnified; *h*, the husk whose outer part is pericarp and whose inner part is testa; *a*, the gluten or aleurone layer whose cells are filled with grains of protein; *b*, a partof the starchy tissue which makes up the body of the grain. — *After Cobb.*

In nearly all seeds both nitrogenous and non-nitrogenous foods occur, but the latter are always present in greater amount than the former. Usually, in any one kind of seed, one kind of non-nitrogenous food is present in large excess over all other kinds of food. On this basis seeds may be classed as *starchy*, *oily*, or *horny* as the case may be. The seeds most extensively used for food, including the grains, are starchy seeds.

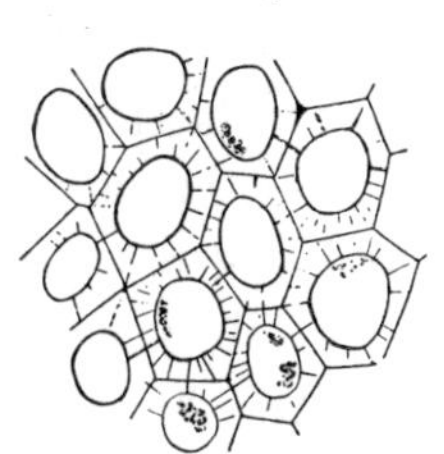

FIG. 155.—Section through part of the endosperm of a persimmon seed. Note the greatly thickened walls of reserve cellulose.

H. Germination of Seeds. — Germination of seeds is, as you noted, the renewal of growth by the embryo. If certain conditions are present, this renewal of growth occurs in all good seeds. The conditions necessary for it are seed maturity, abundant moisture, suitable temperature, and the presence of oxygen.

Light, which is so essential to the later stages of green plants, is not necessary for the germination of their seeds. As you have noted (page 86), the rate of growth is generally greater in the absence of light than it is in its presence, and this applies to the first stages of the seedling as well as to other parts of the plant. For the gradual transformation of a seedling into an independent plant, however, the illumination of its food-making parts is essential. It is their inability to obtain proper illumination which is one of the principal causes of the very high death rate among seedlings. Perhaps you have noted the thousands of tree seedlings which start up each spring in a forest. Hardly one in a thousand lives even into the second season. It is the shading by older plants which is the principal cause of their death.

a. Water Cultures. — Soil is not necessary for germination nor is it even necessary for the later growth of the plant. It is entirely possible to grow plants without soil. They may be brought to maturity, even to the maturity of flowers and fruit, by means of what are called *water cultures*. With their leaves in the light, and their roots in water that contains in solution those substances the plant needs, some plants seem to do nearly as well as when growing under natural conditions. Corn, by this method, has been matured through several generations without coming in contact with a single grain of soil.

b. The Nutrient Salts. — Of those substances that must be placed in the water of water cultures, you have just previously noted (page 344) that nitrogen is one; there must be present some soluble compound of whose molecule nitrogen is a part. Other elements besides nitrogen which are used in the manufacture of protoplasm are sulphur and phosphorus. They, like nitrogen, must be presented to the plant in the form of *available* compounds; that is, as a part of molecules which, by their solubility for one thing, are available for plant use.

You noted (page 344) that the principal kinds of nitrogen-containing compounds in which the nitrogen is available for plant use are compounds of the class called nitrates. Nitrates are compounds of nitrogen, oxygen, and of one other of various elements of a group called *bases*. Calcium, magnesium, potassium, and sodium are the bases of those nitrates which are of common occurrence in the soil, and these other elements are taken into the plant along with the nitrogen.

Nitrates belong to that chemical class of substances known as *salts*, and it is common to speak of the "nutrient salts of the soil," meaning those salts which are necessary to plant life. It is nitrogen- and phosphorus-containing salts that the fertilizers used by farmers principally supply. What general use the plant may make of the calcium, magnesium, potassium, and sodium salts, as well as using them as incidental to getting nitrogen, is not well understood. It appears, however, that for some plants at least they are quite necessary for their own sake as well as for the nitrogen which their compounds contain.

Iron also is necessary to the plant. You have noted (page 232) that in the absence of iron chlorophyll is not manufactured.

c. Digestion in Seeds; Enzymes. — It is in connection with the germination of seeds that digestion in plants may be best considered. You know that digestion is just as universal a process among plants as it is among animals; food must be transformed from solid to liquid before it can be used in the cells of either plants or animals (see page 42).

To understand digestion it is necessary to know something about *enzymes*, and of these you may have heard in your study of the human body. Enzymes are present in the saliva, in the gastric juice, and in the intestines. They are substances that cause other substances to change; they are the agents of digestion, which cannot occur without them; they change insoluble substances to soluble ones, and they may change insoluble ones back into solution again. It is an enzyme, for example, which causes starch to change to sugar, and the same enzyme may reverse its action and change sugar back again to starch.

With all the changes of substances from solid to liquid and back again which are constantly going on in plant life, it is evident that enzymes have a very important part to play. Enzymes are manufactured by the protoplasm.

One of the properties of enzymes is that in causing other substances to change they do not, apparently, change themselves. This being the case, a very small amount of an enzyme is able to cause changes in very considerable amounts of other substances. Many enzymes, so small in amount that they could not be separated out from the tissues in which they work, were recognized first by the changes which they caused. Though they themselves may not be evident, the work which they do is very evident. Thus, in the case of the coconut, during germination much oil is changed to sugar. This is doubtless due to the action of an enzyme, but no one has as yet been able to isolate the enzyme which does this. If any one did so, he would probably have in this enzyme the basis for an enormously useful digestive medicine, for the transformation of fats is one of the things which very frequently causes trouble in human digestion.

The first step in the germination of seeds appears to be the absorption of water. This water causes the cells to enlarge. The seed swells. The next step is the digestion of food; that is to say, it is the action of an enzyme upon the food stored in the seed with the result that this food is made soluble and moves into the growing parts of the embryo.

This process may be readily studied in germinating grains, as in wheat. The aleurone layer (see *Figure 154*) first shows signs of life. The protoplasm in it produces an enzyme called *diastase*, which is the kind of enzyme most commonly found in seeds. It is one which has the power to transform starch to sugar. It is the starch next to the aleurone layer which, in wheat, is the first to be changed. Digestion after that proceeds rapidly, and liquid food is supplied to the seedling as fast as needed.

d. Assimilation.— In germinating seeds the dissolved foods pass by osmosis from cell to cell until they reach those cells which are growing. Here the protoplasm is very active, and here the foods must be built up into new protoplasm. It is this transformation of foods by protoplasm into new protoplasm, or into other substances, which is called assimilation. Assimilation involves numerous and intricate chemical changes very few of which are understood. One of the simplest of them is the transformation of liquid food into cellulose, the cell-wall material, which, as you noted, is a carbohydrate. The most intricate of all kinds of assimilation is the transformation of food into protoplasm itself, and this building up of the living substance is not understood at all.

e. Mechanics of Germination. — The first external evidences of germination are the swelling of the seed, the breaking of the testa, and the emergence of the embryo. Usually small seeds germinate more quickly than large ones; this seems to be due to the fact that water and oxygen reach all parts of them more readily than is the case with larger seeds. Starchy seeds germinate more quickly than oily ones. The digestion of starch is a more simple process than that of fat.

In some seeds there are thin spots in the testa through which the embryo emerges. Germination may often be hastened by cutting away pieces of the testa and so making easier the entrance of air and water and the exit of the young plant.

Usually the tip of the hypocotyl is the first part of the embryo to appear outside the seed. The advantage of this is evident. The hypocotyl is the water-absorbing part and all the water must come from without. Of food, on the other hand, there is enough stored within for some time, so that the emergence of the food-making parts is not so pressing a need.

The cotyledons may appear above the soil or they may not. Germination in which they do appear above the soil is said to be *epigean*, a word meaning *above the soil*. When the cotyledons remain below the soil, germination is said to be *hypogean*. Pumpkin, mustard, beech, and some kinds of beans furnish examples of epigean germination.

The seedlings of all cereals and of many other plants are hypogean.

In many epigean species, as in the beech, squash, and mustard, the cotyledons as well as the rest of the embryo increase in size at germination. They turn green and manufacture considerable new food before they wither and disappear.

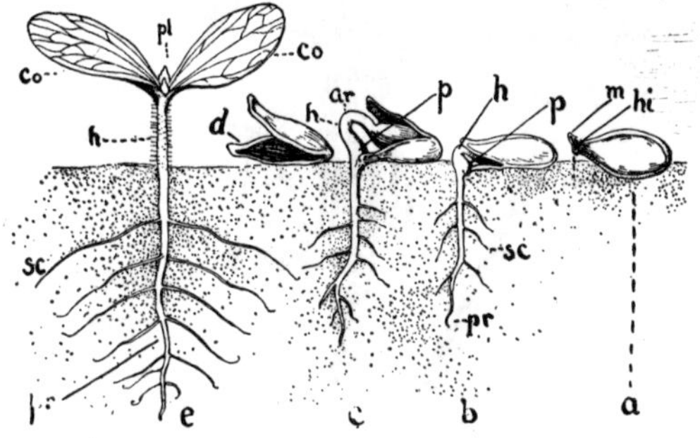

FIG. 156. — Development of a squash seedling; *a*, mature seed; *m*, micropyle; *hi*, hilum; *b*, *c*, and *e*, stages of the seedling; *h*, the hypocotyl; *p*, an outgrowth of the hypocotyl called the peg; *ar*, the arch of the hypocotyl; note in this stage that the peg is pressed firmly against the lower half of the testa furnishing thereby a resistance which enables the arching growth of the hypocotyl to draw the rest of the embryo out from the seed coat; *d*, the empty testa (seed coat); *pl*, plumule; *co*, cotyledons; *pr*, the tap-root; *sc*, branches. — *After Andrews.*

Some seeds have special structures which are of mechanical help in the withdrawal of the embryo from the testa. This is well illustrated in the squash. (Study *Figure 156.*)

68. Agencies of Seed Dispersal. — The principal agencies of seed dispersal are wind, water, and animals. Among animals, man and all the various means of transportation operated by him are to be reckoned.

A. *Wind.* — Of all agencies of seed dispersal, wind is by far the most important for the largest number of plants.

You have noted the various and common structures of seeds and fruits which favor their transportation through the air. (See *Figures 143, 146,* and *157.*) The fruits of many *Compositæ* and of many successful trees furnish the most conspicuous examples of this type of dispersal.

FIG. 157. — Ripe fruits of dandelion. Note the tufts of hair which act like parachutes and enable these fruits to travel far on even the slightest breeze. — *After Kerner.*

The catalpas, the maples, the elms, and most members of the pine family have winged fruits or seeds. (See *Figures 159 A,* and *146.*) The wings borne by seeds and fruits of many tropical trees are of remarkable size, one very successful family of tropical trees receiving its name (*Dipterocarpeæ,* meaning *winged seeds*) from this character. (See *Figure 158.*)

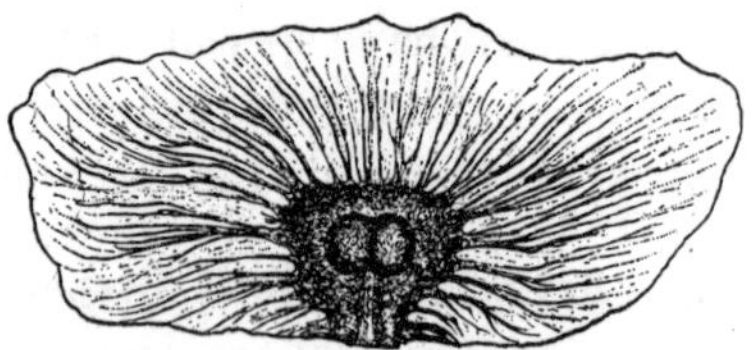

FIG. 158. — Winged seed of *Bignonia,* a tropical tree. — *After Strasburger.*

You have noted the tumbleweeds (page 327); recall also that it is from light seed hairs, structures favorable for wind dissemination, that cotton is obtained.

The dispersal of seeds by wind, like the dispersal of pollen by wind, involves enormous waste. The chance that the wind will drop seeds on "good ground" is, of course, better than the chance that it will drop pollen on a stigma, yet the wind-blown seeds that perish far exceed in number those that survive.

Despite its wastefulness, wind is by far the greatest sower of seeds in nature. It is also by far the greatest sower of spores, which are one-celled reproductive bodies. Spores are especially conspicuous in ferns and mosses. The first plants to appear on the new islands which resulted from the great volcanic eruption of Krakatao in 1883 were wind-sown plants. They were forms of ferns and of other non-seed plants whose abundant, light spores the wind carries far better and farther than it carries even the lightest seeds.

B. Water.—Lakes and oceans and running streams are all agencies by which wide scattering of seeds is accomplished. Some seeds are especially resistant to the entrance of water and so are able to remain in water for a long time without sprouting.

In such a great drainage basin as that of the Mississippi Valley the dispersal of seeds by the streams has much to do with the spread of the same kinds of plants over wide areas. Plants growing near the headwaters of the Ohio, Missouri, or Mississippi may gradually spread, by the water-carriage of their seeds, through the entire system traversed by the water which flows down from their original home.

Even more far reaching is the dispersal brought about by ocean currents. Seeds are carried along the coasts, from island to island, and even from continent to continent, by the ocean currents. Darwin estimated that about one kind of seed in ten is able to retain its vitality in sea water as long as a month. This, at the usual rate of movement of ocean currents, might mean a successful journey of over a thousand miles.

There are many floating fruits, however, which soon decay in water, especially in salt water. The coconut, often seen floating in tropical seas, loses its vitality in the water in a few days. There are some plants of the tropical beaches, however, whose seeds have been found to be uninjured after floating in rough, salt water for 143 days.

C. Animals. — The distribution of seeds and fruits by animals may be said to be of two kinds, voluntary and involuntary. Of the fruits distributed involuntarily, those with hooked appendages are most familiar. (See *Figure 159.*) Besides burdock and cocklebur, many other common weeds use this method of distribution.

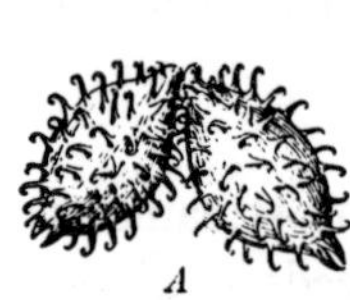

FIG. 159. — The fruits of cocklebur (*A*) and burdock (*B*) showing the hooked appendages by which they become attached to the coats of passing animals. — *After Kerner.*

Of the edible fruits, some, on account of their edibility, are of much service in placing seeds favorably for germination, while for others edibility appears to be a loss rather than a gain. It all depends upon whether the seeds as well as the pulp of the fruit are digested.

The seeds which pass from the body of animals uninjured are quite likely to be deposited in conditions favorable to the germination and later growth of the young plant.

FIG. 159 *A*. — The seasonal changes in a twig of elm. The uppermost picture shows the condition of the buds early in April. The second picture shows the clusters of flowers. The flowers of elm are monoclinous, but the pistil is not conspicuous at the time the anthers are ripe; in the picture, the prominently exposed stamens are quite evident. Pollination is by wind. The third picture shows the condition of the green, winged fruit late in May. The young leaves have not yet reached full size. In the fourth picture (September) the leaves are mature. The last picture shows the twig late in October; note the condition of the winter buds. — *Redrawn from Bailey.*

Animals of the same kind tend to frequent similar regions. Seed is likely to be dropped in the kind of place in which its parent grew. This applies to seeds carried involuntarily as well as to those eaten. Thus, for example, wading birds (see page 64) fly from swamp to swamp, and grazing animals keep largely to one kind of land.

The most useful animals in seed dispersal (not considering man and his planted crops) are such birds as the robins, thrushes, and blackbirds. They eat fleshy fruits in abundance and through their alimentary tracts the seeds pass uninjured.

QUESTIONS AND SUGGESTIONS

Section 65. 1. State the general functions of fruits and seeds. 2. Explain why seed "germination" is not true germination. 3. Contrast, as to advantages, small seeds with large seeds. 4. Upon the basis of the part of the plant transported, what are three kinds of seed dispersal? Give examples. 5. Describe akenes and grains, defining pericarp and testa.

Section 66. 1. Discuss the general nature of fruits.

A. 2. Explain three principal ways in which fruits aid in the dispersal of seeds.

B. 3. Explain the difficulties in classifying fruits.

C. 4. Define dehiscent fruits, giving two examples. 5. Define indehiscent fruits, giving six examples. 6. Define the term berry, and give examples.

D. 7. Compare dry and fleshy fruits as to commercial importance.

E. 8. Explain what is meant by inflorescence fruits and give examples.

F. 9. Describe the structure and behavior of the fruit of porcupine grass.

G. 10. Discuss seedlessness in fruits.

Section 67. 1. Discuss the general nature of seeds.

A. 2. Discuss the importance of seeds to man.

B. 3. To what class of plants are seeds of chief importance? Why? Explain why they are of less importance to plants of other kinds.

C. 4. Discuss the means of protection afforded by seeds and fruits.

D. 5. Discuss vitality of seeds, indicating the minimum of vitality desirable in seeds used for planting.

E. 6. Describe the structure of seeds. 7. Describe the structure of seed-plant embryos, defining hypocotyl, plumule, and seedling.

F. 8. Contrast the embryos of monocotyledons, dicotyledons, and pine.

G. 9. Discuss the source of nitrogen and its use by plants. 10. Contrast the two classes of non-nitrogenous foods. 11. Describe the gluten of wheat. 12. Describe the occurrence of starch in plants. 13. Discuss cellulose and its occurrence. 14. Classify seeds on the basis of food content and give examples.

H. 15. Present the conditions necessary to germination, and the relation of light to the life of seedlings. 16. Discuss water cultures. 17. What are the common bases of the nitrogen salts which plants absorb? Discuss the uses of these bases. 18. Describe enzymes and discuss their use, giving examples. What is diastase? 19. Discuss assimilation. 20. Describe the visible steps in germination. 21. Distinguish between epigean and hypogean germination, giving examples.

SECTION 68. *A.* 1. Discuss wind as an agency of seed dispersal, and structures which favor such dispersal.

B. 2. Discuss the value of water in dissemination, giving examples.

C. 3. Give examples of dissemination by animals, and of structures favorable to it.

CHAPTER IX

THE NON-VASCULAR PLANTS

69. **Introductory.** — We have been considering seed plants exclusively. It is time to consider also the plants which do not bear seeds. In doing so we shall follow what is called the evolutionary sequence; that is, we shall begin with plants whose structure is very simple and proceed from them, as evolution proceeded, to those whose structure is complex. We shall go from the *lowest* plants to the *highest*.

All plants may be grouped into those which have a vascular system and those which do not. It is quite certain that those without a vascular system preceded those which have it; those which have it evolved from forms which were without it. So we will begin by considering the *non-vascular* plants.

Since wood is a part of the vascular system, the non-vascular plants are, of course, not woody. They are all soft plants of little height. Mosses and mushrooms are examples of the non-vascular plants, while ferns and seed plants are examples of those which are vascular.

70. **The Four Great Divisions.** — The plant kingdom is divided into four great groups. Those plants which compose the two lower or simpler of these divisons have no vascular system; those which compose the two higher or more complex divisions are plants in which we find a vascular system.

The great majority of familiar plants belong to the highest of these four divisions. Because these are the most familiar plants, the first part of this book deals exclusively with plants of this division, the division of seed plants. But you would have a very imperfect idea of plant life if you knew nothing about the plants belonging to the other divisions. Also you would have a very imperfect idea of the seed plants themselves if you understood nothing of the lower plants, for the seed plants evolved from ancestors which were like some of the lower plants of to-day.

The four great divisions are the following: —

(1) *Thallophytes.* — These are the simplest plants and so are regarded as the lowest from the standpoint of evolution. The word means *thallus plants.* Thallus means a simple kind of plant body. It is not differentiated into roots, stems, and leaves as are the bodies of the higher plants. Seaweeds, toadstools, and the lichens you have seen on rocks are all thallophytes, and there are many others which are still more simple. The green threads you have seen in stagnant water and in running brooks, the green you have noticed on the bark of trees and where water falls on stone or soil, this green is the green of simple thallophytes.

(2) *Bryophytes.* — The word means *moss plants.* Most bryophytes are mosses, but some of them are liverworts. The liverworts are often mistaken for mosses. The difference between them will be explained.

(3) *Pteridophytes.* — The word means *fern plants* and most pteridophytes are ferns. There are certain other plants, however, which do not look like ferns at all, yet which also belong to this division.

(4) *Spermatophytes.* — The word means *seed plants.* You

are already familiar with them. The first three groups may be called the non-seed plants. One of our main objects in studying the non-seed plants is to see how, from plants like them, the seed plants have been evolved. In doing this, we shall also gain an acquaintance with these interesting groups for their own sake.

THALLOPHYTES

These plants fall into two great groups, the *algæ* and the *fungi*. Algæ contain chlorophyll; fungi do not. Mushrooms are examples of fungi. They, like animals, do not possess the power of photosynthesis.

71. **Algæ.** — It is believed that all the other groups of plants evolved from forms like algæ. The lowest of the algæ show us the simplest kind of plant life. They are evidently of much scientific importance. It is from forms like the simplest algæ that the higher plants evolved. The larger and more complex algæ, such as grow in salt water, also evolved from simple algæ, but are believed not to have given rise to forms more complex than themselves. The great majority of algæ are water plants; they grow either in the water or where water is abundant.

Some of the marine algæ (seaweeds) are of much economic importance. They are much used in Japan for food. In France seaweeds are collected for the extraction from them of iodine. Recently, our own government has begun the collection of seaweeds in the Pacific to use them as a source of potassium for fertilizers. The structure of land plants appears to be largely determined by the need to protect against loss of water.

But plants surrounded by water do not need such protection; neither, since water is all about them, do they need special structures for conveying water from one end of the plant to the other, as is the case with most land plants. So it does not surprise us to find that the structure of algæ is very simple.

A. Unicellular Forms. — The simplest algæ are those in which the individual plant is just one cell. *Pleurococcus* is such a plant. Often you have seen it growing on the north side of trees. The north side is more favorable than the other sides because the sun does not strike it, and hence the bark is more likely to remain moist. You have seen it also on old damp boards and on wet rocks. It looks like green paint. A little of it as seen under the microscope is represented in *Figure 160.* Each cell is a separate plant. Lacking the pressure of surrounding cells, such cells tend to assume the form of a perfect sphere. The pressure of the water upon them is practically equal on all sides. They illustrate perfectly the effects of osmosis upon an elastic cell wall. (See page 102.) Reproduction is accomplished, as the picture indicates, by the division of the whole body of the parent. There is no such thing as the differentiation of a part of the body for reproduction, such as we have seen in the seed plants.

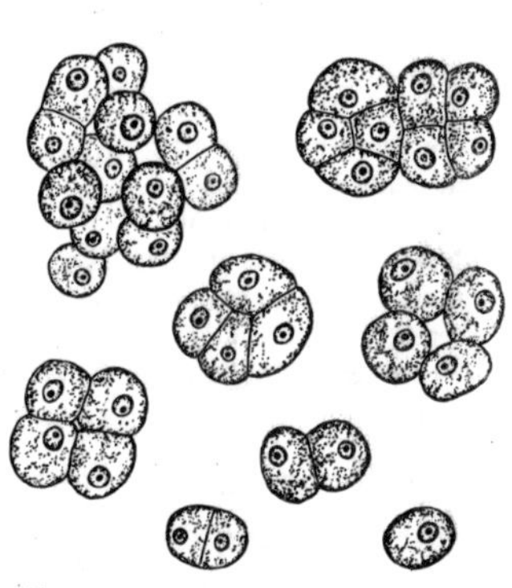

FIG. 160. — *Pleurococcus.* Clusters of the cells showing the manner of reproduction by simple division (*fission*). Highly magnified.

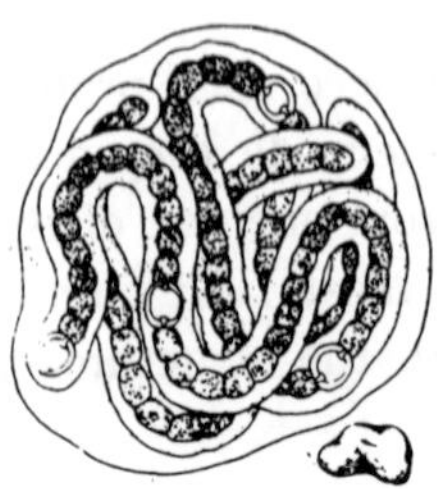

FIG. 161.—*Nostoc.* The small figure at the right indicates the jelly-like mass formed by a sort of mucilage excreted by the cells of this plant. Such masses, brown in color, are often found on damp rocks in shady places. Within them the filaments of *Nostoc* lie embedded as shown in the larger figure. Note four cells of the filaments which are different from the others. These are called *heterocysts*.

B. Colonial Forms. — In algæ of this kind the cells are individual plants, but they occur in groups called *colonies*. They are held together by a substance which is manufactured and excreted by the cells. *Nostoc* and *Rivularia* are examples of such forms. *Nostoc* is found in jelly-like lumps; *Rivularia* in blue-green patches. Both are found in damp places. (See *Figures 161* and *162*.) It will be noted that the cells are not all alike. In *Rivularia* there is a difference between the cells at the base and those at the apex of the chain, while in *Nostoc* occasional cells are much larger than their neighbors. The reproduction of these plants is by cell division only, as in *Pleurococcus*. *Oscillatoria* is a blue-green form which is often found on wet soil. In it the cells are so closely pressed together that they lose their spherical shape. They have the shape of round pill-boxes. (See *Figure 163*.) Yet they

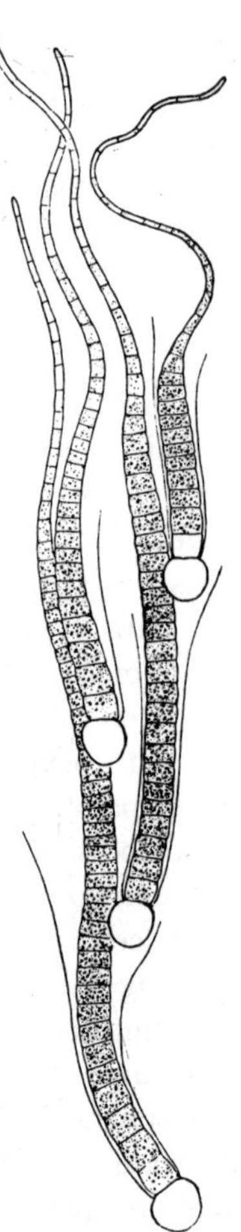

FIG. 162. — *Rivularia.* A colonial alga found in running water. The filament tapers into a whip-like extension.

are also able to live independently. Such a chain of cells as we find in *Oscillatoria* is called a *filament*. The filaments of this plant have the peculiar power of rotating and of oscillating from side to side.

Fig. 163. — *Oscillatoria.* Portions of three filaments very highly magnified.

C. Many-celled Fresh-water Forms. — The largest many-celled forms are some of the seaweeds, but just now we shall consider only those which grow in fresh water. In these the individual cells do not exist independently; except temporarily as spores. The individual plant is composed of a group of cells, usually in the form of a filament. Probably you have noticed such filaments. They are green, thready growths which are common in brooks and ponds and lakes. In brooks you may have seen them fastened to rocks with their long green threads spread out in the water. They make the rocks slippery. In ponds you have seen them floating on the surface and under the water. When they are not attached to sticks or stones, the bubbles of gas they give off often lift them to the surface. Perhaps you have noticed these bubbles in the light-green pond scum which appears on stagnant water. Unattractive though this scum may be to the naked eye, under a microscope it is a thing of beauty. It is made up of thousands of filaments whose chloroplasts have graceful and symmetrical forms. (See *Figure 168.*)

These many-celled algæ have cells of more than one kind, especially in connection with reproduction. Differentiation of cells begins among the many-celled algæ and this means differentiation of function. *Ulothrix*, a form not uncommon in brooks, shows very well this differentiation of cells. (See *Figure 164*.) You note that the lowest cell, which is quite different from the others, serves as an anchor. Other cells serve especially in reproduction. *Any single cell which reproduces without fusion with another cell is called a spore.* Multicellular algæ reproduce chiefly by spores. In *Ulothrix* certain cells of the filament, under conditions which favor this process, begin to divide into spores. What was a vegetative protoplast becomes a considerable number of spores. An opening appears in the wall, and these spores swim out. Their swimming is accomplished by the movement of their *cilia* (singular, *cilium*). A cilium is a thread-like extension of protoplasm. The spores of *Ulothrix* have four cilia at the top. The swimming spores of other algæ have other

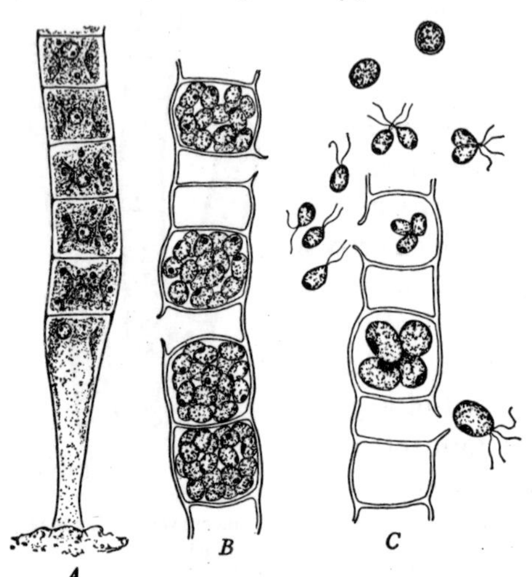

FIG. 164. — *Ulothrix*. *A*, the base of a filament showing the manner of attachment, and the differentiation of the lowest cell. *B*, portion of a filament whose protoplasts have formed spores; note the two cells from which the spores have already escaped. *C*, portion of a filament showing zoöspores of two sizes both before and after their escape into the water; at the top are shown two spores each of which has resulted from the fusion of two of the smaller zoöspores; these are *oöspores*. See context.

arrangements of their cilia, while some spores of algæ have no cilia at all. Swimming spores are called *zoöspores* (*animal-spores*).

The most interesting fact about *Ulothrix* is that in it we find the *simplest form of sex reproduction.* The number of spores produced by a single protoplast is not a fixed number. Sometimes the protoplast goes on dividing until it forms as many as thirty-two spores. The greater the number of spores, the smaller they are. As the spores become smaller, their power to produce new plants appears to become reduced. Though the larger spores germinate directly into new plants, the smaller spores have been observed *to unite in pairs before germination occurs.* It appears that these spores, too feeble to produce new plants alone, unite their forces, and the cell resulting from their union accomplishes what the separate cells could not. United, such cells can reproduce; separate, they cannot reproduce. Such cells, having lost the power to reproduce independently have thereby lost the right to be called spores. They are called *gametes. Cells which cannot reproduce independently, but which can by fusion produce a cell that reproduces, are gametes.* Evidently, wherever there is a sex process there are gametes. (Study *Figure 164.*) *Ulothrix* produces both spores and gametes. The eggs and sperms of which you have already heard are evidently gametes. The fusion of the gametes of *Ulothrix*, just as the fusion of sperm and egg, is fertilization. The cell which results from the union of the gametes is called the *oöspore* (*egg-spore*) or fertilized egg.

Thus we see that gametes are derived from swimming spores, and that this evolution of a spore into a gamete is the evolution of sex. Remember that cells which must fuse before reproduction occurs are gametes, and cannot properly be called spores; cells which reproduce directly are spores, and cannot properly be called gametes. Thus the fertilized egg is properly called a spore; a sex process is responsible for its origin, but it is not a sex cell itself; only gametes are sex cells.

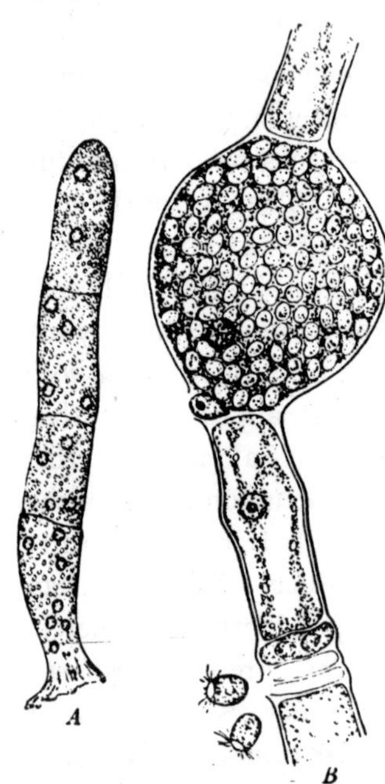

FIG. 165. — *Œdogonium*. *A*, a young filament. *B*, portion of a mature filament. The largest cell is an *oögonium*, that is, a cell which contains an egg. The oögonium of this plant contains just one large egg which is richly supplied with food, indicated by the numerous light oval bodies. Near the bottom of this picture are three *antheridia* (cells which produce sperms). From two of these antheridia sperms have escaped. A third sperm, without its cilia, may be noted at the bottom of the oögonium. It has reached the egg and is about to fuse with it.

In *Ulothrix* we have the appearance of sex, but we do not have its differentiation; that is to say, although the sex act is performed, there appears to be no difference between the gametes, no differentiation into male and female. But in *Œdogonium*, another common fresh-water alga, we have distinct male and female gametes. (See *Figure 165*.) In this form certain cells become enlarged and rounded. The protoplast is transformed into an egg. Other vegetative cells subdivide into smaller cells and these produce sperms. The sperms escape into the water. They swim actively and appear to be attracted to the egg. Presently one of them reaches the egg and fertilizes it. The cell which produces the egg is called an *oögonium;* cells which produce sperms are called *antheridia.* Oögonia and antheridia are sex organs.

Another very significant thing is to be noted in connection with *Œdogonium.* It is the oöspore which, protected by a thick wall, survives through the winter. When this oöspore germinates in the spring it does not grow directly into a plant like the one which produced it. *It subdivides into four cells, and each one of these four cells may*

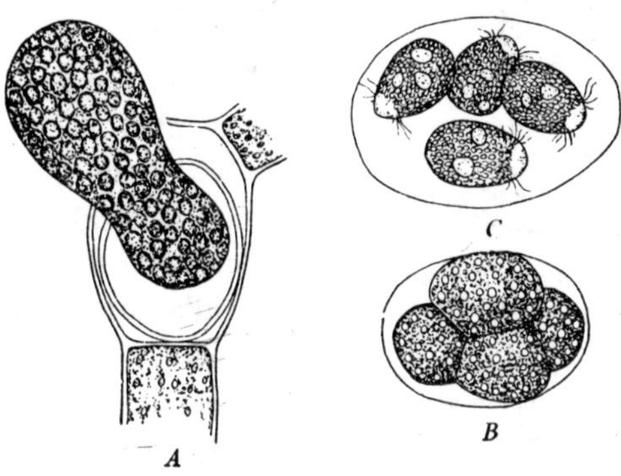

FIG. 166. — *Œdogonium.* *A*, the oöspore (fertilized egg) escaping from the heavy wall which surrounded it. *B*, the oöspore after germination divided into four large zoöspores. *C*, the zoöspores completed and ready to escape.

then produce a new plant. (See *Figure 166.*) In this matter the fertilized eggs of plants generally show a distinct difference from the fertilized eggs of animals. The eggs of animals in nearly all cases directly produce forms like their parents. The egg of *Œdogonium* does not, however, directly produce a form like its parent. It first produces a number of spores and each one of these may become a new plant. This is important for the reason that this habit of the egg, the habit of producing something different from its parent, becomes more and more pronounced in the more complex plants.

Two other common green algæ are *Vaucheria* and *Spirogyra.* (See *Figures 167, 168,* and *169.*) *Vaucheria* forms dark-green, velvety masses. It is much more coarse in texture than *Spirogyra,* which is the pale-green, slippery pond scum.

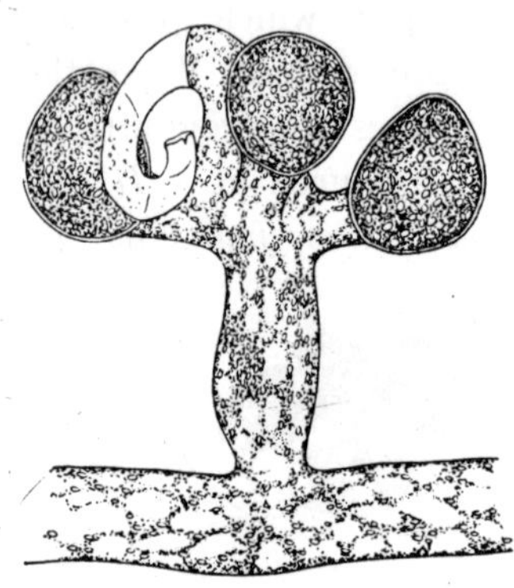

FIG. 167. — *Vaucheria.* Portion of a filament bearing a branch on which are three oögonia each containing a fertilized egg. The antheridium, now empty of sperms, appears as a drooping, curved cell arising from the top of the branch.

Vaucheria has no cross partitions in its filaments. Its protoplasts appear to be merged into one great protoplast with many nuclei. Such a structure is called a *cœnocyte.* In *Vaucheria* the oögonia and antheridia appear on special branches.

Spirogyra gets its name from its large and beautiful spiral chloroplasts. In it the sex differentiation appears to be a differentiation between individuals rather than between cells. Filaments lying side by side produce short tubes which unite the cells lying opposite each other. (See *Figure 169.*) The protoplasts of one filament pass through these tubes and fuse with the protoplasts of the other filament. From this behavior the migrating protoplasts may be regarded as sperms and the passive ones as eggs. Structurally, however, there appears to be no difference between them. The filament which loses its protoplasts may be regarded as a male individual, and the receiving filament as a female individual.

D. The Groups of Algæ. The algæ are divided into four groups. The first two are mainly fresh-water forms; the others are mainly salt-water or marine forms.

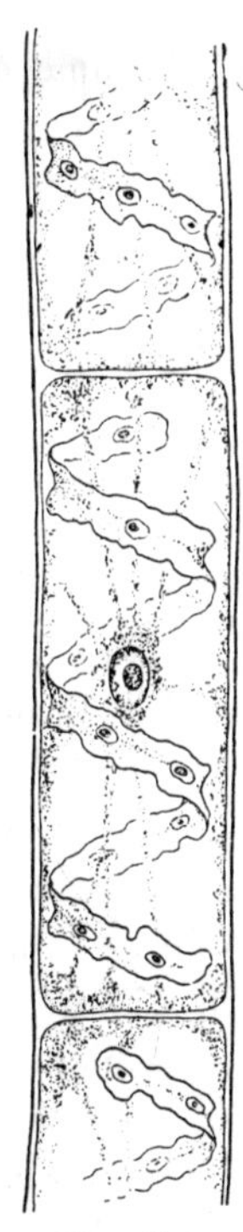

FIG. 168. — *Spirogyra.* Portion of a young filament showing the large spiral chloroplast.

(1) *Cyanophyceæ.* — The word means *blue algæ.* They are so called because a blue pigment (coloring matter) occurs along with the green chlorophyll. *Oscillatoria, Nostoc,* and *Rivularia* belong to this group.

(2) *Chlorophyceæ.* — The word means *green algæ.* Most of the common fresh-water algæ belong to this group, and forms like certain members of this group are thought to be the remote ancestors of the higher plants.

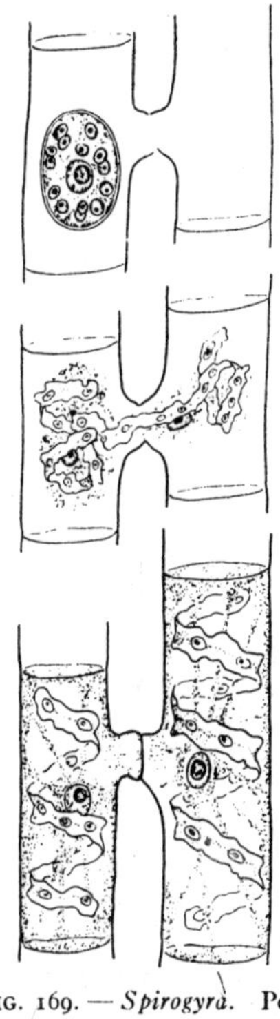

FIG. 169. — *Spirogyra.* Portions of adjacent filaments showing the sex process as described in the context.

(3) *Phæophyceæ.* — The word means *brown algæ.* Although chlorophyll is present, other pigments which are brown and yellow obscure it and determine the color of the plant. These plants are abundant along all seacoasts. They grow anchored by holdfasts

to the rocks, and their tough leathery bodies are not injured by the beating of the waves. The rockweeds, which are members of this group, have air bladders, and swollen tips in which the oögonia and antheridia occur. (See *Figure 170.*) The largest of this group are the *kelps*, some of which attain a length of several hundred feet. (See *Figure 171.*) Iodine is obtained from seaweeds which belong to this group.

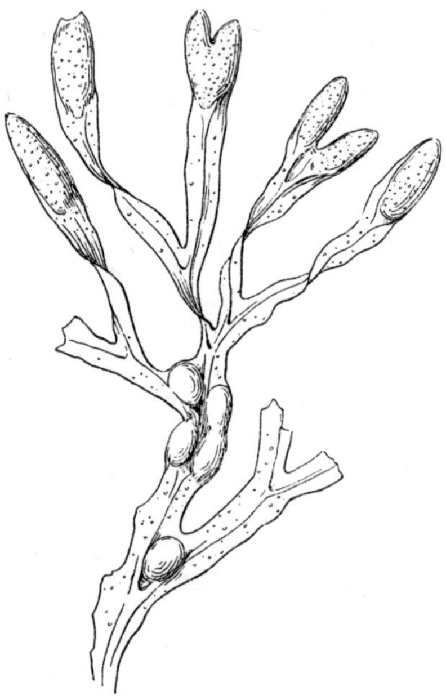

FIG. 170. — Rockweed (*Fucus*). A common kind of seaweed. One of the brown algæ. Note the swollen tips in which oögonia and antheridia occur; also the air bladders which add much to the buoyancy of this plant.

(4) *Rhodophyceæ.* — The word means *red algæ.* These forms are smaller and more delicate than the brown algæ. They get their name, like the brown ones, from the pigment which obscures their chlorophyll. These are the seaweeds which are often dried on cardboard and taken away as souvenirs of the seaside. There are violet, purple, and pink ones, as well as red. They often have very graceful branching forms. They do not grow so far north as the brown algæ, and are most abundant in the tropics. (See *Figure 172.*) Some of the red algæ have rather complex methods of sex reproduction.

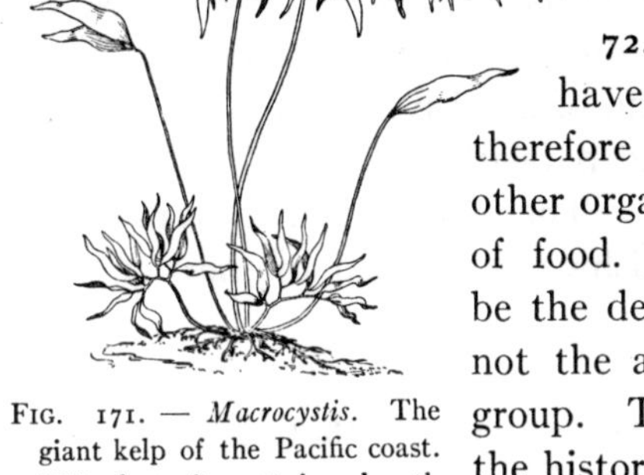

FIG. 171. — *Macrocystis.* The giant kelp of the Pacific coast. This plant often attains a length of several hundred feet. It is one of the brown algæ.

72. **Fungi.** — These plants have no chlorophyll and therefore are dependent upon other organisms for their supply of food. They are thought to be the descendants of algæ, but not the ancestors of any other group. Therefore, in studying the history of plants, they are of less interest to us than algæ. But in studying the relations of plants to man, they are of far more interest to us than are the algæ, as you shall see.

A. Parasites and Saprophytes. — Dependent plants either obtain food directly from other living organisms or use substances for food which were once a part of the bodies of living organisms. Those which use the former method of obtaining food are called *parasites;* those which use the latter are called *saprophytes.* Note that though all fungi are either parasites or sap-

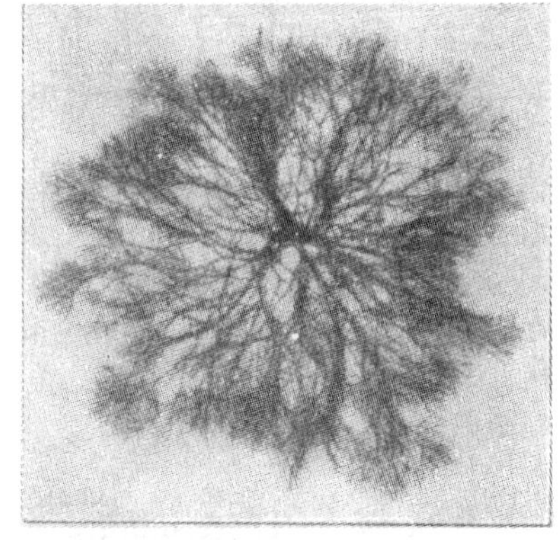

FIG. 172. — One of the red seaweeds showing its finely and gracefully divided body. — *After Hunter.*

rophytes, all parasitic or saprophytic plants are not fungi. Some seed plants do not have chlorophyll; they are as dependent for their food as the fungi are. Indian pipe is a saprophyte, and dodder is a parasite. So the words *parasite* and *saprophyte* do not denote a definite group of plants; they denote a kind of nutritive habit. Some fungi are both saprophytic and parasitic. They take their food wherever they can get it. The plant or animal from which a parasite derives nourishment is called the *host* of that parasite.

B. Economic Importance. — To say that a thing is of *economic importance* to man does not necessarily mean that it helps him. Its importance may be in the way of injury as well as in the way of benefit. Fungi are of huge economic importance to man in both ways. As to injury, certain fungi are the causes of many of the diseases of cultivated crops, of farm animals, and of man himself. As to benefit, they cause the *decomposition* of organic substances, and increase the fertility of soil. In bread making and cheese making certain fungi are essential. Yeast is a kind of fungus.

In the process of decomposition complex organic molecules are changed into simpler ones. These simpler molecules are then available for plant use. The fertility of humus depends chiefly on the activity of bacteria which transform its complex molecules into molecules which plants can use.

C. Bacteria. — These minute, unicellular organisms have by far the greatest economic importance of all fungi. In some of their characteristics they are different from all other plants and animals, and it has been a question whether to consider them plants or animals, but it is customary now to class them among plants as a special group of the fungi.

Bacteria are the direct causes of most of the diseases of men and of other animals. The study of bacteria is a special subject called *bacteriology*, and it is by means of learning the habits of bacteria that our most effective methods of preventing and curing disease have been discovered. They are the smallest known organisms, and their study requires microscopes of very high power of magnification. Doubtless there are some which are too small to be seen even by microscopes of the highest power. The existence of such bacteria has been indicated by the effects they produce.

As to structure, bacteria are of three distinct types. They are spherical, rodlike, or spiral. (See *Figure 173*.) They occur either singly or in filaments. Sometimes they have cilia and swim actively. Under favorable conditions they multiply very rapidly by division like the division of *Pleurococcus*. They appear to be present everywhere, ready to reproduce rapidly under any condition favorable to their growth. The most important principle of modern sanitation is the prevention of conditions which are favorable to the growth and spread of disease-producing bacteria. Many kinds of bacteria are capable of enduring conditions of temperature,

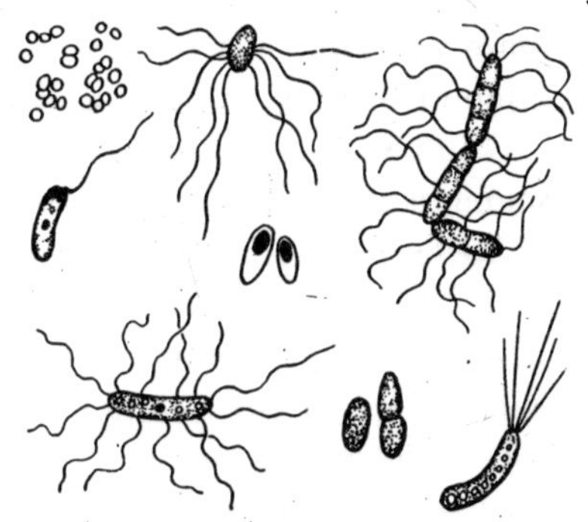

FIG. 173. — Various forms of bacteria, very highly magnified.

of lack of water, and even of lack of oxygen, which would promptly kill all other organisms. Typhoid fever, cholera, tuberculosis, and diphtheria are among the human diseases caused by bacteria.

Bacteria are of great importance in agriculture. Though they cause some diseases which are injurious to crops, this injury is more than offset by other ways in which they are helpful to crops. There are bacteria which inhabit the soil and make it fertile. There are certain substances, common in soil, which contain elements that plants need, but which might as well not be there so far as being of direct use to green plants is concerned. The plant cannot use them any more than we can use sea water for drinking purposes. Certain bacteria, however, are able to act upon these substances and so to alter them that they do become available for plant use. Salt water must be distilled before it becomes available for human use as drinking water. The action of bacteria is nothing like the distilling process, but this comparison may help you understand what is meant by making certain substances *available for plant nutrition.*

Nitrogen is very important in the nutrition of green plants. These soil bacteria, in their own nutrition, act upon certain compounds which contain nitrogen. As a result of this action, certain other compounds which contain nitrogen are produced, and these other compounds may be readily absorbed and used by the plant, whereas the original nitrogen-containing compounds could not be used. Since the fertility of the soil depends largely upon the amount of nitrogen in it which is available for plant use, the benefit of these bacteria is obvious.

It has been demonstrated that soil probably never contains at any one time enough of available nitrogen compounds to produce a good crop. That is, if a good crop is to be produced, the soil bacteria are absolutely necessary. They must be present and actively at work throughout the growing season. *Just as we must depend upon the activity of green plants for our food, so it appears that green plants must depend upon the activity of soil bacteria for some of the materials out of which they make food.*

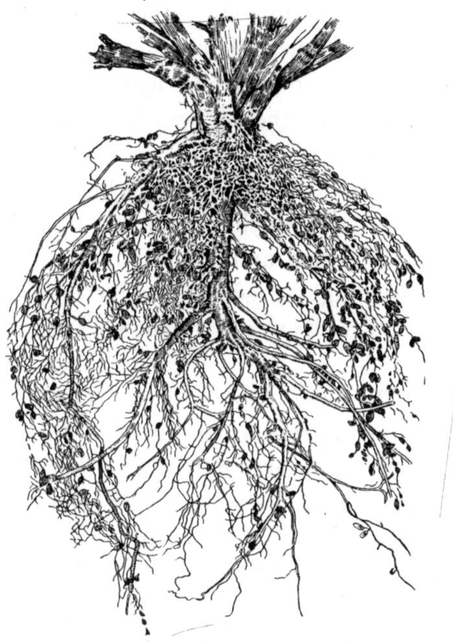

FIG. 174. — Roots of clover showing the numerous nodules. These nodules or tubercles contain bacteria which are of much benefit to the plant that bears them, as described in the context.

That great group of plants, the *Leguminosæ*, to which peas, beans, alfalfa, and clover belong, have a special relation to certain soil bacteria which is of much economic importance. On the roots of these plants are found swellings called *nodules* or *tubercles*. (See *Figures 174* and *147*.) Within these tubercles are found bacteria. The tubercles are caused by the presence of these bacteria. These bacteria of the tubercles have the power to use in their nutrition the nitrogen of the air, a power which green plants do not possess. The activity of the bacteria in the tubercles results in the presence there of certain nitrogen compounds which are very useful to their host. It has been frequently proved that leguminous crops grown on soil rich in the tubercle-forming bacteria do very much better than when grown on soil free from these bacteria. It has been found profitable to *inoculate* soil with these bacteria; that is, to introduce them artificially. Due to the assistance of these bacteria, leguminous crops are especially rich in nitrogen, and leave the soil in better condition as to nitrogen than it was before they were grown upon it. For this reason they improve the yield of other crops, such as corn or wheat, which may succeed them on the land, and which do not form tubercles. Soy beans are often grown and plowed under for the express purpose of enriching the soil for other crops, and their power to do this is due entirely to the bacteria of their tubercles.

D. True Fungi. — We speak of *true* fungi because bacteria are not true fungi. Some of their characteristics make it desirable to class them with fungi, but certain other characteristics make them very different from the true fungi.

There are thousands of different kinds of true fungi and they have great economic importance. We can describe only a few of the most common kinds. Mushrooms and toadstools are the best known of the true fungi. Perhaps you also know of mold and mildew. They also are true fungi. If you live in the country, you have heard of wheat rust and oat rust, and of corn smut. These plant diseases are caused by true fungi.

The true fungi have a plant body called a *mycelium* (plural *mycelia*). A mycelium is a mass of more or less closely interwoven thread-like structures. These threads may be so closely interwoven as to form a sort of mat, or even to form what seems to be solid tissue, as in toadstools. You may have noticed an example of mycelium on the surface of moldy bread.

From a common toadstool we may learn the general habit of growth of all true fungi. Toadstools often make us wonder. We wonder what they grow from and how they grow so quickly. They seem to appear suddenly in the night. They are saprophytes. The part we call a toadstool is the reproductive part. The nutritive part is hidden in the rich soil or in the decaying wood or in whatever it is from which the toadstool arises. This nutritive part is a mass of burrowing and absorbing hairs. These hairs are called *hyphæ.* The mass of them we have already called the mycelium. The behavior of hyphæ suggests the behavior of root-hairs. Like root-hairs they are much concerned with absorption, but they absorb food itself as well as substances used in the manufacture of food. The foods they absorb come mainly from the decay of the dead parts of plants. The fallen leaves and branches with which the floor of a forest is littered decay and become a part of the soil of the forest, the part you have learned to call humus. It is rich in organic material, and this material is food for fungi. The mycelia of toadstools are usually abundant in humus.

The thing we call a toadstool is only a temporary part of the plant. The permanent part is the mycelium. The toadstool is a spore-bearing organ. The spore-bearing organs of fungi are called *sporophores.* The advantage of having sporophores above the soil is that the thousands of light spores they produce may thus be scattered by the wind. Some spores of fungi are light enough to float in the air.

The sporophores of toadstools and mushrooms are much more complex than those of the simpler and less conspicuous fungi. Usually, especially in parasitic fungi, the sporophore is nothing more than a single erect filament which produces spores at the end. (See *Figure 175.*) A toadstool is a compound sporophore. The stem part of it is called the *stipe*. The cap part of it is called the *pileus*. On the under surface of the pileus you will notice the edges of the *gills*. (See *Figures 176* and *177.*) It is on these gills that the spores are borne. If a ripe pileus is laid, gills down, on a piece of paper where the air is still, a *spore print* may be produced. The pileus and the paper must be left alone for a day or two. The ripe spores are usually dark colored. They fall on the paper and reproduce quite accu-

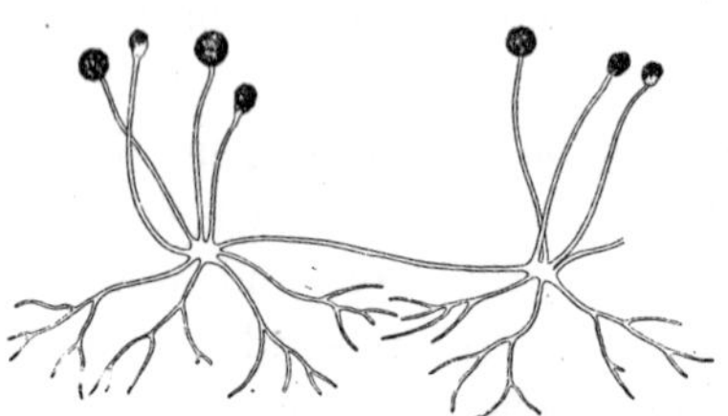

FIG. 175. — *Mucor*. The bread mold, showing *hyphæ* below and *sporophores* above.

FIG. 176. — *Lepiota*. A common edible mushroom. Note that the *pileus* in the youngest stages appears to be merely a swelling at the top of the *stipe*. Presently this swelling becomes ruptured at the base and flattens out revealing the *gills* beneath. Note the gills in the specimen which is tipped.

rately the graceful pattern of the gill edges. (See *Figure 178.*)

The spores of toadstools and of all other fungi which belong to the same group are borne at the ends of very slender branches which arise from a structure, usually club-shaped, which is called the *basidium*. (See *Figure 179.*)

Fig. 177. — A common edible mushroom. The gills show plainly. Note the ring of tissue surrounding the stipe and indicating the former attachment of the edge of the pileus. — *After Gibson.*

When conditions are just right, the sporophores of fungi suddenly appear. Moisture, temperature, and the amount of food present in the hyphæ are some of the conditions upon which the appearance of sporophores depends. This is well illustrated in the growing of mushrooms for market. Good mushrooms sell for as much as a dollar a pound, and thousands of pounds are grown artificially every year.

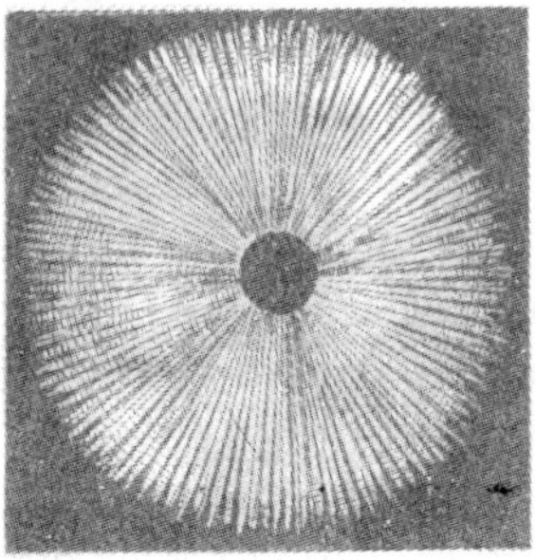

Fig. 178. — The spore print of a mushroom. See context.— *After Andrews.*

The mushroom grower plants what is called *spawn*. The spawn of mushrooms is simply a mass of mycelium in a state of suspended animation. It is the object of the mushroom grower to get this spawn to produce as many sporophores as possible; the spreading of the mycelium is of no value to him unless it produces an abundant crop of the reproductive organs. Mushrooms are usually grown in cellars or in mushroom houses, which are but dimly lighted. Light is not needed for the crop. To get the right temperature, the right moisture, and the right food supply is the problem. The spawn is spread in manure which needs to be at just a certain stage of decay in order to produce the best results. The manure is covered with a thin layer of black earth. Presently, if conditions are right, young mushrooms, called buttons, begin to break through the soil.

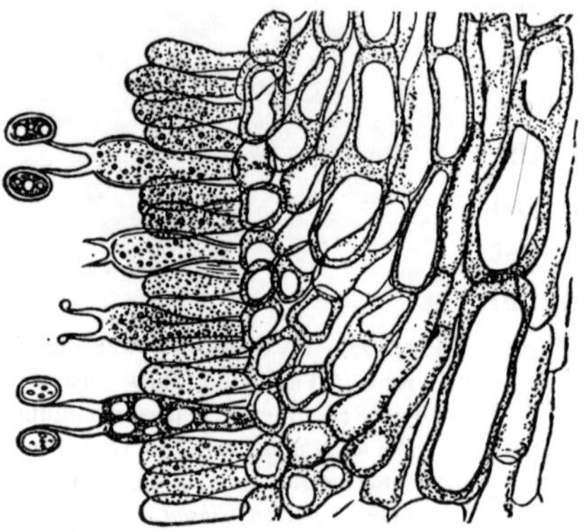

FIG. 179. — A portion of the surface of one of the gills of a common mushroom showing the way in which the spores are borne. Note the club-shaped structures bearing two prongs at whose tips the spores are borne. These club-shaped structures are basidia. See context.

All true fungi follow this general plan of growth and nutrition that we have noted as to toadstools and mushrooms, but their reproductive habits are quite dissimilar. Some fungi have sex methods of reproduction in addition to the non-sex methods. As to nutrition, all true fungi have hyphæ which absorb food. They also assimilate it. The whole work of nutrition goes steadily on in the mycelium, and then, when the conditions are just right, sporophores appear, the spores are scattered, and the sporophores disappear again. They shrivel up and die. The mycelium continues its work as long as it can. When the conditions for growth become unfavorable, mycelia of saprophytic fungi appear to pass into a condition of suspended animation; when the conditions are right again, they renew their activity.

a. Difference between Toadstools and Mushrooms. — These plants belong to the same group of true fungi. The difference between them is simply that mushrooms are edible and toadstools are not. There is no single way by which you always can tell a mushroom from a toadstool. Some of them look much alike. Yet it is not at all difficult to become so well acquainted with a number of the common mushrooms that you are in no danger of mistaking toadstools for them. Mushrooms which are good are very good indeed, and it is a great pleasure to collect them in the woods and fields. The best way to learn them is to go into the field with some one who already knows them. They are most abundant in early fall, especially just after rains, and they are found in pastures as well as in timber. One of the best mushrooms, and one which you can easily recognize, is the *morel*. (See *Figure 180*.) It is usually found in the woods late in April or early in May.

FIG. 180. — *Morchella*. The morel, a common edible mushroom which appears in early spring. The morel does not belong to the same botanical group as true mushrooms, but it is commonly called a mushroom. It has the advantage of being very easily recognized.

Puffballs, and the bracket fungi (see *Figures 181* and *182*) which grow on stumps and logs are relatives of the mushrooms. Puffballs are edible. The giant puffball, which sometimes gets as big as a watermelon, is, when fresh, one of the best of the edible fungi. The puff of dust which rises when you step on an old puffball is a cloud of spores. Some of the large bracket fungi become hard with age and may endure for years. The brackets are sporophores. The spores are borne inside the many pores which are on the under surface.

FIG. 181. — Young puffballs. When in this condition, puffballs are good to eat, the only drawback being the worms which may be found in the firm white meat inside. The wormy parts may be cut away, and sometimes large puffballs are found quite free from worms. The puffball is a sporophore. The white tissue within becomes a mass of spores, and the tough skin turns brown. Trillions of spores may be contained in one puffball. When the ball is kicked the spores rise from it like a puff of dust. — *After Gibson.*

b. Bread Mold. — Moist bread kept where it is fairly warm will soon begin to be covered with a white, furry growth. This is the mycelium of bread mold. Branches of the mycelium grow into the bread and absorb food from it. Hyphæ which branch away from the mass of the mycelium in this manner and absorb food are called ***haustoria***, in this or in any other fungus. (See ***Figure*** *183.*) After the mold has become well developed, little vertical branches appear, and the swollen tips of these branches turn black. These branches are sporophores, and the blackness of their tips is the color of the ripe spores which they produce. (See *Figure 175.*) These light, powdery spores are scattered by currents of air.

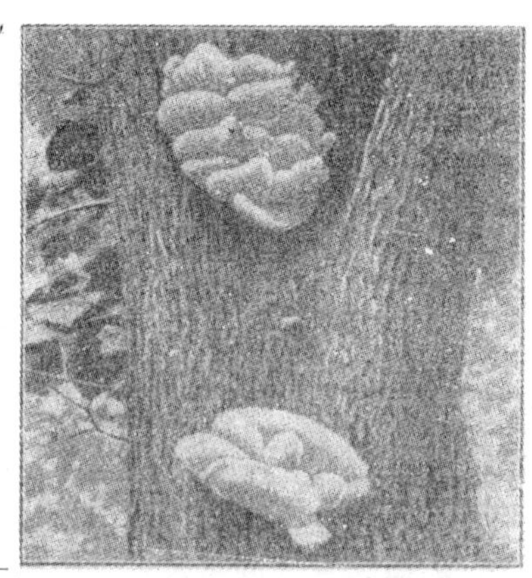

FIG. 182. — A bracket fungus growing on a linden tree. — *After Andrews.*

Bread mold, like some other of the simpler fungi, has a sex method of reproduction. Under right conditions, special hyphæ appear whose tips come together in pairs. These are the sex branches. (See *Figure 184.*) The hyphæ of bread mold, like the filaments of *Vaucheria*, are not divided by cross walls, but a cross wall appears behind the tip of the sex branches. In this way a protoplast is cut off from the rest of the body. After the tips of two sex branches have come in contact, the walls are absorbed, and the two protoplasts fuse. Evidently these protoplasts are gametes and the result of their fusion is an oöspore. This oöspore turns black. It forms a heavy wall about itself. It is capable of surviving under conditions which would cause the death of the spores of the sporophore.

c. Mildews. — If you have ever noticed the leaves of lilac bushes you have probably noticed that some of them often seem dusty. This dusty appearance is due to a parasitic fungus which grows upon them. It is the lilac mildew. Many parasitic fungi grow within the tissues of their host, only the sporophores growing up to the surface; but this one spreads its mycelium over the surface of its host, only the haustoria growing down into its tissues.

(See *Figure 185.*) These haustoria penetrate the epidermal cells of the lilac, and from them absorb food. Since the cells of the mesophyll are not injured by it, however, lilac leaves appear to get along pretty well in spite of the mildew. Numerous little sporophores arise from the mycelium, and it is the abundance of spores which makes the mycelium look like dust.

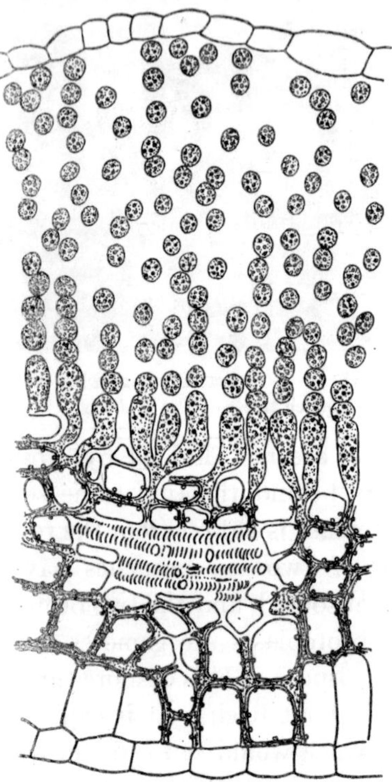

FIG. 183. — *Albugo candida.* A kind of parasite (white rust) which attacks members of the mustard family. The figure shows the appearance of a section of a mustard leaf cut through one of the parts attacked. Note the way in which spores are produced. The top row of cells is the epidermis of the host. It is finally ruptured and the spores escape. Note the way in which the hyphæ (shaded) burrow between the cells of the mesophyll and send little suckers (*haustoria*) into the protoplasts.

This fungus, like bread mold, also has a sex method of reproduction. The sex organs are very small, but the structures which result from the sex process are not small. They are big enough to be seen with the naked eye. They look like dark-brown dots. They are called *ascocarps.* The outside of them is a thick-walled case, and the inside of them is what has resulted from the growth of the oöspore. For this oöspore has a habit we have already noted in *Œdogonium*. It does not directly produce a plant like its parent. It produces a number of thin-walled *sacs* or *asci*, and in each *ascus* a number of spores are found. (See *Figure 186*.) Thus, as a result of the sex process, a considerable number of individuals are produced. It is by means of the ascocarps that this fungus lives over winter.

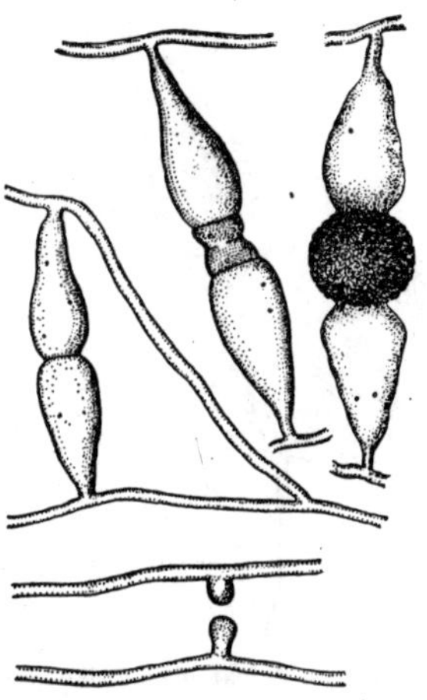

FIG. 184. — Sex reproduction in the bread mold. Note that the ends of two hyphæ which come together are cut off from the rest of the mycelium by cross walls. The parts thus cut off fuse, and a dark colored oöspore results.—*After Hunter.*

d. Wheat Rust. — That parasitic fungus which causes wheat rust produces four distinct kinds of spores. In this it is not entirely different from other fungi of the same group, but it is the one whose peculiar life history was first understood. Its name is *Puccinia*.

Like the rust of oats, wheat rust is first noticed as rust-colored marks on the stems and leaves of its host. The rust-colored areas spread. If you walk through a field of rusted oats or wheat, your clothes become covered with a reddish-brown dust. Under a microscope this dust is found to be nothing but spores. The rusty patches on the plant are nothing but spores. The mycelium, which burrows among the tissues of the leaf and stem, has sent hundreds of sporophores to the surface, and these have borne thousands of these rust-colored spores. This kind of spore is called the *summer spore*. By means of its summer spores this parasite spreads rapidly in a field where it has gained a start. (See *Figure 187*.)

Late in summer another kind of spore appears. This is usually after the wheat has been cut. We find this new kind of spore on the stubble. Stubble is what is left of the wheat after the harvest. This second kind of spore forms black marks on the old stalks; it is sometimes called *black rust.* These are the *winter spores* of the wheat rust. These black spores have heavier walls than the summer spores, and it is by means of them that this parasite lives over winter.

Fig. 185. — A leaf of lilac showing the mildew (*Microsphæra*) upon it. The dark dots are *ascocarps.* They are explained in the context.

The winter spores germinate in early spring and a third kind of spore appears as a result of their germination. These winter spores are well stocked with food and the generation which they produce is a little filamentous structure which is not a parasite at all; its nourishment is drawn from the supply stored in the spore. (See *Figure 188.*) This third kind of spore is borne at the ends of very slender branches of the filaments. The filament from which these branches arise is a basidium, the structure you have already noted in connection with toadstools. This kind of spore is called the *early spring spore.*

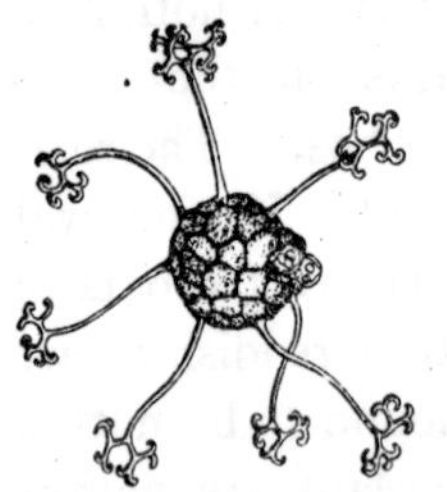

Fig. 186. — An ascocarp of the lilac mildew. Note the peculiar forked outgrowths. The ascocarp has been crushed, and two asci with their contained spores may be seen protruding at the right.

The surprising thing about early spring spores is that usually they do not attack wheat; they attack other plants. It was in England that the life history of wheat rust was discovered, and there it was found that these early spring spores germinate on the leaves of barberry. The barberry is a common shrub. It was thought that if the barberry was destroyed, the wheat rust would disappear. But it was found that these early spring spores could germinate on other plants as well as on barberry. Soon after these early spring spores have produced a mycelium in the tissues of whatever leaf it is they attack, reddish patches appear at the surface of the leaf. These reddish patches form what are called *cluster cups*. In them are found the fourth kind of spore, the *late spring spore*. (See. *Figure 189*.) It is this fourth kind of spore which germinates on the young wheat plants, thus completing the life cycle of this remarkable plant.

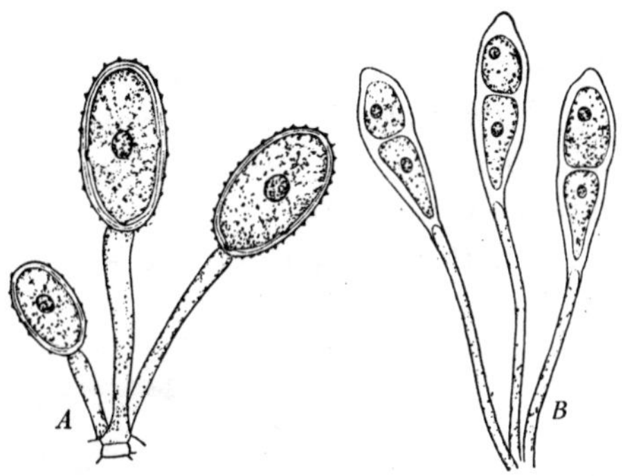

FIG. 187. — *A*, summer spores of wheat rust. *B*, winter spores of wheat rust.

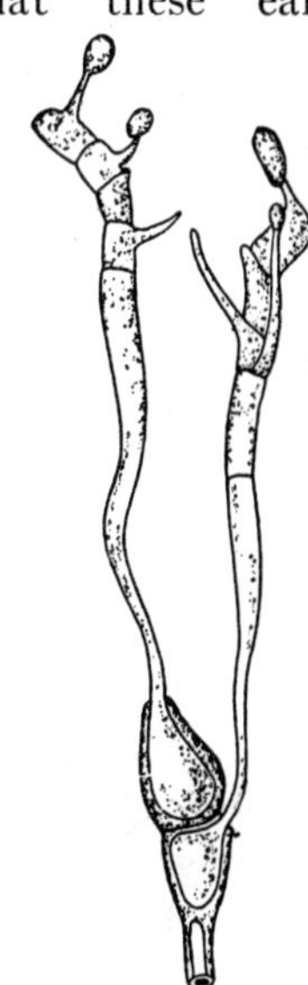
FIG. 188. — The winter spores of wheat rust germinating. Note the short filaments at whose ends short branches appear bearing the early spring spores.

c. Classification of True Fungi. — There are three great groups of true fungi, and they all have been illustrated in the forms we have been considering.

(1) *Phycomycetes.* — The word means *alga fungi.* They are so called because they resemble algæ in structure, especially algæ like *Vaucheria*, and they also resemble algæ in their manner of sex reproduction. These forms

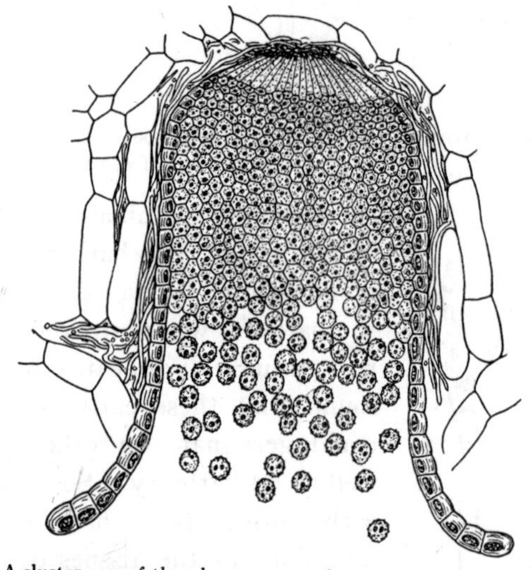

FIG. 189. — A cluster cup of the wheat rust on the under surface of a barberry leaf, very highly magnified. The spores produced in it are "late spring spores."

have no cross walls in their hyphæ; they are composed of cœnocytes. This group includes all true fungi in whose life histories neither asci nor basidia appear. The bread mold is a member of this group.

(2) *Ascomycetes.* — The word means *ascus fungi.* This group includes all the true fungi which produce asci. You learned of asci in connection with lilac mildew. An ascus

is a sac in which spores are borne. (See *Figure 186.*) The asci are not always inclosed in a case as in lilac mildew; in other ascomycetes they form the lining of cup- or funnel-shaped structures. The hyphæ of this group have cross walls. Some of them are known to reproduce sexually, but the sex organs are not like those of algæ.

(3) *Basidiomycetes.* — The word means *basidium fungi.* You have learned what a basidium is in connection with toadstools, and also in connection with wheat rust. Both of these forms are basidiomycetes. Any forms in whose life history basidia appear are put in this group. The hyphæ have no cross walls, and no sex organs have been discovered.

E. Fungi and the Soil. — The importance of bacteria in rendering the soil fertile was described a few pages back. Some of the true fungi are also important in connection with this subject.

Soil is a very complex thing. Except where it is quite bare of plants, like bare sand, it contains more or less of organic material. This organic material comes principally from the bodies of plants which have lived and died in and upon the soil. This organic material is food for fungi, for true fungi as well as bacteria. Rich soil, like stagnant water, is full of invisible life, and this invisible life affects the visible plants which grow from such soil. Humus is filled with fungi. They are as much a part of the soil of the forest as the rotting branches and the decaying leaves. This hidden life in the soil is not well understood, but we do know that the fungi, which principally compose this soil life, produce certain changes which are beneficial to green plants.

Nitrogen, as you know, is necessary to plant life. Though nearly four fifths of the air is nitrogen, green plants get none of their nitrogen directly from it. They are unable to use nitrogen in gaseous form. That which they use comes from the soil. Compounds which contain it are present in the soil, and these, when dissolved in soil water, are the source of nitrogen for green plants. Certain compounds called nitrates are of chief importance to plants as sources of nitrogen. (See page 344.)

The soil on which wild plants grow has its supply of nitrates kept up. The wild plants return their nitrogen to the soil when they die. But soil on which farm crops grow is constantly losing nitrogen. The plants which it produces do not return to it when they die. The farmer is constantly taking away the crops and the crops are rich in nitrogen. He does not allow them to rot upon the ground. For this reason one of the great questions which confronts farmers is how to keep the soil of their fields from becoming too poor in nitrogen. Manure and certain other fertilizers help keep up the nitrogen supply.

Bacteria, as you have noted, are of much importance in connection with this subject. It is the action of bacteria upon manure and other organic materials which results in the production of nitrates, the best available form of nitrogen for green plants.

Probably you already know that it is not good farming to grow the same crop upon the same soil year after year. Farmers avoid doing this either by letting some fields rest (lie fallow) in some seasons, or by rotation of crops. To grow corn for two seasons and then clover for one upon the same land is an example of the rotation of crops. Why should such practice improve the fertility of the soil?

There are a number of reasons why it does, but the principal reason is thought to be that such practice keeps the soil from becoming too poor in nitrates. You remember that leguminous crops, like clover, increase the nitrogen supply on account of the work of the bacteria in their tubercles.

The effects which true fungi produce upon the fertility of soil are not yet well understood, but it is known that their hyphæ penetrate everywhere in soil which contains organic matter, and they produce many chemical changes in such soil. Certain soil fungi become closely attached to the roots of other plants, forming a growth which is called *mycorhiza.* (The word means *root fungus.* See *Figure 190.*)

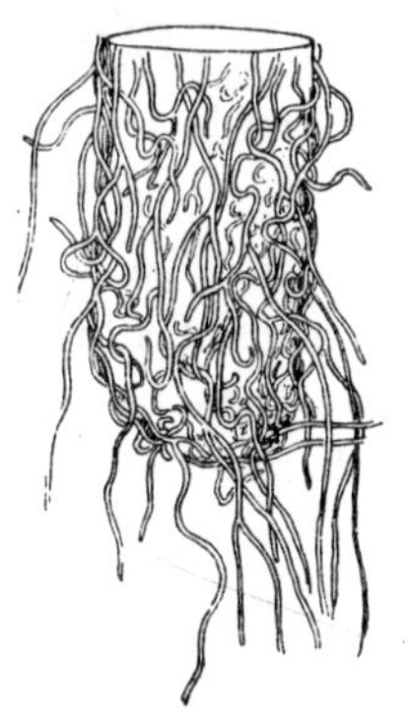

FIG. 190. — Mycorhiza. The tip of a beech root covered with a close mass of fungal hyphæ.

This mycorhiza, which is very common upon the roots of forest plants, is believed to be an advantage both to the fungi and to the roots on which they grow. The fungi are believed to derive nourishment from the roots, and the roots are believed to use the filaments of the fungi somewhat as they use their own root-hairs. On account of this direct connection with true fungi of the soil, the power of roots to absorb water and solutes is greatly increased.

F. Lichens. — Probably you have noticed on rocks and trees brownish and greenish patches which are formed by structures having the texture of very thin leather. They stick closely to whatever they grow upon. These structures are *lichens.* (See *Figure 191.*) Lichens are the other group of plants besides bacteria which are included under fungi but which are very different from ordinary fungi. They are different in that they completely envelop the host from which they draw nourishment. They are composed of fungi and algæ which grow together. The principal part of the structure is formed by the hyphæ of the fungi, and enmeshed among them are found the cells of the alga. (See *Figure 192.*) The fungus of the lichen derives its nourishment from the cells of the alga, while the alga derives its moisture and solutes from the hyphæ of the fungus. This arrangement makes it possible for lichens to live where either algæ or fungi alone could not live. We find lichens on bare rocks. They are the first plants to appear in such places. By their growth and decay they gradually produce a little soil. Then other plants come in and crowd the lichens out. Thus the lichens are the first on rocks of what are called *pioneer plants.* Sometimes you may find upon lichens little saucer-shaped outgrowths in which spores are produced.

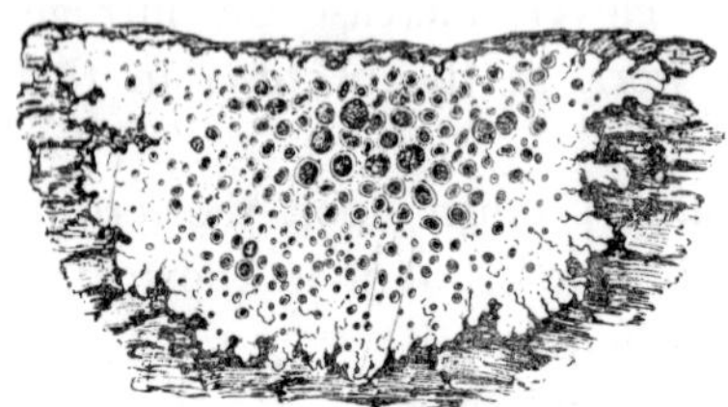

FIG. 191. — A lichen growing upon the bark of a tree.

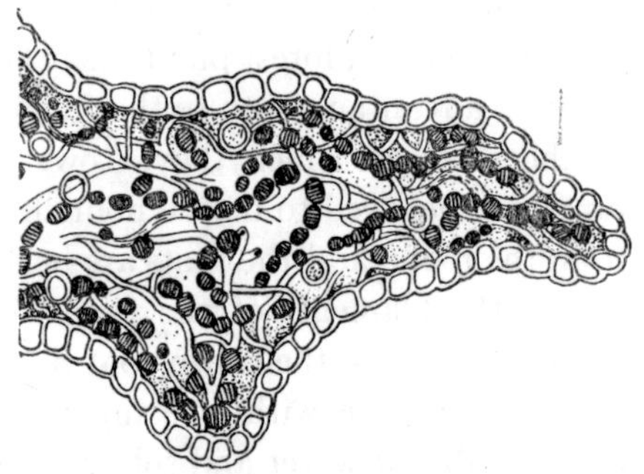

FIG. 192. — Section through the body of a lichen showing the cells of an alga (shaded) within it.

Bryophytes

You know what moss is. You have seen it growing in all sorts of places, and you have probably called things moss which are not moss at all. People are likely to say that any plant whose leaves and stems are not distinct is a "kind of moss." But the stems and leaves of true mosses are distinct, though they are usually very small. True mosses form one of the divisions of bryophytes. The other division is the liverworts. The liverworts are simpler than the mosses and we will consider them first.

73. **Liverworts.** — These plants grow where it is damp and shady. They grow on decaying rock and on decaying logs and sometimes on the soil itself. They have distinct under and upper surfaces. They are prostrate. In this they are different from their relatives, the mosses, which do not have distinct under and upper surfaces, even though some of them are prostrate.

The simplest liverworts are thought to be much like the first plants which ever grew on the land. These simplest liverworts are found on water or on mud which is left after the water has disappeared. It is believed that plant life began in the water. Occasionally pools of water in which this early plant life grew would dry up. It is believed that some of these water forms acquired such structures that they did not die as soon as such pools dried up; they did not need to be completely surrounded by water in order to live; they were able to live because they could absorb water from the damp surface beneath and because they were protected on top by a structure which prevented the prompt loss of this water by evaporation. Their structure enabled them to retain water better than it could be retained by the ancestors from which they sprang. Such forms began the conquest of the land by plants and such forms are found to-day among the liverworts. Such forms are said to be *amphibious*, which means that they are capable of living both in the water and on the land.

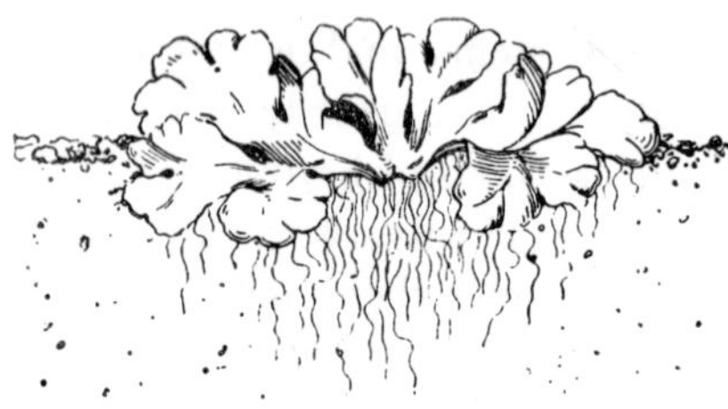

FIG. 193. — *Riccia*, a simple liverwort.

The body of the simplest liverworts is a sort of plate of cells having certain outgrowths beneath. (See *Figure 193*.) The commonest liverworts, however, such as are found on damp rocks in shady places, have a more complex structure. An outline of this structure as it appears under the microscope is shown in *Figure 194*. This is a view which you get by examining a thin slice of the liverwort, cut through from top to bottom. The same kind of liverwort, as it appears to the naked eye, is shown in *Figure 195*. The erect outgrowths are concerned with reproduction. The prostrate part is called the thallus, a word which you met in studying thallophytes; it describes a plant body which is not differentiated into stems and leaves.

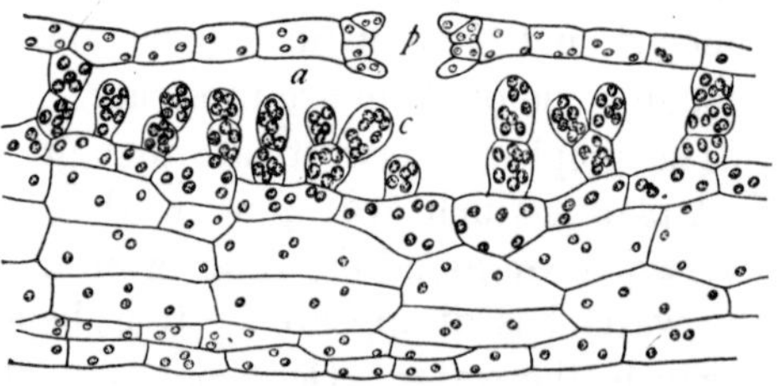

FIG. 194. — Section through the thallus of *Marchantia*. Note the large air chamber, *a*, under the upper epidermis; *p* is a permanently open pore; the chloroplasts are principally contained in the cells of peculiar outgrowths, *c*, which arise from the bottom of the air chambers.

The view of the inside of the body of the liverwort shows that it is built a good deal on the plan of a green leaf. You see the chloroplasts in the cells, more abundant on the upper side. You see a pore in the epidermis which corresponds in function to the stomate of a leaf; it admits air which diffuses freely in the *air chamber* beneath. In the kind of liverwort which is pictured (*Marchantia*) the air spaces are more prominent than in the leaf, and thin-walled cells containing chlorophyll grow up in a peculiar manner from the base of the air chamber. The relations between the air and the thin-walled cells, however, are just the same as in the leaf. The cells of the upper epidermis of this liverwort are waterproof, but the cells of the under epidermis are permeable to water.

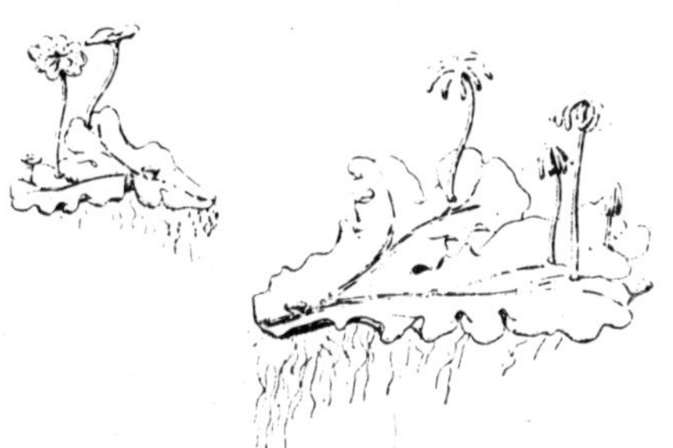

FIG. 195. — Two plants of *Marchantia*. The one at the right bears *archegoniophores*. The one at the left bears *antheridiophores*. Each one bears a single *cupule*. See context.

A. Reproduction. — Liverworts reproduce in three ways. The simplest of these ways is by growing in front and dying behind. Since liverworts have the habit of branching frequently, you can see that this process results in the formation of new individuals. You have noted the same process in connection with the growth of underground stems in seed plants.

Another way of reproduction is by means of reproductive buds which are often borne in little cup- or crescent-shaped structures called *cupules*. (See *Figure 195*.) The reproductive buds which the cupules contain are called *gemmæ*. Gemmæ are bud-like bodies which become detached from the parent plant and reproduce it. They occur in mosses and in some pteridophytes as well as in liverworts. The gemmæ of common liverworts are green bodies a little smaller than the head of a pin; they are plates of cells which become detached from the short stalks on which they are borne, and are carried away by the water which often washes over the surface of the liverwort.

The third way of reproduction in liverworts is much more complex than the other two. Also it is the kind of process from which the sex and seed reproductive habits of the higher plants have been evolved. To understand seeds it is important for you to understand the method of reproduction about to be described. If it were not for seeds, agriculture would be a very different thing from what it is, and if agriculture were different, our lives would certainly be different from what they are. So this process has a real relation to your own life.

Liverworts produce eggs and sperms. Plant eggs and sperms are not new to you; you learned of them in algæ. But in liverworts the egg is produced in a structure called the *archegonium*. Archegonia you have not met before. Neither algæ nor fungi have them, but liverworts and mosses, and all the fern plants bear their eggs in archegonia. An archegonium is a flask-shaped structure which bears an egg in its swollen base or *venter*. (See *Figure 196*.) In *Marchantia*, which is that kind of liverwort shown in *Figure 195*, the archegonia are borne on those structures which bear finger-like branches at the top; the archegonia are found under the fingers, near their base; the necks of the archegonia hang down. This stalked structure with the finger-like branches is an *archegoniophore.* Some liverworts do not have archegoniophores; their archegonia are embedded in the thallus with the necks up and open at the top. As you have already guessed, it is down the neck that the sperms pass to the egg.

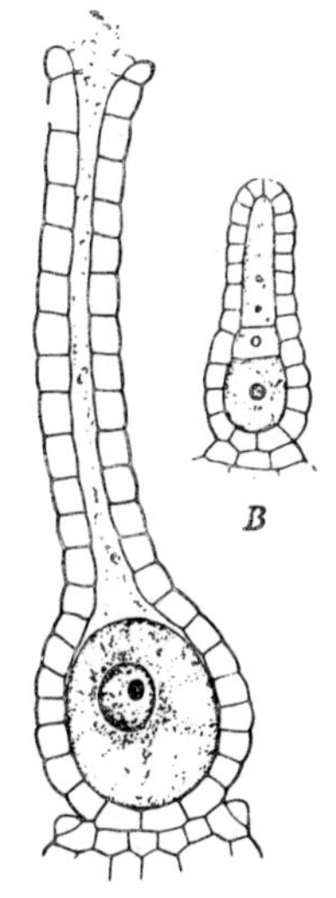

FIG. 196. — Archegonia of *Marchantia*. *A*, a mature archegonium, ready for fertilization. *B*, young stage, showing the row of central cells of which the lowest becomes the egg.

The sperms are borne in organs called antheridia, a term which you used to describe the sperm-bearing organs of thallophytes. The antheridia of liverworts produce great numbers of sperms. (See *Figure 197*.) All the sperms of bryophytes are very small and bear two cilia. In the picture of *Marchantia* (*Figure 195*) you have noticed those branches whose flat tops are scalloped at the margin. These are *antheridiophores* and the antheridia are found embedded in their flat tops. They open on the upper surface. Antheridia are borne in other ways by other liverworts.

Perhaps you have already wondered how fertilization is accomplished in *Marchantia*. You know that in some way the sperms must reach the eggs or else all these structures that we have been describing are of no use to the plant. Sperms are not like pollen; they are not blown by wind or carried by insects. In order to move they need water; they swim actively by means of their moving cilia. They are so small that a film of dew is quite enough for them to swim in, and when the sperms of *Marchantia* are discharged from the antheridia, there is nearly always enough moisture on the surface of the plant to permit them to swim to the archegonia. To reach the eggs they have to swim up, but at the time when swimming occurs the stalks of the archegoniophores are quite short; not nearly so long as they become later. Most of the liverworts do not bear their archegonia upon stalked organs, and of course in such cases the journey of the sperms to the eggs is much simpler than in *Marchantia*.

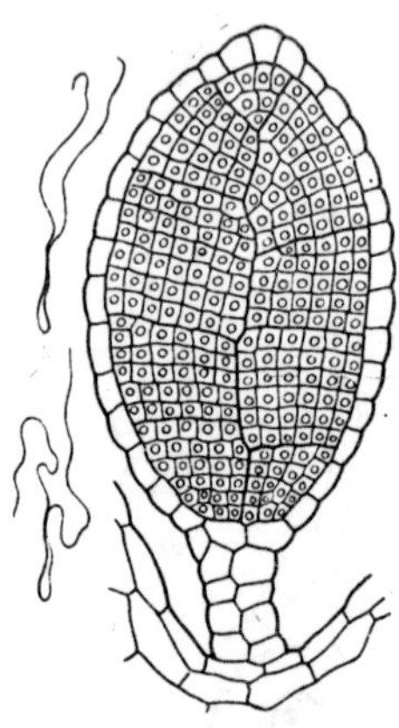

FIG. 197. — A young antheridium of *Marchantia*. Also two of the biciliate sperms from a ripe antheridium.

What happens after fertilization is particularly important for you to note. You remember that in *Œdogonium* the fertilized egg or oöspore does not grow directly into a plant like its parent; it divides into four spores, and each of these is capable of producing a new plant. In liverworts we find this process carried still further. We find the oöspore not only forming a number of spores; we find it growing into a special organ many of whose cells are not spores at all. This organ is called a ***sporogonium***. (See *Figure 198*.) The sporogonium in *Marchantia* is an organ big enough to be seen easily by the naked eye. It contains spores, but it also contains many cells of other kinds. Besides the cells of the *wall* and the *foot* and the *seta*, there are the *elaters*, peculiar spirally-marked cells which appear among the spores and, by twisting as they dry, push the spores out of the ripe sporogonium. The spores are light and may then be scattered by currents of air. Thus you see the advantage which it is to the plant to have its sporogonia borne on stalked organs whose stalks elongate after fertilization has occurred.

This little history of the sex method of reproduction in a liverwort, and of what happens after the egg is fertilized, illustrates what is called the *alternation of generations*, and this is a matter important enough to have a heading of its own.

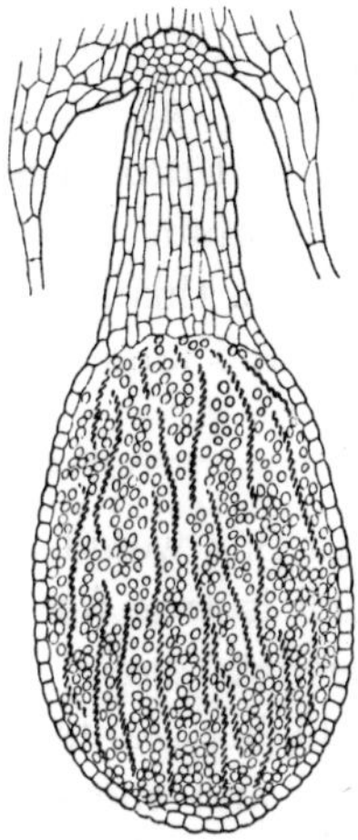

Fig. 198. — Section through the *sporogonium* of *Marchantia*. This structure hangs down from beneath the top of the ripe archegoniophore. Note that it has three parts. The largest part is the *capsule*. It contains the spores. Scattered among the spores the spiral *elaters* are seen. Above the capsule there is the stalk or *seta*, and above that there is the *foot* which is embedded in the tissue of the archegoniophore.

B. Alternation of Generations. — In animals the fertilized egg produces a form like its parent. From the egg of a chicken we get another chicken. There is no alternation of generations. The generations are all alike.

But in plants, at least from liverworts up, the fertilized egg does not produce a form like its parent. It produces a form like its grandparent. One generation is not succeeded by another just about like it; it is succeeded by one which is not like it at all. Then this new, entirely different generation produces in turn, not itself, but the generation which produced it.

Let us see how this is illustrated in the liverwort. You

may recall (see page 268) that we have defined a generation as *an individual which arises from a single cell.* The spore of a liverwort is, of course, a single cell. It gives rise to the liverwort plant. That is one generation. The liverwort plant produces sperms and eggs (gametes) and a fertilized egg or oöspore results; the oöspore is a single cell and it produces a structure called a sporogonium. The sporogonium is therefore another generation. We may illustrate this by a diagram in which large P stands for the plant and large S stands for the sporogonium.

P —sperm / —eggs > oöspore—S—spore—P

All this would not be very important if it were a thing which is true for liverworts alone. But it is not true of liverworts alone. It is true of them and also true of all the higher plants. Ferns and seed plants, as well as liverworts and mosses, all have alternation of generations. We are explaining it now for them as well as for liverworts. It is a thing which it is important for you to understand.

Since two distinct generations occur in all the higher plants, it is desirable to have names by which they may be distinguished whatever their forms may be. Whatever their form may be, these generations are always alike in the reproductive cells which they produce; that is, *one generation always produces spores, the other always produces gametes.* The generation which produces spores is called the *sporophyte;* the one which produces gametes is called the *gametophyte.* Another way of stating the difference between these two generations is to say that sporophytes always come from fertilized eggs, while gametophytes always come from spores. In liverworts the plant is the gametophyte and the sporogonium is the sporophyte.

Note that although a sporogonium is a sporophyte, these two words are not synonymous. A sporogonium is a particular kind of structure which occurs only in bryophytes. Sporophyte has a much broader meaning. *It names the generation which produces spores whatever its form may be.* It may be a sporogonium, as in bryophytes, or a large plant, as in seed plants. You remember that seed plants produce pollen, and that a grain of pollen is a spore. Therefore the generation which produces pollen must be a sporophyte.

C. Leafy Liverworts. — There is a group of liverworts whose bodies are divided into distinct stems and leaves which look much like moss stems and leaves. These liverworts are often mistaken for mosses, but they are different from mosses in that they have distinct upper and under surfaces. They are often found on moist and shaded bark. Such liverworts are called *leafy* or *foliose* in distinction from those, like *Marchantia*, which are *thallose.*

D. Green Sporogonia.—There is a kind of thallose liverwort which produces a sporogonium which contains chlorophyll. This sporogonium grows up from a prostrate thallus and looks a good deal like a blade of grass. (See *Figure 199.*) This liverwort is *Anthoceros.* It is not very common. Its archegonia are not borne on erect archegoniophores as in *Marchantia,* but are found embedded in the thallus.

This green sporogonium of *Anthoceros* is important because it forecasts what we find in the ferns, namely, *the two generations growing as independent plants, each manufacturing its own food.* Although the sporogonium of

Anthoceros absorbs much of its food from the thallus in which its foot is embedded, it is also able to manufacture food itself by means of the chloroplasts which it contains. It is believed that forms like *Anthoceros* were the ancestors of the fern plants.

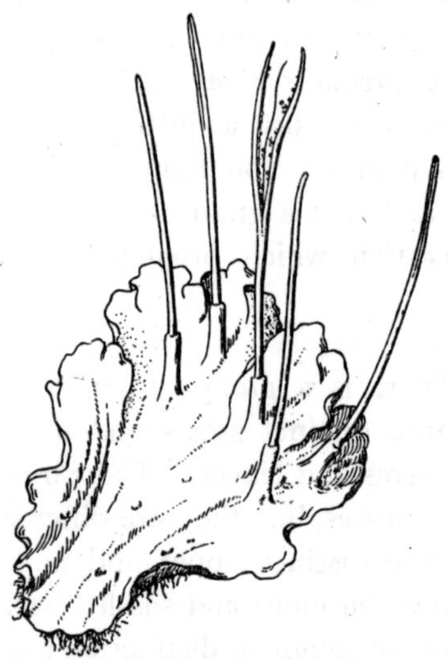

FIG. 199. — *Anthoceros*, a liverwort which produces green sporogonia. — *Redrawn from Bergen.*

E. Liverworts as Pioneers. — You remember that lichens were described as the first plants to grow on rocks. It was said that they are followed by other plants. If the rocks are moist and shaded, liverworts are very likely to be the plants which follow the lichens. You may have seen rocks on which liverworts were gradually growing over the lichens and killing them. Similarly, we often find mosses following liverworts and crowding them out. Mosses do not need as much moisture as liverworts do, so, on dry rocks, they are usually the plants which follow the lichens.

74. **Mosses.** — Mosses grow in patches, in clumps, and in cushion-like masses. If you have pulled up a bunch of moss, you have noticed the brown, dead part below. It is as large or larger than the live, green part above. Moss keeps growing above and dying below. You have noticed the tiny

green leaves and the delicate stems. A moss leaf has very simple structure, far more simple than a liverwort thallus. The leaves of many mosses are but one cell thick; they are so delicate that the light shines through them. (See *Figure 200.*) In spite of their delicate leaves, mosses are able to live in dry places. Clumps of moss hold water even better than a sponge. Even if mosses do become dried out, they are usually able to recuperate when they become moist again.

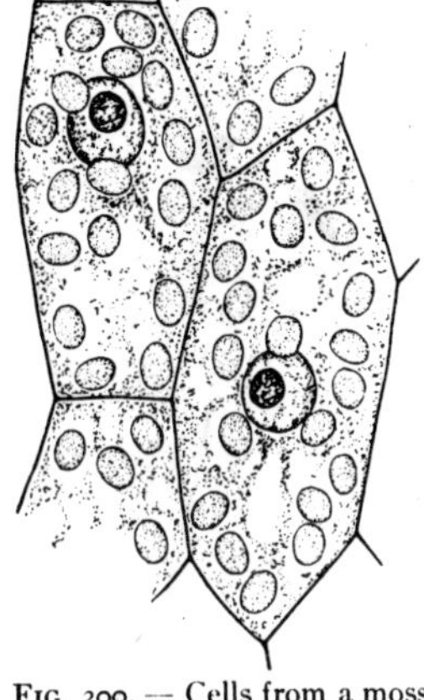

FIG. 200. — Cells from a moss leaf which is but one cell in thickness.

The dead brown parts of moss clumps absorb water as well as the live parts. Young moss plants have filamentous growths at the base called *rhizoids*. (The word means *root-like.*) The rhizoids serve as holdfasts, and probably absorb moisture. (See *Figure 201.*)

FIG. 201. — A young moss plant showing the *rhizoids* at the base.

A Reproduction. — Perhaps you have seen what is sometimes called moss fruit. It grows up from the top of the leafy stems. (See *Figure 202.*) Sometimes all the moss plants in a clump appear to be bearing these structures at the same time. In one common kind of moss these structures are called *pigeon wheat.*

This so-called moss fruit is a sporogonium It bears spores in the capsule-shaped enlargement at the top. This spore-containing enlargement is called the capsule, the stalk which bears it is called the seta, and the third part of the sporogonium, which is hidden in the top of the moss stem, is called the foot. The spores escape from the ripe capsules and are scattered by the air. When they germinate, they produce a filamentous growth which looks like an alga. This growth is called *protonema.* A protonema can always be told from filamentous algæ by the fact that its cross walls are diagonal instead of straight. (See *Figure 203.*) Buds appear upon the protonema, and these buds develop into the erect, leafy moss plants.

FIG. 202. — *Polytrichum,* a common moss bearing what is known as pigeon wheat. Each of these "fruits" of the "wheat" is a sporogonium. Note the long seta and the capsules. The foot is embedded in the apex of the stem of the leafy part.— *After Hunter.*

The archegonia and antheridia of mosses are found at the tops of the leafy stems. (See *Figures 204* and *205.*) Sometimes archegonia and antheridia are found on the same plant, sometimes on separate plants. The archegonia of mosses are longer and more slender than those of liverworts and the antheridia are club-shaped. After fertilization the oöspore develops into the sporogonium, whose three parts have already been noted. The spore-producing part of the sporogonium is much less in proportion to the sterile part than is the case in the sporogonium of *Marchantia*. By sterile part is meant any part which does not produce spores. This increase in the sterile part of the sporophyte is significant, for in higher plants we find that nearly all the cells of the sporophyte are sterile. The sporogonium of certain mosses, like that of *Anthoceros* among liverworts, contains chlorophyll and does some food making. The bulk of its nourishment, however, is probably absorbed by the foot from the tissues of the leafy stem. As in liverworts, *the sporophyte is a parasite upon the gametophyte.* The gametophyte in mosses includes both the protonema and the leafy part.

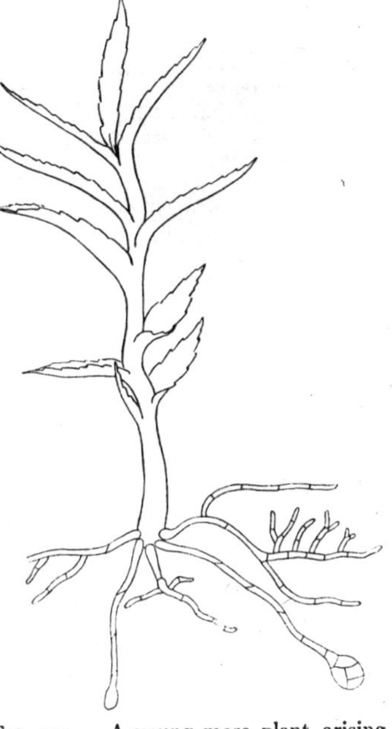

FIG. 203. — A young moss plant arising from the *protonema* and putting out rhizoids. Note the diagonal cross walls of the filaments. At the end of one of these filaments a resting bud appears. A moss plant may arise from this.

B. Increase of the Sporophyte. — Already we have noted that the sporophyte becomes more prominent as we go from the simpler plants to the more complex ones. At first the sporophyte is only a number of spores and a case. Gradually it increases in size and becomes differentiated into several kinds of tissues and organs. It begins to do leaf work. We have seen this in both liverworts and mosses. (See *Figure 206.*)

In ferns you will find that the sporophyte ceases to have any nutritive connection with the gametophyte except when it is very young. It becomes an independent plant which makes its own food. It is the leafy part of the fern, while the gametophyte is an insignificant little structure which is rarely noticed. Finally, in seed plants, we find that *sporophytes are the plants themselves*, while the gametophytes, which began by being hosts of the parasitic sporophytes, end in seed plants by being themselves parasites, and the sporophytes are their hosts. You have already learned something of this in studying the flower. (See page 268.)

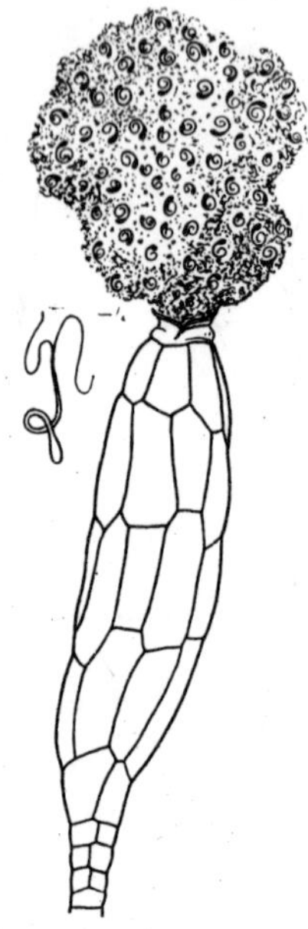

FIG. 204. — An antheridium of moss discharging its contents. At the left a single sperm is shown. Note that it has two cilia.

Since a gradual increase of the sporophyte and a corresponding decrease of the gametophyte is a universal characteristic of the plant kingdom, we should naturally expect to find that there is some advantage to plants in this matter. This advantage is not difficult to perceive. The thing which makes the sporophyte increase and become independent is that the nutritive work of the plant, at first done by the gametophyte, is gradually transferred to the sporophyte, while the thing which makes the gametophyte decrease in importance is that it loses this nutritive work. As you have noted, the sporophyte of lower plants gets its food from the gametophyte, while in the higher plants gametophytes get their food from sporophytes.

The sporophyte produces spores. The gametophyte produces sperms. It is better for food making to be connected with the spore-producing generation than with the sperm-producing generation. Why? The answer to this question is also an answer to the question, "Of what advantage to plants is the increase of the sporophyte and the decrease of the gametophyte?"

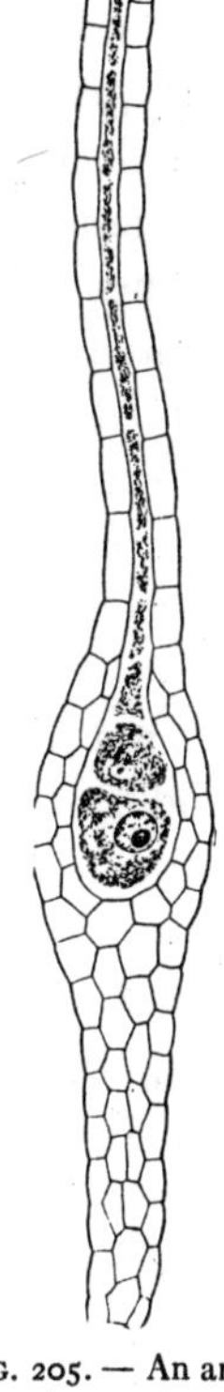

FIG. 205. — An archegonium of moss. Note the long neck and the stalk below the venter.

Spores and sperms must both move away from where they are borne. Spores must be scattered to reproduce the plant, while sperms must move to reach the eggs. Spores move through the air. Sperms always move through fluid. Now in nutrition one of the principal things which a plant body needs is light. If this plant body is a spore-bearing body, the more it grows up into the light, the better are the chances of its spores to be carried away through the air; it can help its leaf work and its spore distribution at the same time. But if this plant body is a sperm-bearing body,

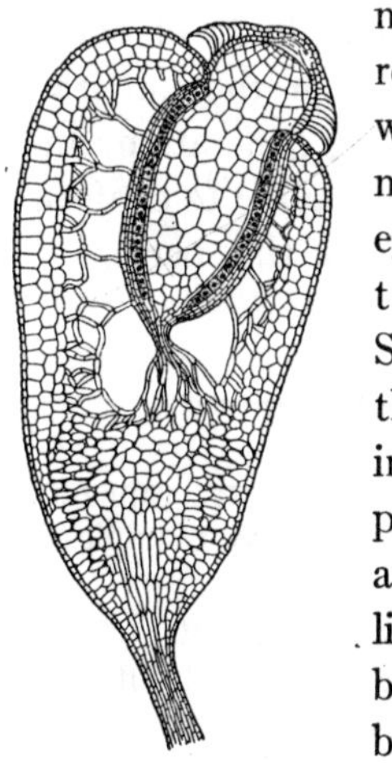

FIG. 206. — Lengthwise section through a young capsule of a moss sporogonium. Only the cells which are shaded produce spores. Note how greatly the spore-producing tissue is reduced as compared with the sterile tissue.

the more it grows up into the light, the poorer are the chances of its sperm to get to eggs on other plants; what is helpful to its leaf work is hurtful to its sperm work. In mosses and liverworts we have leaf work done by the sperm-bearing generation, but mosses and liverworts never become tall; they are lowly plants which are often covered by enough moisture to permit the sperms to get from one plant to another. Possibly mosses indicate just about the limits of height to which it is possible for sperm-bearing stems to ascend if the sperms are to do their work; they may be said to represent the limits in the evolution of a leafy gametophyte. Further evolution in the nutritive body of plants was accomplished only by forms in which the nutritive work was transferred to the spore-bearing generation, and so was not limited in its development by having to conform to conditions necessary to the movement of the sperms.

QUESTIONS AND SUGGESTIONS

SECTION 69. 1. What is meant by evolutionary sequence? 2. What is meant by non-vascular plants? Give three examples of them other than those given in this section.

SECTION 70. 1. Name the four great divisions of plants, and explain the meanings of these names. 2. State a principal difference between algæ and fungi, and give an example of each.

SECTION 71. 1. Describe the places in which you have seen algæ growing. 2. In what ways are the needs of algæ as to structure different from the needs of ordinary land plants?

A. 3. Describe *Pleurococcus*.

B. 4. Describe *Nostoc*. 5. Describe *Oscillatoria*.

C. 6. Describe the general structure of many-celled, fresh-water algæ. 7. Describe the reproduction of *Ulothrix*, defining cilia, gametes, and oöspore. 8. Distinguish between spores and gametes. 9. Describe *Œdogonium*, defining oögonia and antheridia, and explaining a difference between the behavior of its egg and the egg of an animal. 10. Describe *Vaucheria*, defining cœnocyte. 11. Describe *Spirogyra*.

D. 12. Name the four groups of algæ and explain the meaning of the names. 13. Describe the brown algæ. 14. Describe the red algæ.

SECTION 72. 1. State the general characteristics of fungi. 2. Give three examples of fungi which you have seen.

A. 3. Explain the difference between parasite and saprophyte, defining the word host.

B. 4. State the general facts as to the economic importance of fungi to man, defining decomposition.

C. 5. State a fundamental principle of sanitation with reference to bacteria. 6. Give the general facts as to bacteria. 7. What is the principal way in which soil bacteria are helpful to crops? 8. Describe the relationship of certain soil bacteria to *Leguminosæ*, defining tubercles.

D. 9. Describe mycelium. 10. Describe the life of toadstools, defining hyphæ, sporophore, stipe, pileus, gills, basidium, and spawn. 11. Describe the culture of mushrooms. 12. State the difference between toadstools and mushrooms, and tell what you know about edible fungi. 13. Describe bread mold, defining haustoria. 14. Describe lilac mildew, defining ascus and ascocarp. 15. Describe wheat rust, defining its four kinds of spores. 16. Name the three great groups of true fungi, explaining what these names mean. 17. State the chief characteristics of these three groups.

E. 18. Describe mycorhiza. Of what advantage is it believed to be to the plants which have it?

F. 19. Describe lichens. Where have you seen them growing? 20. Why are lichens called pioneer plants? How do they reproduce?

SECTION 73. 1. State a fundamental difference between mosses and liverworts. 2. Describe what is thought to be the way in which plant life came from the water to the land. What is meant by an amphibious plant? 3. Describe the vegetative (non-reproductive) structure of *Marchantia* as described by the text and the pictures.

A. 4. Describe the asexual methods of reproduction in *Marchantia*, defining cupules and gemmæ. 5. Describe the sex method of reproduction in *Marchantia*, defining archegonia and antheridia. 6. Describe the sporogonium, defining foot, stalk, and elaters.

B. 7. Describe alternation of generations in liverworts, defining sporophyte and gametophyte.

C. 8. How are foliose liverworts to be distinguished from thallose ones? How are they to be distinguished from mosses?

D. 9. Describe the sporogonium of *Anthoceros*. Why is it of especial interest?

E. 10. Describe the relations between mosses, liverworts, and lichens as to growth upon rocks.

SECTION 74. 1. Describe the habits of growth of mosses. 2. Describe the structure of a moss plant, defining rhizoid.

A. 3. Describe the sporogonium of moss, defining foot, seta, and capsule. 4. What is a protonema? 5. Describe sex reproduction in a moss.

B. 6. Explain the advantage to plants in having photosynthesis and spore-production performed by the same generation. 7. Why is it that gametophytes are limited in stature?

CHAPTER X

THE VASCULAR PLANTS

75. Introductory. — Vascular plants are plants with a vascular system. A vascular system is composed mainly of large cells called vessels. This system is the principal means of mechanical support and of sap movement for such plants as possess it. The fern plants and the seed plants possess vascular systems. Plants of the lower groups do not.

The greatest gap in the plant kingdom is between the vascular and the non-vascular plants. From the simplest thallophytes to the most complex bryophytes there is a sort of series of forms which gradually increases in complexity. Also from simplest pteridophytes to highest spermatophytes a similar series may be traced. But between bryophytes and pteridophytes there is a great gap. Here there are many "missing links" in the chain of plant evolution.

A vascular system is to a plant all that your blood vessels and bones are to you. It forms the principal paths of movement in the plant, as well as the mechanical support which enables it to stand erect. You have learned of this in your study of stems (Chapter V). The vascular system is composed of xylem and phloëm, two groups of tissue that are always found together, usually in the form of what are called vascular bundles. (See page 82.)

Vascular plant bodies are all sporophytes, the gametophytes of these plants being either insignificant little green bodies, as in the common ferns, or parasites within the sporophyte, as in seed plants. Evidently, then, the evolution of a vascular system occurred after the principal nutritive work of the plant began to be done by the sporophyte generation.

Vascular plants fall naturally into two great divisions, those which produce seeds and those which do not. Those which do not produce seeds are the pteridophytes. We will consider them first. Among the pteridophytes are found plants which are believed to be much like the ancestors of seed plants.

Pteridophytes

There are three divisions of pteridophytes. The common ferns with which you are familiar belong to one of these divisions. You recall that in bryophytes the mosses are the most familiar and abundant members of that group. They are the forms which give the group its name, and yet they are not the forms of that group from which higher plants were evolved. Similarly in pteridophytes, the ferns are the most familiar members of the group and give it its name, yet it is not from ferns like those of to-day that the seed plants were evolved.

The ancestors of seed plants were ferns of ancient days, long since extinct and not at all like modern ferns in appearance. They were more like trees. *Fossils* of these ancient ferns are abundant in those layers of rock from which coal is principally obtained. (See *Figure 206 A*.) Fossils are those evidences of once living things which we find in rock.

76. **Ferns (Filicales).** — The *Filicales* are composed of the *true ferns* and the *water ferns*. Water ferns are small aquatics, sometimes introduced into artificial ponds. They are of slight importance as compared with the true ferns. Of all pteridophytes the true ferns are, to you, by far the

Fig. 206 *A*. — The fossil of an ancient fern.

most important group. They are by far the most abundant and familiar. They are handsome plants. You have admired their large, graceful, much-divided leaves. Perhaps you have seen them growing abundantly in moist woods and have brought them home to plant. (See *Figure 207*.) In conservatories you may have seen the beautiful tree-ferns of the tropics. Though they are common in temperate regions, true ferns are even more abundant in the tropics.

The stems of our common ferns are rhizomes; that is, they are underground. They live through the winter, growing on from season to season, putting up new leaves. These large leaves of the true ferns are sometimes called *fronds*. They have distinct veins. You recall that the leaves of mosses have no veins. Veins are a part of the vascular system. The veins of ferns differ from those of seed plants in that they branch by forking; that is, one vein divides

FIG. 207. — Ferns growing abundantly in moist woods. Note the large, compound leaves. It is evident that they do not require strong illumination.

into two equal veins. (See *Figure 208*.) In spring you may have noticed that the young leaves of ferns are coiled in a roll; they gradually uncoil. (See *Figure 209*.) By this habit as well as by their *venation* (vein arrangement) fern leaves are easily identified.

A cross section of the stem or the petiole of a true fern shows the well-developed vascular system. (See *Figure 210*.) You note that the xylem and phloëm do not occur in bundles such as you saw in seed plants. In the stem which is pictured the vascular system forms a cylinder, but it is not a cylinder with xylem on the inside and phloëm on the outside such as you have seen in seed plants. In this cyl-

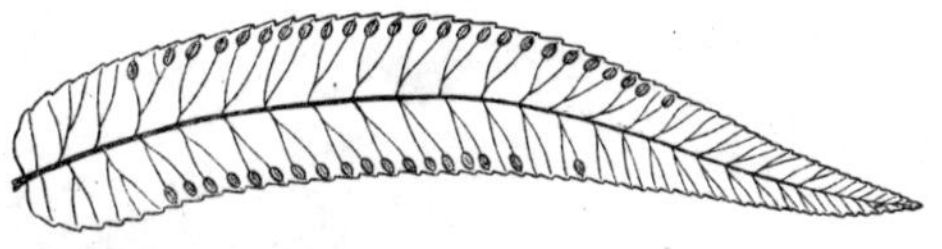

FIG. 208. — A fern leaflet, showing the way in which each vein forks into two branches (*dichotomous venation*).

inder phloëm is present on both sides of the xylem. Other arrangements of the xylem and phloëm are found in the stems of other kinds of pteridophytes.

Reproduction. — On the under side of fern leaves you may have noticed brown dots. (See *Figure 209.*) You may have noticed that what you now know to be spores come from these dots. These brown dots are groups of spore cases and these spore cases are called *sporangia.* The brown groups of sporangia are called *sori.* When young the sorus is usually protected by a covering called an *indusium.*

The appearance of one of these sporangia under the microscope suggests a helmet, such as was worn by the armored knights of old. (See *Figure 209, D.*) The crest of the helmet is a row of thick-walled cells called the *annulus.* When the spores are ripe, the annulus is of service in discharging them. It behaves as shown in *Figure 211.*

You may confuse the word *sporangium* with the word *sporogonium.* Both of these structures contain spores. But sporogonia occur only in bryophytes, and they constitute the entire sporophyte generation of the plants which produce them. Also, the spore-producing part of a sporo-

FIG. 209. — *Aspidium*, a common fern. *A*, the entire plant, showing the rhizome, the coiled young leaves arising from it, and three mature fronds, one of which shows *sori* as dark dots on the under surface. *B*, a leaflet, showing the shield-like coverings (*indusia*) of the sori. *C*, section through a sorus and the leaf which bears it, as seen through a microscope; *i*, the indusium; *s*, the sporangia. *D*, a sporangium, showing the annulus and two escaping spores. — *After Wossidlo.*

gonium is often only a small part of it. Sporangia, on the other hand, are exclusively devoted to producing spores,

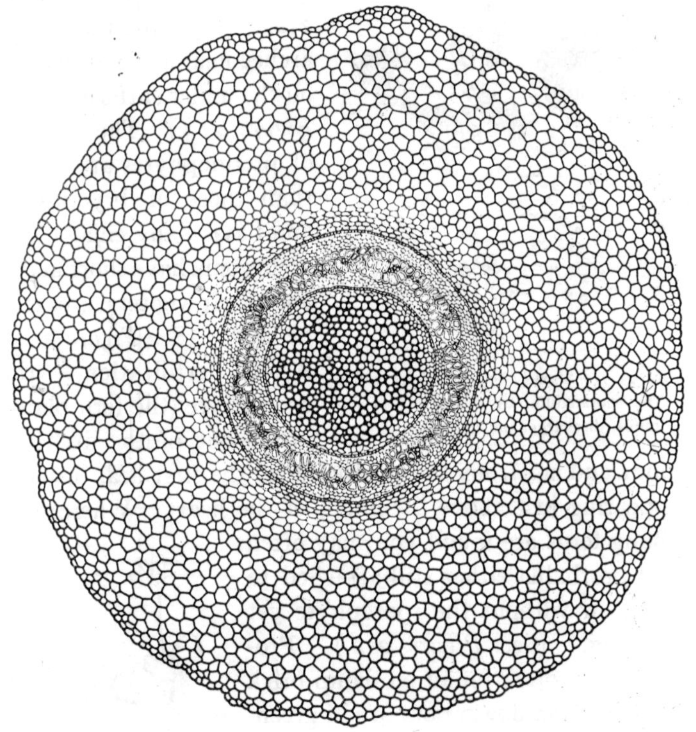

FIG. 210. — *Adiantum*, the maidenhair fern. Cross section of the stem. The fol lowing tissues are evident: epidermis, a thick cortex, outer phloëm, xylem (thick-walled cells), inner phloëm, pith. The phloëm, the xylem, and the pith, taken together, constitute the stele.

and they are only a very small part of the whole sporophyte generation.

When the spores of pteridophytes germinate, they produce a structure called the *prothallium*. Since this structure is produced by a spore, it is, of course, a gametophyte. The prothallia of true ferns are small, heart-shaped, flat, green bodies. (See *Figure 212.*) They produce their archegonia and antheridia on the under surface. (See *Figure 213.*) The sperms of pteridophytes are larger than those of bryophytes and have many cilia. The archegonia have short necks and their venters are embedded. (See *Figure 214.*) After fertilization, the sporophyte (the fern plant) grows out of the archegonium and up from the prothallium, the first leaf usually appearing just above the notch of the heart. Where very young ferns are found you may often find prothallia either still attached to them, or without the sporophyte yet showing. (See *Figure 212.*) Prothallia are often found on damp brick walls and on flower pots in old greenhouses in which ferns are grown.

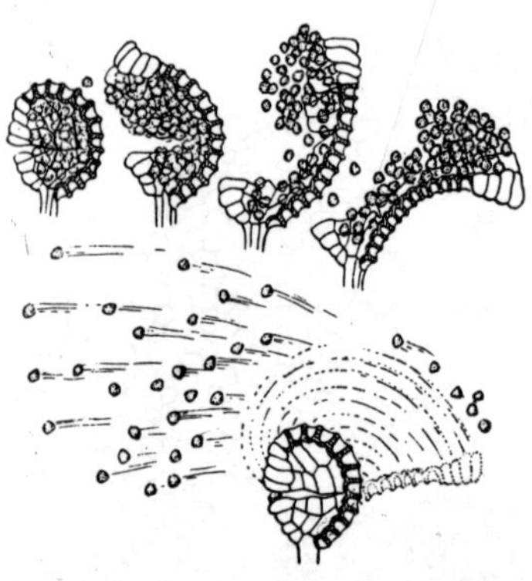

FIG. 211. — Shows the method by which the sporangium, by means of its annulus, discharges the spores.

The growth of new fern leaves from their underground stems is more conspicuous than their growth from prothallia, and when ferns are planted in yards or gardens the underground stems are used. They send up new fronds each season (see page 156).

FIG. 212. — Prothallia (gametophytes) of a true fern. *A*, prothallium with a young sporophyte arising from it. *B*, a group of prothallia without sporophytes.

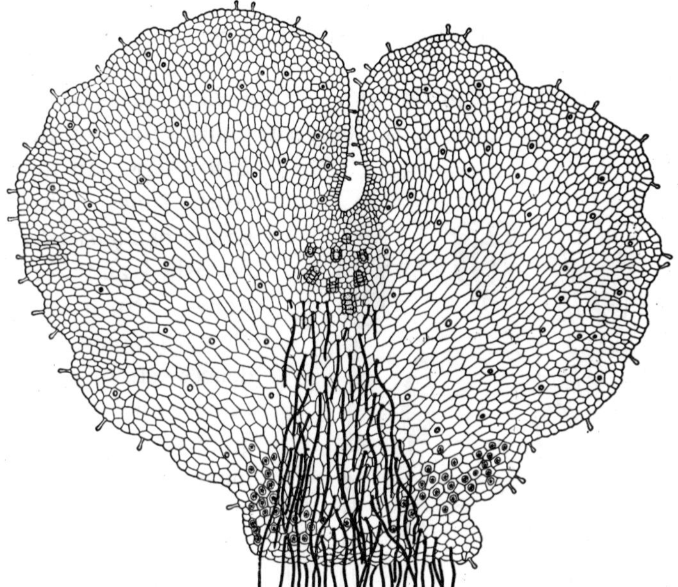

FIG. 213. — Prothallium (gametophyte) of *Aspidium*. The under surface, as viewed through a microscope. The necks of eight archegonia are indicated just below the notch of the heart. The hair-like outgrowths are rhizoids. At the base of the prothallium, among the rhizoids and to the right of them, numerous antheridia are indicated.

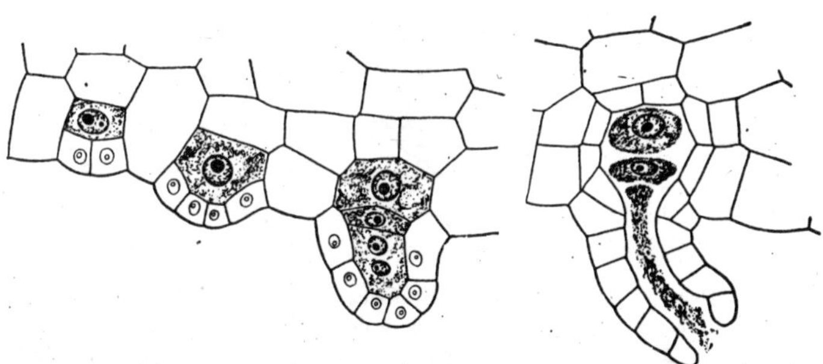

FIG. 214. — Archegonia of a fern, showing the manner of development. At the right is an archegonium ready for fertilization. From its neck is excreted a substance which attracts to it the swimming sperms which come near.

77. Horsetails (Equisetales). — *Equisetum* means *horse-tail.* It is the name of the only surviving group of the *Equisetales.* In ancient times other groups of *Equisetales* flourished and were very abundant. Now the only evidences left to us of these once great groups are fossils found in coal and in other kinds of rock.

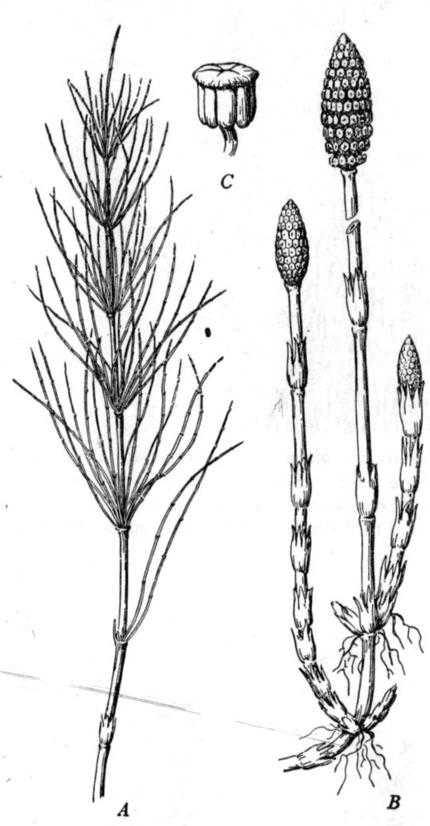

Fig. 215. — One of the horsetails (*Equisetum arvense*). *A*, a vegetative shoot; note the scale-like leaves at the joints. *B*, the fertile shoots which, in this species, appear in spring before the vegetative shoots; note the cone-like strobili at the tops; these are composed of sporophylls which are closely fitted together. *C*, a single sporophyll. See page 424.

The horsetails or equisetums are also called the scouring-rushes. Probably you have seen them. They have slender, round, green stems which are jointed, and they are often much branched in such a way as to suggest a horse's tail. The stems are stiff and rather brittle. They may be easily pulled apart at the joints. A ring of small, brown, scale-like leaves occurs at each joint. (See *Figure 215.*) Equisetums grow in both dry and wet places. They are often found on railroad embankments.

Reproduction. — Equisetums have underground stems as well as aërial stems and their underground stems live over winter. One of the commonest kinds produces unbranched shoots in the spring before the branched shoots appear. These unbranched shoots bear a yellowish, cone-shaped structure at the top. (See *Figure 215.*) This structure is composed of peculiar leaves, each of which is shaped like the letter T. Only the tops of the T's are visible. The bottoms are attached to the stem. The tops fit closely together, forming the surface of the cone. On their under sides these leaves bear spo-

Fig. 216. — *Botrychium.* The moonwort. One of the true ferns which does not bear its sporangia on the under side of the green leaves, but upon specialized parts. In this fern the uppermost part of the frond is devoted to spore production. It is a case of partial differentiation of a sporophyll.

rangia. When the sporangia are ripe, the leaves which bear them separate and the spores escape. Since these peculiar leaves are exclusively devoted to spore production they are called *sporophylls*. You will find the word sporophyll of much importance from this time on. It is applied to *any structure, leaf-like in origin, whose main business is the bearing of spores.*

There are some of the true ferns which also produce their spores on special leaves or special parts of leaves instead of on the under surface of the ordinary leaves. *Botrychium* is an example. (See *Figure 216*.)

The cone-shaped aggregation of sporophylls which we find in equisetum is called a *strobilus*, and this word too you will find to be important from this time on. *Any group of sporophylls growing close together is called a strobilus, and it is from strobili that flowers have evolved.* From what you already know you can see that stamens and carpels are sporophylls; *i.e.* they are leaves as to origin and they produce spores.

78. Club Mosses (Lycopodiales.) — This is the third division of pteridophytes. Some of the members of it are called ground pines. They grow in pine forests and have a resemblance to seedlings of pines. They are often used for Christmas decorations. They have long stems which spread over the ground. Their leaves are small, numerous, and scale-like. (See *Figure 217*.)

The leaves which bear the sporangia of these plants are usually those near the tips of the stems. The sporangia appear on the upper surfaces of the leaves near the axils, and there is *only one tc a sporophyll*. (You recall that in ferns the sporangia are many and on the under side of the sporophylls). Not all of the stem tips are fertile; that is, not all of them bear the little bean-shaped, yellowish sporangia.

Some of the club mosses do not have the strobilus so clearly defined as it is in the equisetums, but in others the sporangia-bearing tips have their leaves more closely set together than those of the sterile tips, and the strobilus is very distinct. It has a club-shaped appearance; hence the name club mosses. (See *Figure 218*.)

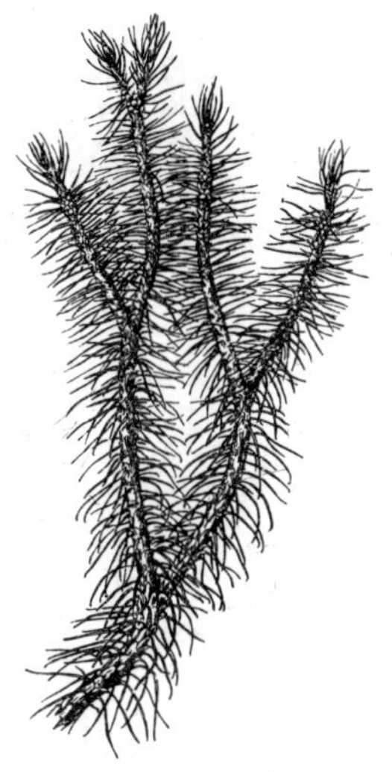

FIG. 217. — *Lycopodium*. Ground pine or club moss. This is a picture of one of the simplest species of this genus. In it the sporangia are not restricted to certain leaves, but a sporangium is usually found in the axil of each leaf. Near the top may be noted some sporangia which have not yet discharged their spores.

Selaginella. — This is the name of a group of club mosses. Some of the *Selaginellas* are quite common in greenhouses. Their general structure and habit of growth are like that of the club mosses already described, but they are more delicate and graceful. They are smaller. Their leaves are both smaller and broader. They are not such hardy plants as the coarser club mosses. (See *Figure 219*.)

In tracing the history of plants the *Selaginellas* are very interesting and important. They have the remarkable habit of producing *two kinds of spores*. Heretofore we have found but one kind of spore. This habit of producing two kinds of spores would be of no great interest to us if it were a habit of the *Selaginellas* alone. But the fact is that *all seed plants also produce two kinds of spores.* It is evident that seed plants are descended from ancestors which had two kinds of spores, and *Selaginellas* are important to us because they are two-spored forms which throw much light upon the evolution of seed plants.

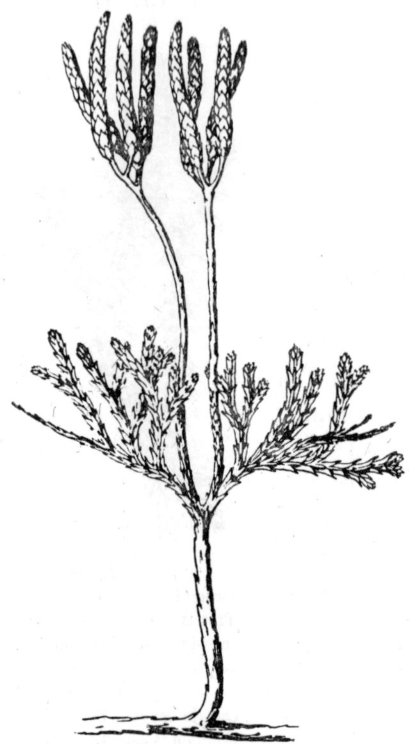

FIG. 218. — *Lycopodium.* A variety which bears distinct strobili of the kind which gave rise to the name club moss.

This habit of producing two kinds of spores is called *heterospory.* (The word means *different spores.*) The habit of producing one kind of spore, the habit with which we are already familiar, is called *homospory.* (The word means *same spores.*)

The spore-bearing leaves (sporophylls) of *Selaginella* are found at the tips of the stems as in the other club mosses. The clusters of sporophylls (strobili) are evident, but the sporophylls themselves are not very different in their general appearance from the ordinary leaves. (See *Figure 219.*)

In addition to heterospory, there are other important

facts about *Selaginella* which forecast the habits of seed plants and throw light upon their evolution. Not only are there two kinds of spores produced, but *these spores are produced separately in two kinds of sporangia.* And not only are there these two kinds of sporangia, but *these sporangia are produced separately on two kinds of sporophylls.* (See *Figure 220.*) And these facts are also true of seed plants.

FIG. 219. — *Selaginella.* A delicate kind of club moss whose reproductive structures and habits are very important in indicating the evolution of seed plants. The cone-like structures at the tips are strobili. Each of the leaves which compose them bears a single sporangium upon its inner face. — *After Coulter.*

In what way are these two kinds of spores different from each other? They are different in two ways. They are different both in size and in what they produce. (Study *Figures 221, 222,* and *223.*)

The larger spores are called *megaspores* and they produce gametophytes which bear eggs, but not sperms. The smaller spores are called *microspores* and they produce gametophytes which bear sperms but not eggs. So heterospory gives us not only two kinds of spores, but also two kinds of gametophytes. The megaspores produce female gametophytes and the microspores produce male gametophytes.

The sporangia which produce megaspores are called *megasporangia*, and the sporophylls which bear them are called *megasporophylls*. Similarly we speak of the *microsporangia* and *microsporophylls*. All these facts are also true of seed plants and the same terms are used to describe them.

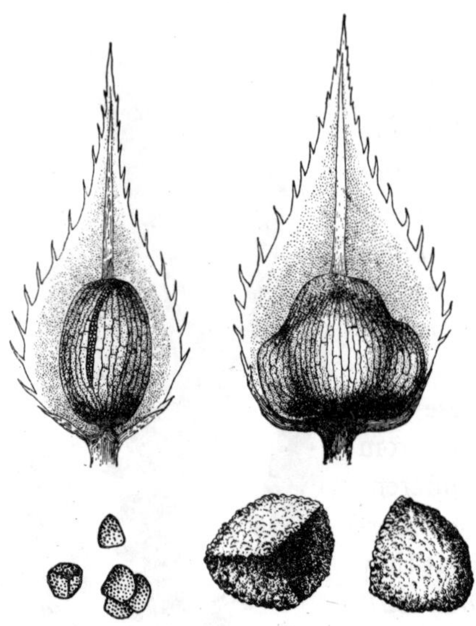

FIG. 220. — The two kinds of sporophylls of *Selaginella*. The one at the left bears a sporangium containing many small spores (microspores); the one at the right bears a sporangium containing just four large spores (megaspores). The megaspores and microspores are shown below the sporophylls which bear them; note the difference in size.

This matter is made easier to understand by studying the pictures. In *Figure 220* you see the two kinds of sporophylls, the two kinds of sporangia which they bear, and the two kinds of spores which are produced. The megasporangium contains only four megaspores.

The life history of *Selaginella* may be indicated by the following formula in which S stands for the sporophyte generation and G for the gametophyte generation.

S — strobilus ⟨ megasporophylls — megasporangia — megaspores
microsporophylls — microsporangia — microspores
— female G — egg
— male G — sperm ⟩ oöspore — S

This formula also indicates the life history of any seed plant, except that in seed plants there may be two kinds of strobili.

Surely *Selaginella* throws light upon the evolution of seed plants! We are not yet done, however. There is yet another feature of *Selaginella* to be considered which is also a feature of seed plants. This other feature is that *the gametophytes do not escape from the spores which produce them.* No longer do we find the spores producing little green independent gametophytes such as we noted in the true ferns. Instead of that, the gametophytes remain largely within the walls of their parent spores. Remember that a spore is but a single cell. When it becomes two cells, it ceases to be a spore; it begins to be a gametophyte. So a structure of even a few cells held within the walls of its parent spore is as truly a gametophyte as a much larger green structure, such as a moss plant, which soon leaves its spore behind it.

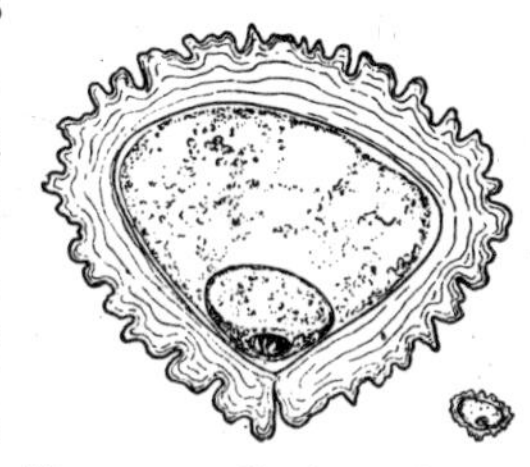

FIG. 221. — Sections through spores of *Selaginella* as seen through a microscope. This shows their actual difference in size.

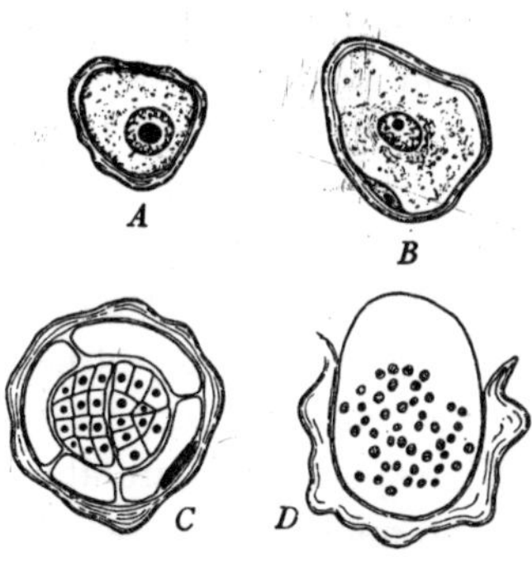

FIG. 222. — *Selaginella.* Sketches showing stages in the germination of the microspore and the development from it of the male gametophyte. *A*, mature microspore. *B*, the microspore containing two cells; it has now ceased to be a spore; the male gametophyte has begun. *C* and *D*, later stages in the development of the male gametophyte; *D* shows the sperms ready to be discharged and shows that the male gametophyte, even when mature, does not escape from the old microspore wall.

This matter of the size of

the gametophytes and their retention within the spores is also easier to understand if you study the pictures. In *Figure 222* you see a microspore and various stages of the male gametophyte to which it gives rise. In *Figure 223* you see the mature female gametophyte and see how the walls of the old megaspore still embrace it. Of course the wall of the microspore must be broken in order that the sperms may escape, and the wall of the megaspore must be broken so that the sperms may enter and fertilization occur. These breakings of the wall are accomplished by the growth of the gametophytes which they inclose. The sperms do not have to go very far in order to reach the egg for the reason that the microsporophylls are above the megasporophylls, and, when the microspores are shed, some of them fall and lodge upon the megasporangia. Thus, evidently, when the sperms are ready to be discharged, they are already near the eggs which they are to fertilize.

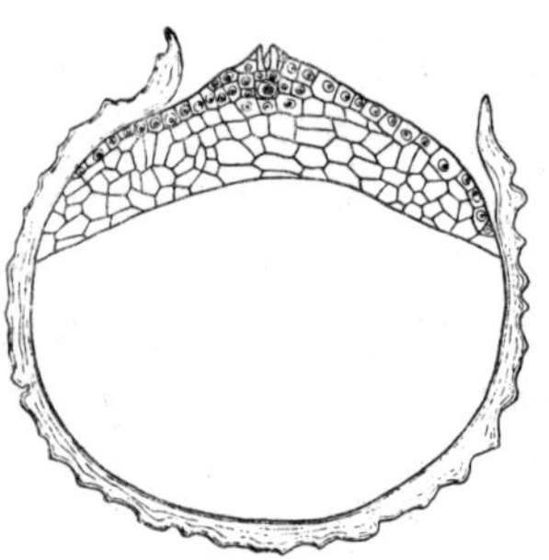

FIG. 223. — *Selaginella.* The mature female gametophyte. Note that it is still largely inclosed by the old megaspore wall. One archegonium, ready for fertilization, may be noted at the top of the gametophyte. The clear space is filled with the food material with which the megaspore was originally supplied.

There is still one more fact about *Selaginella* which needs to be mentioned, not only because it is true of seed plants, but also because it is the fact which more than anything else led to the evolution of seeds themselves, as you shall see. This is the fact that, in the case of the megaspore, not only is the female gametophyte retained within it, but ***the megaspore itself is retained in the mega-***

sporangium which produced it. This fact is referred to as *the retention of the megaspore.* It needs to be borne in mind.

SPERMATOPHYTES

We now come back to that great group of plants with which we began. You are sure to find that what you have learned concerning the lower groups of plants will help you to a better understanding of this, the highest group.

Seed plants fall naturally into two great divisions, gymnosperms and angiosperms. As you have already noted (page 183) the fundamental difference between these groups is that gymnosperms have their seeds exposed, while angiosperms have their seeds inclosed. The seeds of angiosperms are inclosed in that organ called the ovary. Gymnosperms have no ovary. Gymnosperms evolved before angiosperms, and they are simpler. We will consider them first.

79. **Gymnosperms.** — (The word means *naked seeds.*) Pines and other trees like them are the gymnosperms with which you are familiar. You know that they have needle-like leaves which stay on the year round, and for this reason these plants are called *evergreens.* You know that they produce *cones,* and that the seeds are in these cones. You know that pine lumber is a soft, smooth wood which is much used by carpenters.

That division of gymnosperms to which pines and forms like them belong is called *Coniferæ.* (The word means *cone bearers.*) Besides the pines, the cedars, hemlock, spruce, and arbor vitæ also belong to the *Coniferæ.*

Cones are aggregations of sporophylls; that is, they are

strobili. If we examine cones carefully, we shall learn a good deal about the reproductive methods and structures of gymnosperms.

A. Two Kinds of Cones.—Pines, and the other *Coniferæ* as well, produce two kinds of cones; that is, they have two kinds of strobili. The large hard cones which produce the seeds are composed of megasporophylls. They are sometimes called *pistillate* cones, because, like the pistils of flowers, they bear ovules. *Ovules are megasporangia.*

The other kind of cone is much smaller and softer and does not remain on the tree so long; it is composed of microsporophylls; it is called the *staminate* cone because it produces pollen. *Pollen sacs are microsporangia.* The staminate cones are usually found on the trees in spring; they are soft, tassel-like structures, more like catkins than like the hard pistillate cones. After they have shed their microspores (better known as pollen), they wither and die.

B. Pollination. — The pollen of the pines has wings. (See *Figure 224.*) Great quantities of it are produced. It is scattered by the wind. When the pollen is flying in pine woods, the lumbermen speak of "showers of sulphur," for the great quantities of pollen look like that. Some of the pollen falls on the hard cones, the kind which produces seed. To understand what happens then we must understand the structure of the seed-producing cone. (Study *Figure 225.*)

The seed-producing or pistillate cones are composed of hard, woody scales which grow close together. When the pollen is flying in the spring, the young cones, which are then very much smaller than those which contain ripe seeds, spread their scales apart enough to admit some of the drifting pollen. This pollen slides down the scales

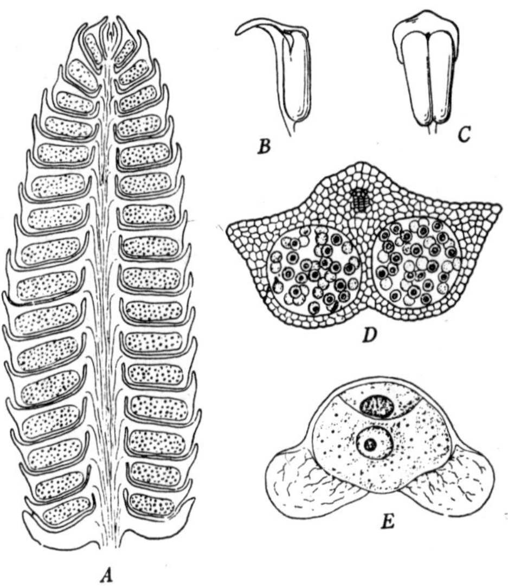

FIG. 224. — *A*, pollen-bearing (*staminate*) pine cone cut through lengthwise and showing the relation of the pollen sacs (*microsporangia*) to the scales (*microsporophylls*). *B* and *C*, two views of a single sporophyll. *D*, enlarged cross section of a single sporophyll, showing young pollen grains (*microspores*) inclosed within the microsporangia. *E*, a single mature pollen grain containing what is now the male gametophyte.

until it lies at their inner ends, where they join the stem part of the cone. This brings the pollen very near to the ovules, those small oval bodies which are borne on the surface of the scale, near its base. (See *Figure 225*.) These bodies, as you have noted, are the megasporangia. It is inside of these megasporangia that the megaspores are produced, but by the time the pollen arrives, the megaspores have germinated as internal parasites and thus the female gametophytes have been produced. One gametophyte matures in each megasporangium. It produces two or more eggs. (See *Figure 226.*) The sperms are in the pollen and the eggs are in the megasporangia. How are they to be brought together?

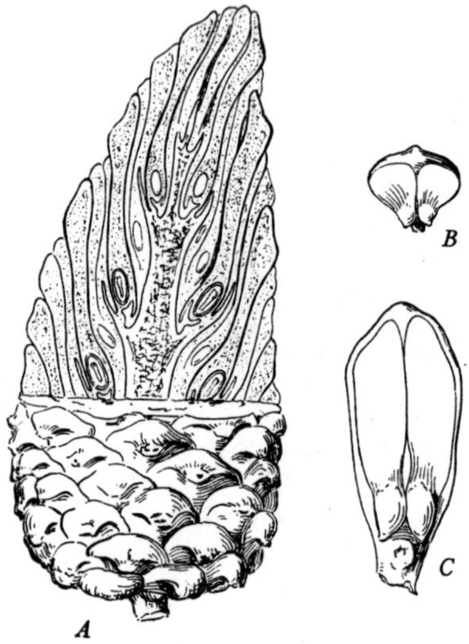

B

C

A

FIG. 225. — *A*, seed-bearing pine cone (*strobilus*), cut so as to show the relation of the seeds to the scales (*megasporophylls*). *B*, a young scale showing the position of the two ovules (*megasporangia*). *C*, the two-winged seeds which develop from the ovules shown as they lie upon the inner face of the scale.

C. The Pollen Tube. — You have learned of pollen tubes in your study of the flower. You know that they grow down through the styles, penetrate the ovary, and that by means of them the sperms finally reach the eggs. But gymnosperms have neither ovaries nor styles. Their ovules are borne on the surface of the sporophylls and the pollen comes directly in contact with them. None the less, pollen tubes are needed and pollen tubes are produced. They are needed to penetrate that tissue of the ovule which separates the sperm from the egg.

D. The Embryo. — After fertilization the pine cone enlarges until it reaches that full size with which you are most familiar. The cones usually hang to the tree for a year or more after they are pollinated, generally shedding their winged seeds before they drop. Meanwhile the embryo grows inside the seeds. The embryos of all our common gymnosperms are different from those of angiosperms in that they have several cotyledons instead of only one or two. (See *Figure 227*.)

FIG. 226. — *A*, diagram of a lengthwise section through a megasporophyll of pine; the shaded portion is the young female gametophyte lying within the ovule (*megasporangium*). *B*, lengthwise section through the ovule only; this ovule is older than the one represented in the other figure; two pollen tubes are seen entering it; they are approaching the two eggs which lie at the lower end of the female gametophyte, which is represented by the long oval.

E. The Seed. — You remember that in discussing *Selaginella* (page 430) we spoke of the retention of the megaspore, and said that this was responsible for the evolution of the seed. You can now see how this is. Evidently if the young pine cones shed their megaspores, there would be no seeds. The megaspores would be cast out from the ovules just as the pollen grains are cast out from the pollen sacs, and seeds would be no more produced than they are in the pteridophytes. Also we can see that this retention of the megaspores within the ovules which produce them is a thing which has involved the evolution of the pollen tube.

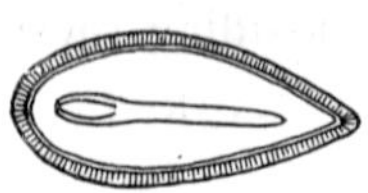

FIG. 227. — Seed of pine. Note that the embryo has several cotyledons. It lies embedded in the endosperm. Note the thick testa.

If the megaspores or the female gametophytes which they produce were cast out into the world instead of being confined within the tissues of the ovules, there would be no need for pollen tubes. This habit of getting the male gamete to the female gamete by means of a tube is called *siphonogamy*, a very expressive term which indicates that the sperm is "siphoned" to the egg.

F. Why Two Sets of Names? — We have said that the megasporangium of gymnosperms is the same thing as the ovule. This is also true of angiosperms. All ovules are megasporangia, though of course all megasporangia are not ovules. (Those which occur among pteridophytes are not ovules and *Selaginella* is not the only member of the pteridophytes which produces megasporangia.)

Also we have referred to pollen as microspores. We may now carry this doubling of terms a little farther. Let us see how it applies to other parts of flowers. You can see that a stamen is a microsporophyll, and the pollen sacs it bears are microsporangia. The carpels are megasporophylls, the megasporangia which they bear being called ovules.

All this is confusing. Surely it would be better to have only one name for each of these kinds of structures. The trouble is that these organs of flowers all received names long before their relationships to organs of lower plants were understood. Pollen, ovule, stamen, and carpel are terms used long before microspores and megaspores were understood. Ovules were thought to be little eggs; that is what the word ovule means. Of course we know now that the ovule, even though it contains the egg, does not belong to the same generation as the egg. In spite of such inconsistencies, this and the other old terms which describe the parts of flowers will continue to be used. They are convenient, and there is certainly no harm in using them as long as they are understood.

G. Pollen Grains not Exactly Microspores. — It is not altogether accurate to call pollen grains microspores. In a young pollen sac which has not yet opened you can find pollen grains which are microspores; that is, they are single cells. But when pollen grains are shed, they have ceased to be single cells. They contain a number of nuclei, and so, strictly speaking, they have ceased to be microspores. They now contain the diminutive male gametophytes. There are two sperm cells. These and a so-called *tube cell* migrate down the pollen tube. They constitute practically all that there is to the male gametophyte.

H. The Female Gametophyte. — In gymnosperms this generation is composed of very many more cells than is the male gametophyte. (See *Figure 226*.) Only one female gametophyte is developed in each ovule. It comes to occupy most of the interior of the ovule. It keeps on growing after fertilization and forms that tissue known as endosperm. The endosperm furnishes nourishment to the embryo, especially when the seed begins to sprout. (See *Figure 227*.)

I. Generations in the Seed. — How many generations are represented in the seed? The hard outer coat known as the testa is, of course, the modified outer part of the ovule; therefore it belongs to the old sporophyte generation. (It is when the testa is formed that the ovule ceases to be an ovule; when the testa appears, we speak of this structure as a seed.) The endosperm is the female gametophyte, while the embryo is the new sporophyte. Thus we find in seeds three generations represented. It is evident that we do not understand seeds if we do not understand these generations, and it is evident that we cannot understand these generations without understanding the lower plants. We see now that seeds are a sort of climax in the evolution of plant reproduction. We see that the gradual differentiation of reproductive structures has led finally to seeds, which are the most familiar and the most successful of all plant reproductive structures. We see more than ever that we cannot understand seed plants without understanding something of the lower plants from which they have evolved. We need to have in mind a clear picture of this evolution.

J. Summary of the Life History of a Gymnosperm. — The sporophytes of gymnosperms are composed of roots, stems, and leaves; commonly they are that form of plant which we call a tree. These sporophytes bear certain short branches called cones. These cones are aggregations of sporophylls; that is, they are strobili. The cones are of two kinds. One kind bears microsporophylls, the other, megasporophylls. On the under surface of the microsporophylls we find structures called pollen sacs; these are microsporangia. On the upper surface of the megasporophylls we find structures called ovules; these are megasporangia. The pollen sacs contain pollen grains which, when young, are microspores; later their contents become transformed into the male gametophyte. The ovules contain megaspores, one of which develops the female gametophyte; the female gametophyte remains within the ovule and produces eggs. By means of wind, pollen grains reach the ovules. By means of pollen tubes the sperms reach the eggs. Fertilization occurs. The fertilized egg germinates. It develops an embryo, one for each ovule. Meanwhile a hard coat called the testa has formed about the ovule. This structure has now ceased to be an ovule. It is a seed. The seeds are winged. They escape from the ripe cones. Under favorable conditions a seed germinates. A new sporophyte sprouts from it.

This life history may also be indicated as follows: —

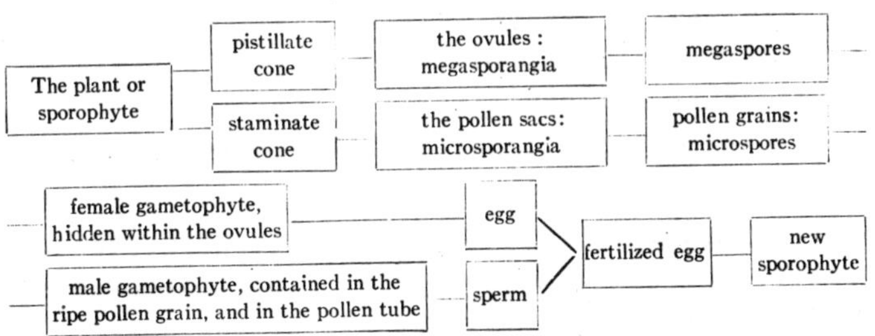

K. General Summary. — At this point it may be well to review by summary the chief features of each of the great groups. This is clearly done by the following quotation from J. M. Coulter's *Textbook of Botany*.

" *Thallophytes.* — Plants with a thallus body, but no archegonia.

Bryophytes. — Plants with archegonia, but no vascular system.

Pteridophytes. — Plants with a vascular system, but no seeds.

Spermatophytes. — Plants with seeds.

Each of these definitions except the last contains a positive and a negative statement. The positive statement distinguishes the group (except thallophytes) from the one below it in order of evolution, and the negative statement distinguishes the group (except spermatophytes) from the one above it.

The chief contributions which each group makes to the evolution of the plant kingdom may be stated as follows:

Thallophytes. — The beginnings of plant structures and of three kinds of reproduction.

Bryophytes. — Growth on land and the alternation of generations.

Pteridophytes. — Appearance of the vascular system, involving true roots, leaves, and stems; appearance of strobili and heterospory.

Spermatophytes. — Appearance of seeds and flowers."

80. Angiosperms. — As has just been stated, seeds and flowers are the two things which chiefly distinguish seed plants from all other groups. Both gymnosperms and angiosperms have seeds, but only angiosperms have those structures which are popularly known as flowers. Strictly speaking, gymnosperms also have flowers (see page 260), but these gymnosperm "flowers" are very different in appearance from that structure which is characteristic of angiosperms, and to which the term flower was first applied. The flower is the final organ to be explained, — explained as to its relationships with simpler reproductive structures which are like those from which it evolved.

A. The Greatest Group of Plants. — There is hardly need to tell you what angiosperms are. You know that all plants which have seeds inclosed in an ovary are angiosperms, and nearly all the plants with which you are familiar have this character. Nearly all of the first part of this book is devoted to angiosperms. Six chapters are given to the description of the sporophyte generation of this group of plants. All that has been said about roots and stems and leaves and about the life which goes on in them, was based upon angiosperms. They are the common, familiar plants. It is an angiosperm which comes at once to mind when we hear the word "plant." The most conspicuous, the most useful, and the most highly differentiated of all plants, are the angiosperms.

There is no need to describe again the nutritive body of these plants, or the structure of the flowers which it bears. With these you are familiar. It remains only to discuss the relationships of the flower and those hidden reproductive parts which you could not understand until after you had studied the lower plants.

B. The Flower. — What is a flower? It is a hard thing to define. We may say that flowers are structures which have evolved from strobili. They, like strobili, are groups of sporophylls. Yet they are more than that. Flowers have other parts than sporophylls. These other parts are around the sporophylls, outside of them. They are usually colored. They compose what is called the perianth. With it you are already familiar. You know that it is commonly composed of sepals and petals. Yet there are some flowers, naked flowers, which do not have a perianth. Such flowers are strobili and nothing more. They differ

FIG. 228. — Cross section of a young ovary of lily. There are two vertical rows of ovules in each of the three carpels which compose this ovary. The diagram indicates a single megaspore within each of the ovules shown. The nucleus of this megaspore divides and redivides until an eight-celled female gametophyte results, such as is shown in *Figure 230*.

from the strobili of lower plants only in that the megasporangia grow within a closed organ, — the ovary.

Thus in flowers we find megasporangia hidden. This is new. Heretofore we have found them borne upon the surface of the megasporophylls. It was not necessary to cut open any inclosing structure in order to find them. But now, among all angiosperms, we find that the carpels (megasporophylls) do not bear the ovules (megasporangia) upon their surface; they inclose them completely in a structure called the ovary. (See *Figure 228*.) A new task has been given to the pollen tube. Its journey has been extended. It must now penetrate more than the tissue of the ovule in order to reach the egg; it must penetrate also the tissues of the carpels which inclose the ovules. The carpels are usually united at the base into a compound ovary and extend upwards in the form of style and stigma, and all these tissues must be penetrated. The pollen tubes grow down from the stigma, through the style, and penetrate into the ovules at last. (See page 269.)

FIG. 229. — An ovary of lily shortly after fertilization has occurred. *Figure 228* represents a cross section of such an ovary when younger.

C. The Male Gametophyte. — What was said about the male gametophyte of gymnosperms applies here as well. This generation in angiosperms is composed of only three cells, two of which are sperm cells; it is hidden within the pollen grain and later in the tube which grows from the pollen grain.

D. The Female Gametophyte. — Here we find a difference. What was said about this structure as to gymnosperms will not do as to angiosperms. Though very small among gymnosperms, the female gametophyte is even smaller among flowering plants. When ready for fertilization it is composed of only eight cells. To observe it we must examine under the microscope thin sections taken through young ovules. Look at *Figure 228.* In it you see how the young ovules lie in a cross section of a lily ovary. You see a little oval near the center of each ovule. That indicates the female gametophyte. Now study *Figure 230.* It shows one ovule greatly enlarged. In the female gametophyte which it contains you can count eight cells. There is a group of three at the lower end, the end nearest the opening or micropyle

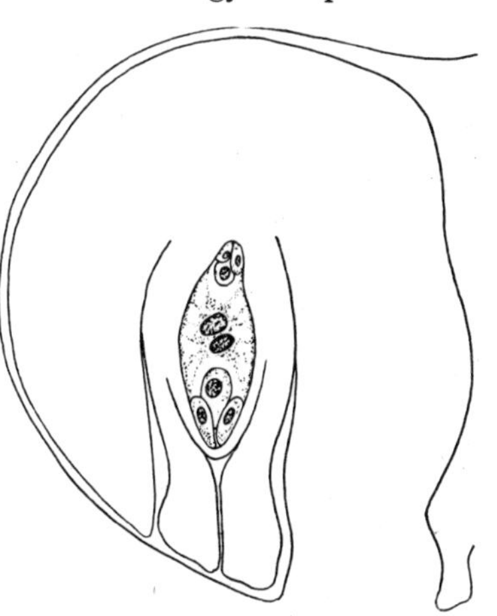

FIG. 230. — Longitudinal section through a mature ovule, showing a mature female gametophyte as it lies within the *nucellus.* This ovule has an outer and two inner *integuments.* See context. Note the *micropyle* between the two inner integuments.

which leads through the coats of the ovule. (These coats of the ovule are the integuments and the tissue they inclose, the tissue in which the female gametophyte lies embedded, is the nucellus. See page 276.) One of this lower group of three cells, the largest one, is the egg. Only one egg is produced by this gametophyte. Usually the pollen tube enters the ovule by way of the micropyle and then reaches the egg by burrowing its way through the nucellus.

Near the center of the female gametophyte you note a pair of nuclei. The peculiar thing about this pair of nuclei is that, about the time that a sperm fuses with the egg, they fuse with each other. Thus they form what is called the *endosperm-nucleus*. The endosperm-nucleus produces the endosperm.

The three nuclei at the other end appear to have no special function; often after fertilization they disappear. Sometimes they persist and contribute nourishment to the embryo.

This thing which we have been calling the female gametophyte, like other parts of the flower, received another name long before its real nature was understood. Botanists noted that in very young seeds the tiny embryo appeared to be lying in a sort of sac. This sac was of course the female gametophyte. They called it the *embryo-sac*, and this name is still commonly used to describe the female gametophyte of angiosperms.

E. Double Fertilization. — You recall that there are two sperms in the pollen tube. Until quite recently it was supposed that after one of these sperms fused with the egg the other simply disappeared. It was thought that it had no special function, save possibly to be of service in case anything happened to its mate.

But a few years ago a surprising discovery was made. It was discovered that this second sperm moves on into the embryo-sac. It reaches that pair of nuclei at just about the time that they are beginning to fuse to form the endosperm-nucleus, and it *joins in their fusion.* In other words, the endosperm-nucleus is the result of the fusion of three nuclei, a *triple fusion.* Thus we have what is described as double fertilization, — one sperm contributing to the formation of the embryo, the other to the formation of the endosperm.

It is hard to explain this double fertilization. There is nothing to explain it by. Nothing has been discovered among the lower plants which throws any light upon the evolution of this habit. So far as we know it originated among angiosperms themselves. It is one of the many unexplained things in plant life.

A plain illustration of the effects of double fertilization is found in corn. Here we find characters of the pollen parent appearing in the endosperm as well as in the embryo, which seems natural enough now that we understand how to explain it. Have you ever noticed red kernels in a white ear of corn? It is not uncommon to find them, and double fertilization explains how they happen. It is presumed that the pollen whose sperms reached these red kernels came from a plant with a tendency to produce red ears. It is the endosperm which is red. Now that we know that one of the sperms fuses with the endosperm-nucleus, we can readily see why the quality of redness which comes from the pollen parent may appear in the endosperm.

F. Plant Breeding. — Fertilization is the basis of plant breeding. What is in sperm and egg limits the nature of the individual which results from their fusion.

Breeding means the effort by man, through control of reproduction, to obtain a progeny better for his purposes than its progenitors were. It is commonly described as effort *to improve the stock*.

Breeding can be understood only if certain fundamental facts are understood and their application appreciated. With some of these facts you are already familiar. One of them is the fact of *variation*, the fact that no two individuals are precisely alike. Remember that nature produces multitudes of individuals, but that it was man, seeking to interpret nature, who determined which groups of these multitudes of individuals we shall call kinds or species. Often it is impossible to tell "where one species stops and another begins" so complete are the *intergrades* between one kind and another. This universal tendency to variation makes it possible for man to select those forms preferred. If the selected forms remain *true to type*, that is, if they produce individuals much like the selected parent forms, an improvement has been made.

But variation hinders as well as helps. The progeny of selected forms may not remain true to type. They are as likely to resemble their inferior grandparents as their superior parents. For this reason *thoroughbred* forms are more valuable than those forms which, though themselves superior, have an unknown ancestry. Thoroughbred or *pedigreed* stock is stock whose ancestry for several generations is known by records to have been of superior grade. Now that plants are being scientifically bred as well as animals, we are beginning to hear of "thoroughbred" corn and wheat just as we have heard of thoroughbred horses. Scientific farmers recognize the superior value of seed corn whose ancestors through many seasons have shown a high record of bushels per acre. That, rather than the superficial appearance of an ear, is the real test of seed corn as to its probable crop-producing capacity.

The scientific breeding of animals and decorative plants has a much longer history than the scientific breeding of agricultural plants. You know of the many kinds of dogs, horses, pigeons, and pigs and of the varieties of cultivated flowers whose existence is due chiefly to the control by man of the operations of nature. By similar methods man seeks now to obtain kinds or *strains* of crop plants which will be even better than those kinds which have resulted from the less scientific agriculture of the past. (See pages 15, 16, and 220.)

Another great fact of life which lies at the basis of breeding is the fact that the sex process of reproduction necessarily results in an individual which, in combining the traits of both its parents, is necessarily different from either. The more dissimilar the parents are, the greater is the likelihood of striking variation from both parents on the part of the progeny. The crossing of different species is called *hybridization* (see page 305), and it is by this means that the most striking forms of cultivated plants and animals have been obtained. Luther Burbank of California is the best-known plant breeder in America, and it is chiefly by hybridization that his most surprising results have been obtained. People have exaggerated his work as to its value in agriculture, but it indicates what results may be obtained by the skilled breeder. Scientific plant breeders in government and state employ are now at work, with many evidences of success, in producing strains of crop plants superior to the present ones with respect to yield, drouth resistance, ability to withstand diseases, etc.

You have learned enough of the nature of plants to realize something of what the problems and the possibilities of plant breeding are. You know that for *many kinds of places there are many kinds of plants* (page 149), you know that these plants *vary in their power to vary* (page 206), and you know that heredity and environment are two forces always at work in determining what the outcome shall be (page 135). The whole nature of any living thing is due to its heredity plus its environment. Its heredity is the sum of the characteristics it inherits. Its environment is the sum of the stimuli to which it responds, and these largely determine the ways in which its hereditary possibilities are fulfilled.

All this is true of ourselves as well as of plants. It suggests wonderful and yet unmeasured possibilities of development in ourselves as well as in plants, and it urges men on to reveal yet hidden possibilities in plant life, possibilities which, once realized, may solve forever the great food problem of the human race. He who reveals such possibilities ranks as a benefactor of mankind along with the greatest of inventors or of statesmen.

QUESTIONS AND SUGGESTIONS

SECTION 75. 1. What are the two great divisions of vascular plants? 2. Describe the functions of the vascular system. 3. What were the ancestors of seed plants? Describe them.

SECTION 76. 1. Describe the structure of a true fern, defining frond and venation. 2. Describe the reproduction of a true fern, defining sporangium, sorus, indusium, annulus, and prothallium.

SECTION 77. 1. Describe the structure of *Equisetum*. 2. Describe its reproduction, defining sporophyll and strobilus.

SECTION 78. 1. Describe the ground pine. How does it differ from ferns in the manner of bearing sporangia? 2. Describe *Selaginella*, defining heterospory, megaspore, megasporangium, and megasporophyll. 3. Illustrate by diagram the life history of *Selaginella*.

SECTION 79. 1. Give examples of the *Coniferæ*. 2. What sort of organ is a pine cone? Explain its structure. *A*. 3. Distinguish between the staminate and pistillate cones of *Coniferæ*. *B*. 4. Describe the pollen of pine and the process of pollination. *C*. 5. Compare the pollen tubes of gymnosperms with those of angiosperms as to the tissues which they penetrate. *D*. 6. Describe the embryo of pine. 7. Describe the habit of pine cones as to attachment to the trees which bear them. *E*. 8. Explain why the retention of the megaspore is responsible for the formation of seeds. 9. Define siphonogamy. *F*. 10. Explain why two sets of terms are applied to flower parts. 11. Show the correspondences between the new terms and the old ones. *G*. 12. Why are pollen grains not microspores? *H*. 13. Describe the female gametophyte of gymnosperms. *I*. 14. Tell how many generations are represented in a seed, and explain your answer. *J*. 15. Give a summary of the life history of a gymnosperm by writing and explaining a diagram of it. *K*. 16. Summarize the chief distinguishing features of the four great groups.

SECTION 80. 1. State the characteristics which distinguish gymnosperms and angiosperms from one another. *B*. 2. Locate and describe the megasporangia of angiosperms. *C*. 3. Describe the male gametophyte of angiosperms. *D*. 4. Define micropyle, integument, and nucellus. 5. Describe the female gametophyte of angiosperms, defining endosperm-nucleus and embryo-sac. *E*. 6. Describe double fertilization, and one of its effects in corn. *F*. 7. Define breeding. 8. Discuss variation and give examples of it which you have noted. 9. What is meant by *true to type?* 10. What is meant by *pedigreed* stock? 11. What is the best test for seed as to its possibilities of yield? 12. Look up in government bulletins or elsewhere the methods used in corn or wheat breeding and report thereon. 13. Report on the history of the seedless orange or of some other improved fruit.

KEY TO INDEX AND GUIDE IN REVIEW

The following tables furnish a key to the relative importance of topics. It is believed that they may be of service to the student in review work, and otherwise help him to differentiate between the more and the less fundamental ideas which the course presents. All terms included occur also in the index, and by use of the latter an explanation of them may readily be found in the text.

Table I is composed of 230 terms which stand for ideas of fundamental importance. An understanding of them constitutes a considerable part of that knowledge which should become the permanent possession of the student. Eighty-five of these terms are printed in bold faced type, not because familiarity with the others is of less importance, but because the eighty-five represent ideas which are foundational to the others. Satisfactory comprehension of the others depends upon comprehension of the emphasized eighty-five. Knowledge of the others without understanding of the eighty-five is superficial. The eighty-five stand for things fundamental in the life of practically all familiar plants, and over half of them stand for things fundamental in all life.

Table II (42 terms), in case Chapters IX and X are given equal attention with the rest of the book, is of equal importance with Table I. The only reason for making a separate table of these terms is that it is not uncommon to give less attention to these topics than to those which compose Table I.

Table III (205 terms) is made up of terms chosen from the entire book, and so important as entirely to justify questions about them in a final examination upon the course. But they are not terms of equivalent importance with those which precede them. The good student at the end of his course will be as familiar with the ideas represented by Table III as he is with those represented by Table I, yet he will recognize readily the more fundamental nature of the knowledge suggested by Table I.

All three tables, however, *by no means indicate all the knowledge which the student should acquire in the course.* They represent chiefly what is common to all courses in elementary botany, but they do not represent that very important part of such courses which is not common to all. They do not represent the part of the course that depends upon the locality in which it is taught, the part whose nature is determined by what plants are of chief local importance and interest. This is a part which is of primary importance, but its nature and content must be determined by the individual teacher. It cannot be dealt with satisfactorily in a basic text.

Thus it will be noted that Table I includes no names at all of kinds of plants. Table II names only those which are essential to a comprehension of the evolutionary sequence, while Table III names only a few of the chief groups, and a few of the most familiar families of seed plants. Important economic plants, and many of the important practices in plant culture, are not listed at all, solely because the study of such should be locally determined. Trees are listed as one entry but those " trees and flowers which every boy and girl should know " are not listed, for the list depends on the locality.

The danger that some students may use these tables simply as a guide to " cramming " is recognized, yet it is believed that this danger is overweighed by the service such tables may render to the more serious student. It is also recognized that this danger is no danger at all in the presence of a skillful teacher, one able readily to unmask sham knowledge.

TABLE I

Accessory parts of flowers, aëration, algæ, angiosperm, **annuals**, **anther**, **assimilation**, **atom**, autogamy, axil, bacteria, **bark**, biennial, blade, bract, bryophytes, buds, bulb, calcium, calyx, **cambium**, **carbohydrate**, carbon, **carbon dioxide**, carpel, **cells**, cellulose, **chlorophyll**, **chloroplast**, cork, corolla, **cortex**, **cotyledons**, crossing, cross- and close-pollination, cutin, cytoplasm, cylindrical arrangement, deciduous, **decomposition**, diclinous flowers, dicotyledons, differentiation, **diffusion**, **digestion**, diœcious, dissemination, **egg**, **element**, **embryo**, endosperm, **energy**, **environment**, **enzyme**, epidermis, epigyny, essential parts of flowers, **evaporation**, **evolution**, **excretion**, family, fat, ferns, fertility, **fertilization**, fibers, fibrous roots, **flower** (structure and function), **food**, forestry, **fruit**, **function**, fungi, geitonogamy, genus, **germination**, growth rings, gymnosperm, **heredity**, host, humus, hybrid, **hydrogen**, hypogyny, improvement of crops, inflorescence, **inorganic**, insect-pollination, internodes, iron, latex, **leaves** (structure and function), lenticel, lichen, **light**, loam, magnesium, membrane, **meristem**, **mesophyll**, mineral, **molecule**, monoclinous flowers, monocotyledons, monœcious, mosses, mushroom, mycelium, nectar, **nitrates**, **nitrogen**, node, nodule, nucleus, nutrient salts, **nutrition**, oil, **organ**, **organic**, **osmosis**, ovary, overproduction, ovule, **oxidation**, **oxygen**, palisade, parasite, parenchyma, **perennial**, perianth, **permeable**, petal, petiole, **phloëm**, phosphorus, **photosynthesis**, pistil, pistillate flowers, **pith**, pod, **pollen**, pollen tube, **pollination**, potassium, primary growth, **protection**, protein, **protoplasm**, protoplast, pteridophytes, radial arrangement, **reproduction**, **respiration**, **response**, rhizome, **root** (structure and function), root cap, root excretion, **root-hair**, rotation of crops, **sap**, saprophyte, scattered arrangement, **scientific method**, secondary growth, **seed**, seed dispersal, seedling, **seed plants**, seed testing, seed vitality, self-pollination, sepal, sex organ, **sex reproduction**, shoot, sodium, **soil**, **solute**, **solution**, **solvent**, **species**, **sperm**, spermatophyte, spongy mesophyll, spore, stamen, staminate flowers, **starch**, **states of matter**, **stele**, **stem** (structure and function), stigma, **stimulus**, **stomate**, storage, style, **sugar**, sulphur, tap-root, tendril, testa, thallophytes, **tissue**, toadstool, tracheary vessel, **transpiration**, trees, tropism, tuber, tubercle, **turgidity**, underground stems, vacuole, **variation**, vascular bundle, vascular cylinder, vascular plants, vascular system, **veins**, venation, vernal habit, **water in plants**, wind dispersal, wind-pollination, wood, xenogamy, **xylem**, zygomorphy.

TABLE II

Alternation of generations, andrœcium, antheridium, *Anthoceros*, archegonium, club mosses, cone, *Coniferæ*, gamete, gametophyte, generation, gynœcium, heterospory, homospory, horsetails, integument, liverworts, *Marchantia*, megasporangia, megaspore, megasporophyll, micropyle, microsporangia, microspore, microsporophyll, nucellus, *Œdogonium*, oögonium, oöspore, prothallium, protonema, retention of megaspore, *Selaginella*, *Spirogyra*, sporangia, sporogonium, sporophyll, sporophyte, strobilus, thallus, *Ulothrix*, unicellular forms.

TABLE III

Actinomorphy, akene, aleurone, annual rings, annulus, apocarpous, archegoniophore, *Ascomycetes*, ascus, autumnal colors, bases, *Basid-*

iomycetes, basidium, berry, blue algæ, bread mold, breeding, brown algæ, budding, capsule, carnivorous plants, catkin, chlorenchyma, *Chlorophyceæ*, cilia, cleistogamy, closed bundles, coal, cœnocyte, *Compositæ*, corn, corymb, cupule, *Cyanophyceæ*, cyme, determinate inflorescence, diastase, dichogamy, dichotomous, double fertilization, drupe, elater, embryo-sac, endogen, endosperm nucleus, epicotyl, epiphyte, *Equisetales*, evergreens, exogen, fertilizers, fig and wasp, *Filacales*, filament, fission, flaccid, foot, fossil, frond, gemma, geotropism, gills, glaciation, glutin, grafting, grass family (*Gramineæ*), green algæ, ground pine, guard-cells, haustoria, head, heartwood, hilum, horny seeds, hydrotropism, hyphæ, hypocotyl, indeterminate inflorescence, indusium, inferior ovary, inoculation of soil, involucre, iodine, layering, leaf fall, leafy liverworts, *Leguminosæ*, liana, lime, *Lycopodiales*, marine algæ, medullary rays, mestome, mildew, morel, mulch, mycorhiza, nerves of leaves, open bundles, ovate leaves, panicle, pedicel, pedigreed stock, peduncle, peg, pericarp, perigyny, *Phæophyceæ*, phototropism, *Phycomycetes*, pigment, pileus, pinnate leaves, pioneer plants, pith-rays, pitted vessels, placenta, plant breeding, planting of trees, plumule, pollen basket, pollen sac, polygamous flowers, polypetalous, pome, primary roots, prop roots, pubescent, *Puccinia*, puffballs, pulvinus, raceme, radicle, *Ranunculaceæ*, receptacle, red algæ, region of elongation, regular corolla, rhizoid, *Rhodophyceæ*, rock phosphate, root branches, root stock, *Rosaceæ*, rosette, runners, salts, samara, sapwood, scalding of leaves, scales, scion, seaweeds, secondary roots, seed coat, seed habit, self-planting, serrate leaves, sessile, seta, shrub, sieve vessel, silk of corn, siphonogamy, slip, sorus, spadix, spathe, spawn, spike, spiral vessel, sporophore, spring wood, stereome, stipe, stipule, stock, stone fruit, stool, suberin, sucker, superior ovary, sympetalous, syncarpous, tassel of corn, thorn, triple fusion, true ferns, true fungi, umbel, *Umbelliferæ*, unisexual flowers, valve, venter, villus, vitality of seeds, water dispersal, water roots, wheat rust, whorl, winter buds, xanthophyll, yeast, yucca and pronuba, zoöspore.

INDEX

(*All numbers refer to pages; those in heavy black type indicate illustrations.*)

D

E

N

O

P

Q

R

T

U

V

W

X

Y

Z